FOOD MICROBIAL AND MOLECULAR BIOLOGY

From Fundamentals to Applications

FOOD MICROBIAL AND MOLECULAR BIOLOGY

From Fundamentals to Applications

Edited by

Saher Islam, PhD

Devarajan Thangadurai, PhD

Jeyabalan Sangeetha, PhD

Zaira Zaman Chowdhury, PhD

AAP | APPLE ACADEMIC PRESS

First edition published 2024

Apple Academic Press Inc.
1265 Goldenrod Circle, NE,
Palm Bay, FL 32905 USA
760 Laurentian Drive, Unit 19,
Burlington, ON L7N 0A4, CANADA

CRC Press
6000 Broken Sound Parkway NW,
Suite 300, Boca Raton, FL 33487-2742 USA
4 Park Square, Milton Park,
Abingdon, Oxon, OX14 4RN UK

Library and Archives Canada Cataloguing in Publication

Title: Food microbial and molecular biology : from fundamentals to applications / edited by Saher Islam, PhD, Devarajan Thangadurai, PhD, Jeyabalan Sangeetha, PhD, Zaira Zaman Chowdhury, PhD.
Names: Islam, Saher, editor. | Thangadurai, Devarajan, 1976- editor. | Sangeetha, Jeyabalan, editor. | Chowdhury, Zaira Zaman, editor.
Description: First edition. | Includes bibliographical references and index.
Identifiers: Canadiana (print) 20220449236 | Canadiana (ebook) 20220449317 | ISBN 9781774910863 (hardcover) | ISBN 9781774910870 (softcover) | ISBN 9781003300601 (ebook)
Subjects: LCSH: Food—Microbiology. | LCSH: Food—Quality. | LCSH: Food poisoning. | LCSH: Food industry and trade—Safety measures.
Classification: LCC QR115 .F66 2023 | DDC 664.001/579—dc23

Library of Congress Cataloging-in-Publication Data

Names: Islam, Saher, editor. | Thangadurai, Devarajan, 1976- editor. | Sangeetha, Jeyabalan, editor. | Chowdhury, Zaira Zaman, editor.

Title: Food microbial and molecular biology : from fundamentals to applications / edited by Saher Islam, Devarajan Thangadurai, Jeyabalan Sangeetha, Zaira Zaman Chowdhury.

Description: First edition. | Palm Bay, FL : AAP, Apple Academic Press, 2023. | Includes bibliographical references and index. | Summary: "The ever-increasing globalization of the food industry demands new interventions and new prevention technologies to improve the safety and quality of food. This multidisciplinary new book addresses the major concerns of the food industry by discussing fermentation, spoilage, mycotoxins, pathogenic bacteria, prions, parasites, viruses, preservation methods, and food quality. The volume provides descriptions of the most advanced systems for identifying, analyzing, tracking, and monitoring microbial contaminants in food. Key features: Focuses on approaching complications in food science related to food safety and traceability Highlights emerging and re-emerging foodborne microorganisms and their virulence characteristics Includes recent approaches for food quality assurance and risk management Describes the practicality of molecular biology and microbial technologies for effectual control of foodborne infections Presents a detailed overview of the utilization of recent molecular techniques in food microbiology With expert contributions from experienced academics involved in food microbiology and molecular biology research, this book offers indispensable guidance and a contemporary update for educators and scholars who are trying to keep informed with latest developments in food microbial and molecular biology"-- Provided by publisher.

Identifiers: LCCN 2022049287 (print) | LCCN 2022049288 (ebook) | ISBN 9781774910863 (hardback) | ISBN 9781774910870 (paperback) | ISBN 9781003300601 (ebook)

Subjects: LCSH: Food--Microbiology. | Molecular biology.

Classification: LCC QR115 .F637 2023 (print) | LCC QR115 (ebook) | DDC 664.001/579--dc23/eng/20221223

LC record available at https://lccn.loc.gov/2022049287

LC ebook record available at https://lccn.loc.gov/2022049288

ISBN: 978-1-77491-086-3 (hbk)
ISBN: 978-1-77491-087-0 (pbk)
ISBN: 978-1-00330-060-1 (ebk)

About the Editors

Saher Islam, PhD
Postdoctoral Researcher, UT Southwestern Medical Center, Dallas, TX, USA

Saher Islam, PhD, is a Postdoctoral Researcher at UT Southwestern Medical Center, Dallas, USA. She worked at Cornell University, New York, and West Virginia State University, West Virginia, in the USA as visiting scholar. She has keen research interests in genetics, molecular biology, biotechnology, and bioinformatics, and has ample hands-on experience in molecular marker analysis, whole-genome sequencing (WGS), and RNA sequencing. She has edited five books, including *Sequencing Technologies in Microbial Food Safety and Quality*, *Nutriomics*, and *Global Food Safety*. She is the recipient of the 2016 Boehringer Ingelheim Fonds Travel Grant from European Molecular Biology Laboratory, Germany.

Devarajan Thangadurai, PhD
Professor, Karnatak University, Dharwad, India

Devarajan Thangadurai, PhD, is Professor at Karnatak University, Dharwad, India. He did his postdoctoral research at the University of Madeira, Portugal; University of Delhi, India; and ICAR National Research Centre for Banana, India. He is the recipient of a Best Young Scientist Award with Gold Medal from Acharya Nagarjuna University, India, and the VGST-SMYSR Young Scientist Award of the Government of Karnataka, India. He has a keen interest and expertise in the fields of biodiversity and biotechnology, genetics and genomics of food crops, and beneficial microbes for crop productivity and food safety towards sustainable agricultural development. He has authored and edited more than 35 books, including *Genetic Resources and Biotechnology (3 Vols.)*, *Genes, Genomes and Genomics* (2 Vols.), *Genomics and Proteomics*, *Biotechnology of Microorganisms*, and *Sequencing Technologies in Microbial Food Safety and Quality* with international scientific publishers in the USA, the UK, Canada, Switzerland, and India. He has extensively traveled to many universities and institutes in the Middle East, Africa, Europe, and Asia for academic work, scientific meetings, and international collaborations. He is also a reviewer for several journals, including *Comprehensive Reviews in Food Science and Food Safety*, *Food Science and Technology International*, *Journal of Agricultural and Food Chemistry*, *Journal of Food Science*, *Journal of the Science of Food and Agriculture*, and *International Journal of Plant Production*.

Jeyabalan Sangeetha, PhD
Assistant Professor, Central University of Kerala, Kasaragod, India

Jeyabalan Sangeetha, PhD, is an Assistant Professor at the Central University of Kerala, Kasaragod, India. She earned her BSc in Microbiology (2001) and PhD in Environmental Sciences (2010) from Bharathidasan University, Tiruchirappalli, Tamil Nadu, India. She holds an MSc in Environmental Sciences (2003) from Bharathiar University, Coimbatore, Tamil Nadu, India. Between 2004 and 2008, she was the recipient of a Tamil Nadu Government Scholarship and a Rajiv Gandhi National Fellowship of University Grants Commission, Government of India, for doctoral studies. She served as a Dr. D.S. Kothari Postdoctoral Fellow and UGC Postdoctoral Fellow at Karnatak University, Dharwad, South India, during 2012–2016 with funding from the University Grants Commission, Government of India, New Delhi. She has carried out her doctoral and postdoctoral research in the fields of microbiology and metagenomics, respectively. She has authored a textbook, *Fundamentals of Molecular Mycology,* and edited 15 books, including *Genomics and Proteomics*, *Industrial Biotechnology*, *Biotechnology of Microorganisms*, and *Sequencing Technologies in Microbial Food Safety and Quality*. She has trained in next-generation sequencing (NGS) and data analysis at the Malaysia Genome Institute, Malaysia.

Zaira Zaman Chowdhury, PhD
Assistant Professor, Nanotechnology and Catalysis Research Centre (NANOCAT), University of Malaya, Malaysia

Zaira Zaman Chowdhury, PhD, is working as an Assistant Professor at the Nanotechnology and Catalysis Research Centre (NANOCAT), the University of Malaya, Malaysia, where she has been appointed as Senior Research Fellow and Post-Doctoral Research Fellow. She received her PhD in Analytical Chemistry from the University of Malaya. She had training in Environmental Analytical Chemistry at TAFE, Sydney, Australia. She has also received a Cambridge International Diploma for Teachers and Trainers (CIDTT) from Cambridge University, UK, with distinction. She is highly experienced in the antimicrobial treatment and the development of food additives using nanotechnology. She has also worked on the synthesis of micro- and nano-structured carbon and biomass processing using catalysts. She has contributed more than 100 papers in top-ranking journals. Currently, she is supervising 10 PhD and 11 MSc scholars at the University of Malaya, Malaysia. She has received several esteemed awards, such as Elsevier's Atlas Award and Best Paper Award on Fourth Generation Nano-antioxidant Synthesis from New Zealand.

Contents

Contributors

Samreen Ahsan
Department of Food Science and Technology, Khwaja Fareed University of Engineering and Information Technology, Rahim Yar Khan, Pakistan

Faiza Anum
ORIC, Lahore Garrison University, Lahore, Pakistan

Priya Ashrit
Department of Biotechnology, M.S. Ramaiah Institute of Technology, Bangalore, Karnataka, India

Saadia Andleeb
School of Applied Biosciences, National University of Sciences and Technology, Islamabad, Pakistan

Sayed Muhammad Ata Ullah Shah Bukhari
Department of Microbiology, Quaid-i-Azam University, Islamabad, Pakistan

Muhammad Farhan Jahangir Chughtai
Department of Food Science and Technology, Khwaja Fareed University of Engineering and Information Technology, Rahim Yar Khan, Pakistan

Chandan Dharmashekara
Department of Biotechnology and Bioinformatics, School of Life Sciences, JSS Academy of Higher Education and Research, Mysuru–570015, Karnataka, India

Gangaraju Divyashri
Department of Biotechnology, M.S. Ramaiah Institute of Technology, Bangalore, Karnataka, India

Rabia Iqbal
Department of Food Science and Technology, Government College Women University, Faisalabad, Pakistan

Bushra Ishfaq
National Institute of Food Science and Technology, Faculty of Food, Nutrition, and Home Sciences, University of Agriculture, Faisalabad, Pakistan

Anisha S. Jain
Department of Microbiology, School of Life Sciences, JSS Academy of Higher Education and Research, Mysuru–570015, Karnataka, India

Muhsin Jamal
Department of Microbiology, Abdul Wali Khan University, Garden Campus, Mardan, Pakistan

Thotawaththage Sanjeewa Prasad Jayaweera
Department of Livestock Production, Faculty of Agricultural Sciences, Sabaragamuwa University of Sri Lanka, Belihuloya, Sri Lanka

Meenatai Kamble
Department of Food Science and Technology, National Institute of Food Technology Entrepreneurship and Management, Sonipat, Haryana, India

Adnan Khaliq
Department of Food Science and Technology, Khwaja Fareed University of Engineering and Information Technology, Rahim Yar Khan–64200, Pakistan

Saira Khan
ORIC, Lahore Garrison University, Lahore, Pakistan

Usman Mir Khan
National Institute of Food Science and Technology, Faculty of Food, Nutrition, and Home Sciences, University of Agriculture, Faisalabad, Pakistan

V. B. Chandana Kumari
Department of Biotechnology and Bioinformatics, School of Life Sciences, JSS Academy of Higher Education and Research, Mysuru–570015, Karnataka, India

Atif Liaqat
Department of Food Science and Technology, Khwaja Fareed University of Engineering and Information Technology, Rahim Yar Khan–64200, Pakistan

Deneththi Nimesh Madushanka
Department of Livestock Production, Faculty of Agricultural Sciences, Sabaragamuwa University of Sri Lanka, Belihuloya, Sri Lanka

Shanthala Mallikarjunaiah
Center for Applied Genetics, Department of Zoology, Bangalore University, Bangalore, Karnataka, India

Tariq Mehmood
Department of Food Science and Technology, Khwaja Fareed University of Engineering and Information Technology, Rahim Yar Khan–64200, Pakistan

Lakshmanan Nandhini
Department of Food Process Engineering, School of Bioengineering, SRM Institute of Science and Technology, Kattankulathur, Tamil Nadu–603203, India

Muhammad Asif Nawaz
Department of Biotechnology, Shaheed Benazir Bhutto University, Sheringal, Dir (Upper), Pakistan

Ankur Ojha
Faculty of Food Science and Technology, National Institute of Food Technology Entrepreneurship and Management, Sonipat, Haryana, India

Shashank M. Patil
Department of Biotechnology and Bioinformatics, School of Life Sciences, JSS Academy of Higher Education and Research, Mysuru–570015, Karnataka, India

Mahesh Pattabhiramaiah
Center for Applied Genetics, Department of Zoology, Bangalore University, Bangalore, Karnataka, India

Sushma Pradeep
Department of Biotechnology and Bioinformatics, School of Life Sciences, JSS Academy of Higher Education and Research, Mysuru–570015, Karnataka, India

Ashwini Prasad
Department of Microbiology, School of Life Sciences, JSS Academy of Higher Education and Research, Mysuru–570015, Karnataka, India

Kollur Shiva Prasad
Department of Sciences, Amrita School of Arts and Sciences, Amrita Vishwa Vidyapeetham, Mysuru Campus–570026, Karnataka, India

M. N. Nagendra Prasad
Department of Biotechnology, Sri Jayachamarajendra College of Engineering, JSS Science and Technology University, JSS Technical Institutional Campus, Mysore–570006, Karnataka, India

Radhakrishnan Preetha
Department of Food Process Engineering, School of Bioengineering, SRM Institute of Science and Technology, Kattankulathur, Tamil Nadu–603203, India

Muralidharan Priyadharshee
Department of Food Process Engineering, School of Bioengineering, SRM Institute of Science and Technology, Tamil Nadu, India

Syed Junaid Ur Rahman
Department of Food Science and Technology, Khwaja Fareed University of Engineering and Information Technology, Rahim Yar Khan–64200, Pakistan

Ramith Ramu
Department of Biotechnology and Bioinformatics, School of Life Sciences, JSS Academy of Higher Education and Research, Mysuru–570015, Karnataka, India

Sana Raza
Department of Microbiology, Abdul Wali Khan University, Garden Campus, Mardan, Pakistan

Redaina
Department of Microbiology, Abdul Wali Khan University, Garden Campus, Mardan, Pakistan

Raveendran Reshma
Department of Food Process Engineering, School of Bioengineering, SRM Institute of Science and Technology, Kattankulathur, Tamil Nadu–603203, India

Hettipala Arachchige Darshanee Ruwandeepika
Department of Livestock Production, Faculty of Agricultural Sciences, Sabaragamuwa University of Sri Lanka, Belihuloya, Sri Lanka

Bindu Sadanandan
Department of Biotechnology, M.S. Ramaiah Institute of Technology, Bangalore, Karnataka, India

Kanza Saeed
Department of Food Science and Technology, Khwaja Fareed University of Engineering and Information Technology, Rahim Yar Khan–64200, Pakistan

Amna Sahar
Department of Food Engineering, Faculty of Agricultural Engineering and Technology, University of Agriculture, Faisalabad, Pakistan

Nimra Sameed
Department of Food Science and Technology, Khwaja Fareed University of Engineering and Information Technology, Rahim Yar Khan–64200, Pakistan

Aysha Sameen
National Institute of Food Science and Technology, Faculty of Food, Nutrition, and Home Sciences, University of Agriculture, Faisalabad, Pakistan

Liloma Shah
Department of Microbiology, Abdul Wali Khan University, Garden Campus, Mardan, Pakistan

Prithvi S. Shirahatti
PG Department of Biotechnology, Teresian College, Siddhartha Nagara, Mysuru–570011, Karnataka, India

Chandan Shivamallu
Department of Biotechnology and Bioinformatics, School of Life Sciences, JSS Academy of Higher Education and Research, Mysuru–570015, Karnataka, India

Anurag Singh
Faculty of Food Science and Technology, National Institute of Food Technology Entrepreneurship and Management, Sonipat, Haryana, India

Soumya Singh
Department of Food Science and Technology, National Institute of Food Technology Entrepreneurship and Management, Sonipat, Haryana, India

Chandrashekar Srinivasa
Department of Biotechnology, Davangere University, Shivagangotri, Davanagere–577007, Karnataka, India

S. Sujay
Department of Biotechnology and Bioinformatics, School of Life Sciences, JSS Academy of Higher Education and Research, Mysuru–570015, Karnataka, India

Arifa Tahir
Environmental Science Department, Lahore College for Women University, Lahore, Pakistan

Farwa Tariq
National Institute of Food Science and Technology, Faculty of Food, Nutrition, and Home Sciences, University of Agriculture, Faisalabad, Pakistan

Tayyaba Tariq
National Institute of Food Science and Technology, Faculty of Food, Nutrition, and Home Sciences, University of Agriculture, Faisalabad, Pakistan

M. Tejaswini
Department of Biotechnology and Bioinformatics, School of Life Sciences, JSS Academy of Higher Education and Research, Mysuru–570015, Karnataka, India

Devarajan Thangadurai
Department of Botany, Karnatak University, Dharwad, Karnataka, India

Abbreviations

ABPP	active-based protein profiling
ACE	accessory cholera enterotoxin
AE	alveolar echinococcosis
AFB_1	aflatoxin B1
AFs	aflatoxins
AGE	acute gastroenteritis
AIDS	acquired immuno-deficiency virus
AQAP	Allied Quality Assurance Publication
BIOHAZ	biological hazards
BOD	biological oxygen demand
BRC	British retail consortium
BSE	bovine spongiform encephalopathy
CAC	Codex Alimentarius Commission
cAMP	cyclic AMP
CCP	critical control points
CDC	communicable disease center
CE	cystic echinococcosis
CNS	central nervous system
CPE	*Clostridium perfringens* enterotoxins
CPS	capsular polysaccharide
CT	cholera toxin
DAEC	diffusely adherent *E. coli*
DEC	diarrheagenic *Escherichia coli*
DGGE	qualitative/semi-quantitative analysis
dsRNA	double-stranded RNA
EAggEC	enteroaggregative *E. coli*
EC	European Commission
EEC	*Escherichia coli* enteritis
EHEC	enterohaemorrhagic *E. coli*
EIEC	enteroinvasive *E. coli*
ELISA	enzyme-linked immunosorbent assay
EPEC	enteropathogenic *E. coli*
ESPs	excretory-secretory products
ET	exfoliative toxins

ETEC	enterotoxigenic *E. coli*
EU	European Union
FAO	Food and Agricultural Organization
FBDs	foodborne diseases
FBP	foodborne pathogens
FDA	Food and Drug Administration
FMI	Food Marketing Institute
FSMS	Food Safety Management System
FSQ	Food Safety and Quality
FSSAI	Food Safety Standards Authority of India
GAP	good agriculture practice
GDocP	good documentation practice
GDP	good distribution practice
GFSI	global food safety initiative
GHP	good handling practices
GI	gastrointestinal
GLP	good lab practices
GM	genetically modified
GMP	good manufacturing practice
GTP	good trading practice
GU	genitourinary
GWP	good working practices
HACCP	hazard analysis and critical control point
HAdV	human adenovirus
HAV	hepatitis A virus
HEV	hepatitis E virus
HIV	human immunodeficiency virus
HPP	high-pressure processing
HRV	human rotavirus
HTST	high-temperature short time
HUS	hemolytic uremic syndrome
IA	intestinal amoebiasis
IEM	immune electron microscope
IFS	International Food Standard
INFOSAN	International Food Safety Authorities Network
ISO	International Organization for Standardization
LAB	lactic acid bacteria
LLO	listeriolysin O
LPS	lipopolysaccharide

MHC	major histocompatibility complex
MRL	maximum residue levels
MSG	monosodium glutamate
NASA	National Aeronautics and Space Administration
NFLs	nutrition facts labels
NGO	Non-Governmental Organization
NGS	next-generation sequencing
NLV	Norwalk virus
NMEC	neonatal meningitis *E. coli*
NMR	nuclear magnetic resonance
NoV	norovirus
NTS	non-typhoidal salmonellosis
OMV	outer membrane vesicle
ONT	Oxford Nanopore Technologies
ORFs	open-reading frames
PAMPs	pathogen-associated molecular patterns
PCB	polychlorinated biphenyls
PCR	polymerase chain reaction
PFOS	perfluoro-octane sulfonate
PONT	point need test
POPs	persistent organic pollutants
PRP	prerequisite program
PTM	protein translational modifications
QA	quality assurance
QMS	quality management systems
qPCR	quantitative PCR
RFEF	radio frequency electric field
RNA	ribonucleic acid
RNS	reactive nitrogen species
ROS	reactive oxygen species
RoV	rotavirus
RTE	ready to eat foods
RT-PCR	reverse transcriptase-polymerase chain reaction
SCV	salmonella-containing vacuoles
SEl	staphylococcal enterotoxin-like
SEs	staphylococcal enterotoxins
SIFs	salmonella-induced filaments
SNPs	single nucleotide polymorphisms
SOPs	standard operating procedures

SPI	salmonella pathogenicity islands
SPS	sanitary and phytosanitary
SQF	safe quality food
SSA	sub-Saharan Africa
SSCP	single-stranded conformation polymorphism
SSOPs	sanitation standard operating procedures
STEC	Shiga toxin-producing *E. coli*
Stx	Shiga toxins
SVFs	street-vended foods
T3SS	type-3 secretion system
TACCP	threat assessment critical control point
TBT	technical barriers to trade
TCP	toxin-coregulated pili
TCR	T-cell receptors
TDH	thermostable direct hemolysin
TLR	toll-like receptor
TNF	tumor necrosis factor
TNFα	tumor necrosis factor α
TQM	total quality management
TRFLP	terminal restriction fragment length polymorphism
TRH	TDH-related hemolysin
TSS	toxic shock syndrome
TSST	toxic shock syndrome toxin
TTGE	temporal temperature gradient gel electrophoresis
UHT	ultra-high temperature
UN	United Nations
UPEC	uropathogenic *E. coli*
USD	United States Dollar
USSR	Union of Soviet Socialist Republics
UV	ultraviolet
VACCP	vulnerability assessment critical control point
vCJD	variant Creutzfeldt-Jakob disease
WFP	World Food Program
WGS	whole-genome sequencing
WHO	World Health Organization
WTO	World Trade Organization
ZEN	zearalenone
ZOT	zona occludin toxin

Preface

The advancements in DNA amplification techniques have quietly revolutionized several domains of biological sciences, including food microbial and molecular biology. Information from DNA, RNA, and protein analyzes can be interpreted through bioinformatics tools, thereby developing our understanding of the involvement of biotic factors, physiology of fermentation starter, spoilage, and foodborne pathogenic microorganisms. New tools in molecular biology are immensely useful in investigating food microbial landscape with molecular perspective. Exploiting the versatility and power of molecular approaches, food microbial and molecular biology greatly improves the characterization of microbial communities, contributing to improving the diagnostic sensitivity and specificity for foodborne intoxications and leading to better prophylaxes and therapeutics against food-related infections. Improvement in food safety and quality (FSQ) assessment protocols has made the food biology discipline stronger for future developments containing the exploitation of single-cell sequencing technologies and gene networks. This book reviews all such developments, delivers an integrated analysis of the study, and assists in recognizing areas of imminent development.

This book provides cutting-edge microbial and molecular practices appropriate to food science. It offers a sophisticated and detailed exploration of the intensive science behind food microbial and molecular analyzes. The book balances the significance of functional applications of food microbial and molecular biology with intrinsic needs for the scientific evaluation of fundamental questions of genetics, survival, growth, control, and prevention of foodborne microbes. This book has three sections that mainly discuss the food safety assessment and management practices; emerging and reemerging food microbes; and the last section comprises principles of existing molecular technologies and deliberates their practicality in solving dilemmas that food science is facing at present.

The book starts with illustrating the worldwide scenario of global food safety and quality in Chapter 1. International regulations and consumer preferences towards food quality and the latest technologies involved in the assurance of food safety and quality are well explained in Chapters 2 and 3. Various food hazards, their causes, preventive measures, and risk management practices are elaborately discussed in Chapters 3 and 4. In Chapters 6 to 8, the authors review the food safety regulations, general characteristics of foodborne organisms, and their clinical manifestations. Chapters 9 to 11 focus on in-depth molecular and virulence studies of selective microbial pathogens and mycotoxins, along with classical examples to overcome and prevent their entry into food systems. The last two chapters of this book review the latest molecular approaches, including sequencing, various bioinformatics tools, and microbial databases to deal with food quality management issues.

This advanced book includes expert contributions from distinguished practitioners and international scientists engaged in food microbial and molecular research. The book provides a thorough knowledge base to the readers. This book is a crucial resource for microbiologists, molecular biologists, food scientists, and graduate students of food science, food biology, food microbiology, and food safety courses.

Finally, the editors immensely thank contributors and publisher for their interest and extraordinary commitment in the preparation and publication of this book.

—Editors

PART I

Food Quality, Safety, and Traceability: Assessment and Management

CHAPTER 1

Global Food Safety and Quality: A Worldwide Scenario

SHASHANK M. PATIL, RAMITH RAMU, PRITHVI S. SHIRAHATTI, M. N. NAGENDRA PRASAD, V. B. CHANDANA KUMARI, S. SUJAY, and M. TEJASWINI

ABSTRACT

Food safety and quality (FSQ) have been emphasized as one of the concerning issues of the 21st century, with factors like inflating population and unstable global economy fueling the deprecation of the same. Despite the unmatchable and extraneous efforts by international and national governing bodies, requisite standards of FSQ have not been achieved by many of the nations. On the other hand, inadequate food is causing nutritional aberrations, of which billions have been suffering since ages. Several factors have been listed out, including physical, chemical, and biological safety taken together with public and private hygiene that determine the safety and quality of the available food. In this regard, many of the diagnostic tools have been used at the cellular and molecular levels to determine the source and type of infection caused. However, despite achieving considerable progress in this arena, issues revoking safety and quality etiquette continue to occur globally to this date. With this under concern, the chapter aims to provide fundamentals of FSQ with respect to the global standards followed in the past and present. The chapter also deliberates on the role of governing bodies, management systems, public, and education that ensure the supply of good quality foods. Highlighting the worldwide scenario of FSQ, the chapter rolls out a prospect on global FSQ standards currently being followed and suggests projections that could likely modulate the concept for improving the lifestyle.

Food Microbial and Molecular Biology: From Fundamentals to Applications. Saher Islam, Devarajan Thangadurai, Jeyabalan Sangeetha, Zaira Zaman Chowdhury (Eds.)

1.1 INTRODUCTION

The population is likely to reach 9 billion by 2050 and is expected to rise to 11 billion by 2100 (FAO 2014). With the growing population come steep increasing requirements like water, shelter, and food supply. Statistics direct an increase in food production by 70% between 2005–2007 to 2050 to feed 9 billion people (FAO, 2009). Malnutrition rates have not diminished despite increasing the agricultural output during the past two decades. On the contrary, overnutrition is also becoming a major global issue though many of the low-to-middle income countries sustaining undernutrition (FAO, 2015; Alders et al., 2017; Mathur and Pillai, 2019). To provide more food, communities around the world must use resources in an appropriate manner to maximize production. Thus, food has been playing an important role in the health efficiency of the world (WHO, 2008).

Wholesome food is much needed because it constitutes a major deciding factor of the nutritional status and productivity of any population. Therefore, its safety is of primary concern (Ayala and Meier, 2017; Damayanti and Wahyati, 2019). Unsafe and adulterated foods are being supplied in the market results in several deaths and, at times, leading to ill health among the population (Uçar et al., 2016). According to the FAO, food quality is "a complex characteristic of food that determines its value or acceptability to consumers" (FAO, 2020a). Nutritional value and safety are the two important elements of food. While nutritional value is attributed to quality, organoleptic, and functional properties, safety is determined by the absence of constituents that might confront one's health (Di Renzo et al., 2015). Ensuring optimal food quality has been extremely challenging owing to a drastic increase in the demand over supply of food (Gorris, 2005; Kearny, 2010; Oldewage-Theron et al., 2016).

In modern times, the growth of food production reached new heights with the rise of the food industry acquired new technologies to produce more hygienic food (Hsieh and Ofori, 2007; Fryer and Versteeg, 2008). As globalization became a popular trend in industries, consumers were exposed to a wide range of products as a result of the extensive growth of international trade (Fukuda, 2015). Meanwhile, such trades can result in the development of new or unfamiliar safety risks related to food (Keiichiro et al., 2015), including microbial, chemical, personal, farm animal-related, and environmental hygiene. This chain of responsibility finally comes down to the government to run things with ease (Fung et al., 2018). In order to be on par with a lengthy food supply chain, it becomes important to improvise the food safety management systems (FSMS) on a regular basis to provide people with safe and hygienic food products (ISO, 2018; Panghal et al., 2018). Despite all the efforts, foodborne diseases continue to be rampant worldwide and are of great concern to prevent such diseases using early detection of the pathogen as well as ensuring food safety. A wide range of methods are being used in this process (Priyanka et al., 2016; Bintsis, 2017).

Apart from the scientific effort, it also becomes essential for the non-scientific community to be aware of food safety. Food producers, vendors, distributors, handlers should bear primary responsibility while consumers must remain vigilant and literate. Government agencies should enforce policies in order to optimize food safety measures. Healthcare workers also need to advise the importance of a safe and healthy diet for optimal healing from diseases. The integrated approach involving all the stakeholders will ultimately ensure food safety (Fung et al., 2018).

The chapter highlights the significance of food safety along with the historical milestones laid in food safety worldwide. Followed by comprehensive approaches worldwide to ensure food safety, current risks are also dealt with. It explains the challenges to food safety and the diagnosing advances. The role of government and the public along with the role of education in food safety is been depicted.

1.1.1 SIGNIFICANCE

Food safety refers to the assurance by certified authority that it will not harm the customers adversely after consumption. At the household level, it is important to ensure food safety and quality (FSQ) but become critical in large and commercial-scale production and processing, at setups that serve fresh food (McFarland et al., 2019). Unlike homes in the past, large scale production of food targets larger per capita incomes, better purchasing power and increased customer demand (Lem et al., 2014). This leads to the manufacture of an array of processed foods and functional foods that should strictly adhere to food safety guidelines. In the recent years, safety challenges faced globally have changed considerably, and issues associated with FSQ have gained significant importance (Kearney, 2010; Radovanovic, 2011; Uyttendaele et al., 2016).

After production comes the food supply. A safe food supply is of great importance in terms of the country's political and economic standpoint. Mead et al. (1999) reported more than 1 billion episodes of food poisoning-related diarrhea to occur each year, resulting in the death of about 3 million children every year. Foodborne illnesses related with microbial pathogens or any other source of contamination pose potential health threat in both underdeveloped and developing countries (Grace, 2015; Odeyemi, 2016). In a recent report, WHO stated 600 million foodborne illness and 420,000 deaths in 2010 (Havelaar et al., 2010). Most of the frequent contaminations occurred due to the presence of norovirus (NoV), *Campylobacter* spp., *Salmonella typhi*, *Taenia solium*, Hepatitis A and mycotoxins (Fung and Clark, 2004; Bintsis, 2017).

Children are said to be suffering disproportionately from the food safety issues and are victims of more than half of such illnesses, being the most susceptible and vulnerable class of the society (Lund and O'Brien, 2011). This is because of the immature immune system that is still in the developmental stage. Owing to this, the pathogenicity level required to make them sick also reduces concomitantly along with kids having limited control over their diet and lack of maturity to judge the food safety risks. Pregnant women and the elderly are also equally vulnerable to foodborne illnesses. It is advisable that pregnant women should be more careful about their diet because foodborne illnesses may also harm the growing fetus.

From an economic perspective, food supply should match up with the growing demand. According to Scharff (2012), it costs around 1,626 United States Dollar (USD) (enhanced model) and 1,068 USD (basic model) of a foodborne illness case. Together it makes 77.7 billion USD for the enhanced and 51.0 billion USD for the basic models. Owing to these factors, the availability of food with higher nutrition value that meet the safety standards has become a growing concern, which resulted in the greater expansion of scope and career opportunities in this sector. Before learning in detail about the challenges faced by the food industries, an insight on the dimensions of food quality and safety, the important essentialities of food quality and safety would be useful.

1.1.2 THE FOUR DIMENSIONS

FAO of the United Nations (UN) states that the safety standards of food are met when the entire population has access to safe and nutritious food sources in order to meet the dietary requirements (FAO, 2006). The four aspects that are concerned for food security also represented as the dimensions include the following (Gross et al., 2000):

1.1.2.1 AVAILABILITY

It is the first dimension of food safety and security. World Food Program (WFP) defines it as "the amount of food that is present in a country or area through all forms of domestic production, imports, food stocks and food aid." It refers to net commercial imports of food material of a nation that is available for people. As the target is to abolish malnutrition and hunger, most of the initiatives undertaken by researchers, practitioners, and teachers during the recent years prove that food safety is not simply a question of availability of food. It depicts that increasing food production will not result in securing food products from damaging. Although the last 50 years of agriculture production yielded more amounts of food commodities, it does not still meet the growing demands of food. Thus, it also becomes essential to secure the food along with the enhanced production (Napoli et al., 2011).

1.1.2.2 ACCESS

Though access as a parameter was first introduced by Amartya Sen in the early 1980's as an important aspect, in recent years, it is not regarded as an important factor. WFP states it as "a household's ability to acquire adequate amount of food for the family, which includes purchases, food assistance programs, gifts as well as borrowing." In fact, there are three elements present in the access of food. These include physical, financial, and socio-economical.

The physical aspect is almost a logical dimension. To meet the demand-supply chain the food must be delivered to all parts of the country. In this case, food security is available at the location where people need it. The financial aspect deals with the financial ability of people to meet their food requirements routinely. When the food commodities are available but cannot be acquired by households due to financial problems is referred to as an in security. During the least years, however, this has been increasingly recognized. Therefore, a paradigm shift has occurred towards food security among the analysts. This is common in rural population, living in almost complete autarchy with diminished amount of available food for consumption. Finally, the third element known as the socio-economic dimension deals with the socio-cultural barriers limiting the access to food, to some ethnic groups in society. This may be developed due to discrimination based on caste, race, or gender. In this case, though the food commodities available physically near the consumers, these barriers restrict them from buying the food. For example, people with human immunodeficiency virus (HIV)/acquired immuno-deficiency (HIV/AIDS) are erroneously kept at isolation from receiving food in public (FAO, 2012).

1.1.2.3 UTILIZATION

This is the third dimension of food safety. WFS states that it refers to "safe and nutritious food that provides adequate nutrition to fulfill the body's functional requirements." It depicts the unavailability of safe and nutritious diet to people even if they can access it. This means even if financially well-settled people are present in a region, they cannot afford food commodities because of the low quality of the food substances. It is also estimated that even people from traditional background can disturb the chains of food commodities. This disturbance can also be expected from refugees and victims of shock. Training may be required to help people optimize their use of the food that is available and to which they have access. Food utilization is also associated with usage of clean water, sanitization, and health care. Thus, utilization also deals with conservation, processing, and preparation of the food commodities and shows how nutrition is closely linked with food safety (FAO, 2020a).

1.1.2.4 STABILITY

The final dimension of food safety is stability. The WFS depicts that it is perennial availability. It applies in the first instance to the above mentioned three dimensions. It does not occur at any moment; instead, it stays permanently with sustainability. Absence of this dimension results in two types of food insecurities, known as chronic food insecurity and transitory food insecurity. Chronic food insecurity is a perennial inability to acquire adequate food requirements, whereas transitory food insecurity refers to be the temporary or short-term food deficit. There is also the possibility of occurrence of cyclical food insecurity like seasonality.

World Food Summit in 1996 deduced that the elements excluded from four dimensions of food security are to be collectively defined as "food requirement to lead a healthy lifestyle." The sovereignty of food at country level can be affected by the factors associated with food preferences, as well as the research at household and individual level (FAO, 2006, 2008).

1.2 HISTORICAL PERSPECTIVES

Every nation in the world has its own record of maintaining food safety measurements. The chapter does not aim at providing an elaborate chain of events that happened throughout the history of the mankind but rather sum up the perspectives in order to mark the milestones that affected hunger, malnutrition, and food insecurity at the global level. It depicts the world food situation in four periods by putting things in chronological order and highlights a few links, which may improvise the present and future approaches to food security. Though it revolves around only circle of league of nations highlighting food security concerns, it does not intend to leave behind the hunger and malnutrition faced by mankind since the ages.

1.2.1 WORLD WAR-I AND LEAGUE OF NATIONS (1930–1945)

It is reported that the League of Nations dealt the things worldwide during 1930's and World War-I. In the early 1930's Yugoslavia proposed that health agencies across the globe should present details of the food positions to assess the world food problems on a global scale. This is believed to be the first national report that introduced the world food problem into the international arena of politics. In 1935, a survey conducted by the health division of the League of Nations reported that acute shortage of food in developing countries first described malnutrition. This resulted in the enforcement of policies to meet all the four dimensions of food safety and security across the world. The United States was the first country to develop and implement policies regarding all four dimensions of food safety in 1896. Though it supplied more than 6 million tons of food commodities to Europe between 1919 and 1926, it resulted in excess stock accumulation, which in turn reduced market prices. To resolve this issue, the agricultural adjustment act was passed in 1933 in order to regulate the production in respective area. Within the framework, a green stabilization and commodity credit corporation were also established to provide financial aid to food handlers. It was then in 1943, where FAO was established, as decided by all the nations of the world (Shaw, 2007).

1.2.2 POST WORLD WAR-II, UN, FAO, AND SURPLUSES (1945–1970)

Canada implemented agricultural price support legislation in 1944. In 1948, the right to food was added in the international covenant on economic, social, and cultural rights, adopted by the UN General Assembly in 1966 and came into force in 1976. After the Second World War, Britain rejuvenated food production, supporting the industrial sector. In 1967–1968 India started to import food commodities due to harvest failure. Indian policymakers became greatly concerned with the US refusing to sell food commodities following India's opposition to involve in Vietnam War. Most of the European countries adopted food safety standards after the Second World War concerning large farmer community of their populations. The first World Food Survey conducted by FAO in 1946 reported that 1/3rd of the population suffers from malnutrition and lack of energy. In the meantime, FAO also studied about overproduction, a major concern of many countries.

In 1949, US established laws to enforce complete utilization of surplus food commodities. In 1952, FAO established a committee on commodity problems that performed an elaborate study on disposal of surplus commodities disposal and recommended a set of principles to deal with it. In October 1960, the UN general assembly passed 27th resolution to provide surplus food commodities to food-deficit regions. In 1961, WHO and FAO together created a food management system known as Codex Alimentarius Commission (CAC) to effectively deal with international standards about processing, labeling, sampling of analysis, hygienic requirements of food commodities. From 1962 to 1967 US passed several domestic and aided the international laws to benefit the food production. In 1999, the food aid convention became one and only legal management system to contribute to providing food without aiming for food security (Shaw, 2007).

1.2.3 FOOD CRISIS AND OTHER EMERGENCIES (1970–1990)

The world food production during the 1950's and 1960's raised by more than 50% and production per capita by more than 20%, resulting in the production of surplus 25 million tons of food commodities every year. Though surplus produces were witnessed during the late 1960s in the cereal markets, it continued efforts from US and Canada brought down the cereal production. But in 1972, cereal production came to an unexpected diminished level, producing 55 million tons less than expected. The same problem hit the Union of Soviet Socialist Republics (USSR), which made it to become a food importer. The Middle East raised petroleum prices at the same time, fueling the food crisis. Though many developing countries augmented their production by 2% per year, they were still dependent on the imports. Considering the crisis, the UN organized a conference in 1974 to focus on resolving food insecurity. The world first witnessed food security as one of the common issues among all the nations. The 1983–1985 and 1992 food crisis due to unfavorable climatic conditions significantly reduced food production in Africa. FAO conference in 1983 prepared a resolution to warrant basic food access to the entire population. World food surveys conducted by FAO in 1977 and 1987 reported a remarkable increment in the production of agricultural products, which disproved Malthusian theories linking increased population growth and food production.

1.3 BRIGHTER YEARS IN FOOD SECURITY (1990–2005)

Followed by the food crisis that hit Africa, 15 major conferences were held to deal with, and approved plans to revive food security. This period witnessed the growth of multidisciplinary dimension as an important approach to food security. International conference on nutrition jointly organized by FAO and WHO in 1992 adopted to eliminate hunger and to reduce all forms of malnutrition. This conference focused on dealing with all dimensions of food security with a positive approach. Significant progress was reported in 1996, 22 years after the 1974 world food conference, at the occasion of the 1996 world food summit organized by FAO in Rome. In the same year, Europe witnessed a new development where new food aid regulation came to act, which financially helped to carry out major scientific projects related to food security. Again in 2002 and 2006, the world food summit and committee on food security had their conferences, respectively, in which it was reported that 150 developing and transitions countries were able to produce national food security strategies. But in 2005, Niger faced severe food shortage, where it was reported that famine was the primary cause. Niger was also hit by inappropriate economic decisions taken by a neighboring country, Nigeria resulting in unexpected increase of prices of basic food commodities. Since then, many acts and laws are passed to ensure proper food security. Many of the scientific reports prepared, stating different conditions, initiated responses from global food security management bodies (Shaw, 2007; FAO, 2012).

1.4 GLOBAL SCENARIO OF FOOD QUALITY AND SAFETY

The global scenario of food quality and safety is greatly affected by foodborne illness all over the world. According to WHO, contaminated food contributes to 1.5 billion cases of diarrhea in children each year, with more than 3 million premature deaths (Mead et al., 1999). These deaths occur equally in both developed and developing nations. In addition to contamination, adulteration with toxic elements can also result in serious health hazards (Rather et al., 2017) and affect the human, economic as well as political health of a nation (DeWall et al., 2005). These effects are described as follows:

1. **Human Costs**: As many countries have not yet adopted adequate surveillance or reporting systems to trace and identify the causes of foodborne illness, it is essential to establish advanced laboratories and other safety measures for quality assessment of food substances. The foodborne illness may range from mild to life threatening. Thus, depending on the specificity of the organism, climatic conditions and hygiene, different age groups can be affected. Though all age groups are susceptible, infants, young children, pregnant women, and elderly are vulnerable more often, resulting in severe consequences (Lund and O'Brien, 2011). In developing countries, poverty, hunger, and malnutrition are the chief sources of foodborne illnesses. Malnourishment among children, including newborn babies, is a major cause of an array of ailments that have some devastating effects (Pal et al., 2015).
2. **Economic Costs**: Issues concerning food security leads to huge economic consequences. Consumer cost includes legal, medical, and other expenses, with absenteeism at school and work. Poverty can hit consumers living at subsistence level. Increased medical expenses, outbreak investigations, handling of medical personnel, food recall and loss of customer costs to national governments. With the globalization of food trade, governments also suffer economic consequences during unsafe food deals done. When the governments face loss, the burden is set on the people in the form of tax. Thus, sometimes entire country may face the consequences (Scharff, 2012; Hussain and Dawson, 2013).
3. **Political Consequences**: Issues related to food quality and safety can have huge political implications. Outbreaks must be dealt carefully with minimal life costs. The failure of health ministries in controlling pandemics results in people losing faith in governments (Lake and Barker, 2018). Governments should provide essentialities to people during outbreaks. The global scenario also deals with different food safety concerns in respective regions of the world. This is followed by the policies and plans of action implemented by the governments. Tables 1.1–1.7 provide a detailed view on global aspects of food safety and security (De Wall et al., 2005)

TABLE 1.1 Food Safety Along with Policies and Plans of Action Followed in Western Pacific Region

Region	Countries in the Region	Food Safety Concerns	Policies and Plans of Action
• Western Pacific Region • Food safety issues are rampant in this region. Though some countries in this region have adopted food safety programs, others do not possess written policies. Many of them lack essential resources to ensure food safety. • The economic development in Australia and New Zealand depend on agricultural exports, so their programs tend to be better developed.	• American Samoa, Australia, Brunei Darussalam, Cambodia, China, Cook Islands, Fiji, French Polynesia, Guam, Hong Kong, Japan, Kiribati, North Korea, South Korea, Republic of Lao PDR, Macao, Malaysia, Mariana Islands, Marshall Islands, Micronesia, Mongolia, Nauru, New Caledonia, New Zealand, Niue, Palau, Papua New Guinea, Philippines, Samoa, Singapore, Solomon Islands, Tokelau, Tonga, Tuvalu, Vanuatu, Viet Nam, and Wallis and Futuna.	• Intensive farming practices, using slaughter by-products and animal waste as feed, over-administration of pesticides in farming, antibiotics, and growth hormones in animal husbandry, increased slaughtering for meat usage. • Unhygienic aquaculture and rearing including fishing from contaminated ponds and use of forbidden chemical agents. • Widespread distribution of contaminated foods in larger processing operations with lack of proper packaging. • Failure of some Pacific Island communities to address marine toxin issues associated with the harvesting of fish species, lack of proper water disposal and other sanitization measures. • Limited access to education and decision-making power at the household level for women, who are often the food handler. • Migration among the people of different countries for international trading. • Lack of adequate food labeling regulations regarding both locally produced and imported food products.	• Many countries have no specific written policies on food safety. Strategies and plans of action are frequently developed from a general health perspective. • Food safety policies are more often not structured with a goal within the context of government policy. Additionally, different government agencies often claim the same jurisdiction over matters of food safety. • Insufficient and inadequately trained food inspectors to implement the national programs effectively, which affects the system monitoring. • Many countries possess advanced programs for export certification that elevate the standards required for them to be exported than if they were sold domestically. • Laboratory analysis capacity varies across the countries. Some of them possess advanced technology and some cannot isolate or identify common foodborne microbes and chemical hazards. • Only a few countries support integrated food safety programs with costumers and industries in order to promote food safety.

TABLE 1.2 Food Safety Along with Policies and Plans of Action Followed in South East Asian Region

Region	Countries in the Region	Food Safety Concerns	Policies and Plans of Action
• Southeast Asian Region • The major nutritional advancements in the last decade may improvised the situation, but it is not adequate to face the new challenges related to food safety. • Access to safe food remains an obstacle to a large percentage of people of the southeast Asian region. Large-scale production in urban areas intensify food safety issues. A regional strategy has been developed to address food safety issues in the region.	• Bangladesh, Bhutan, India, Indonesia, Myanmar, Nepal, North Korea, the Republic of Maldives, Sri Lanka, Thailand, and Timor-Leste.	• Major concerns include inadequate access to clean water, extensive use of pesticides and chemicals in farming and food processing. • Lack of education in both producer and consumer education. • Urban population growth results in conditions like extreme poverty, filth, overcrowding, and poor sanitation. This is believed to aggravate food safety issues. • Street food vendors and foodservice premises play a significant role in the maintenance of food safety. Due to inadequate number of health inspectors, governments fail to enforce health and food safety regulations. • It is rare to find laboratories that can detect common foodborne pathogens and chemical hazards in South East Asia Region. Even if they exist, they are economically not affordable. • Several loopholes are spotted in some countries in terms of policy enforcement and monitoring food security primarily because of the lack of both infrastructure and staff.	• All the countries appointed themselves with a 10-point strategy to reduce the burden of foodborne diseases in 1998. But progress is inconsistent. Though countries have a lead agency to deal with food safety issues within the Ministry of Health, not all of them have developed food safety policies. • Programs have been implemented by the countries to resolve illness and mortality related to food safety. With few exceptions, these programs have failed to address the target and lack information. • Non-official and non-mandatory programs have also been developed by some countries to improve food safety. For example, in Thailand, training programs are used for retailers, restaurants, and other food services to improve food safety and hygiene.

TABLE 1.3 Food Safety Along with Policies and Plans of Action Followed in Eastern Mediterranean Region

Region	Countries in the Region	Food Safety Concerns	Policies and Plans of Action
• Eastern Mediterranean Region • Food safety is regarded as an alarming concern of public health issue in this region. Most of the countries have undertaken extensive reviews of their food safety systems. Some of the countries have also implemented legislative measures. Despite these efforts, full evaluation of food safety systems is not achieved due to lack of information.	• Afghanistan, Bahrain, Cyprus, Djibouti, Egypt, Islamic Republic of Iran, Iraq, Israel, Jordan, Kuwait, Lebanon, Libyan Arab Jamahiriya, Morocco, Oman, Pakistan, Palestine, Qatar, Saudi Arabia, Somalia, Sudan, Syrian Arab Republic, Tunisia, United Arab Emirates, and Yemen.	• Scarcity of suitable land for farming. Food imports are vital to meet up the needs. Thus, to ensure food safety, they depend on systems to control their safety and quality are vital for public health. Food exports generate foreign exchange. Thus, it is essential to have effective food safety systems, also critical to expanding market shares in exports. • Food safety reforms are facing extreme challenges to check on animal diseases. Legislation is outdated, inflexible or fragmented. • Food hygiene, sanitation, and foodborne disease surveillance, are dealt with by the health authorities at central and local/municipal levels. Food quality, safety, and animal origin food issues are often operated by the ministries of agriculture.	• Morocco and Tunisia have developed a national strategy for food control and several countries have drafted new food legislation. Iran, Sudan, and the Syrian Arab Republic have reviewed and updated their food standards and regulations. Yemen has finalized its food safety country profile. • Egypt, Jordan, Morocco, and Tunisia have harmonized their food-safety standards with the Codex Alimentarius and are moving towards an approach based on risk management. • Jordan has established a Food and Drug Administration where all stakeholders in food safety coordinate their efforts. • The United Arab Emirates has adopted the use of customized software for food inspection to monitor and control the safety of food.

TABLE 1.4 Food Safety Along with Policies and Plans of Action Followed in African Region

Region	Countries in the Region	Food Safety Concerns	Policies and Plans of Action
• African Region • Diarrheal diseases among newborns and young children suggest unhygienic food handling in African region. • Frequent outbreaks of poisoning initiated limited implementation of surveillance systems. • The major driving force for these is poverty that compels the population to consume unhygienic food and water along with poor access to adequate healthcare. There is also a rise in communicable diseases, trade pressure, and poor environmental conditions exacerbate the situation.	• Angola, Benin, Botswana, Burkina Faso, Burundi, Cameroon, Cape Verde, Central African Republic, Chad, Comoros, Democratic Republic of Congo, Republic of the Congo, Côte d'Ivoire, Equatorial Guinea, Eritrea, Ethiopia, Gabon, Gambia, Ghana, Guinea, Guinea Bissau, Kenya, Lesotho, Liberia, Madagascar, Malawi, Mali, Mauritania, Mauritius, Mozambique, Namibia, Niger, Nigeria, Rwanda, Sao Tome and Principe, Senegal, Seychelles, Sierra Leone, South Africa, Swaziland, United Republic of Tanzania, Togo, Uganda, Zambia, and Zimbabwe.	• Foodborne diseases affect women and children more often. Unsanitary conditions, lack of clean water, unhygienic transportation and storage of processed food, lack of knowledge of hygiene among food handlers and less stringent policies that are easily deviated because of corrupt officials. • Mycotoxins contaminate various agricultural commodities. Climate poor harvesting practices, improper storage and transport and marketing contribute to fungal growth and increase the risk of mycotoxin production. • Natural disasters such as floods, droughts, and earthquakes, or intense civil war or border conflicts result in contamination of food supplies.	• Failure to incorporate food laws, enforcement of outdated and inadequate policies into the legislation. • Improper coordination between surveillance, food laboratories, and food inspection. Inadequate systematic monitoring and recordkeeping can affect food safety. • Food safety education for industry and consumers is limited. Shortage of trained personnel in laboratories also affects food safety and quality. • Financial support for food safety programs is often meager. Lack of resources at the local and national levels has triggered collaboration with international organizations and NGO's.

TABLE 1.5 Food Safety Along with Policies and Plans of Action Followed in European Region

Region	Countries in the Region	Food Safety Concerns	Policies and Plans of Action
• European Region • Food safety policies has been reshaped over the past 10 years as a result of a few crises such as bovine spongiform encephalopathy (BSE) crisis in Britain, followed by dioxin contamination of animal feed, resulting in the contamination of numerous meat products in northern Europe. • The European Union (EU), comprising 25 nations gives an integrated approach to food safety issues. A set of food safety regulations have been designed to harmonize existing national requirements.	• Albania, Andorra, Armenia, Austria, Azerbaijan, Belarus, Belgium, Bosnia and Herzegovina, Bulgaria, Croatia, Cyprus, Czech Republic, Denmark, Estonia, Finland, France, Georgia, Germany Greece, Hungary, Iceland, Ireland, Italy, Kazakhstan, Kyrgyzstan, Latvia, Lithuania, Luxembourg, Malta, Monaco, Netherlands, Norway, Poland, Portugal, Republic of Moldova, Romania, Russian Federation, San Marino, Serbia, and Montenegro, Slovakia, Slovenia, Spain, Sweden, Switzerland, Tajikistan, The former Yugoslav Republic of Macedonia, Turkey, Turkmenistan, Ukraine, United Kingdom of Great Britain and Northern Ireland, and Uzbekistan.	• Outbreaks of communicable diseases like bovine spongiform encephalopathy (BSE) and Creutzfeldt-Jakob Disease (vCJD) in Britain proved that diseases could spread even in hygienic conditions, without proper safety measurements. • Genetically engineered foods have been highly controversial in the European Region. The use of antibiotics in food production has grown, and microbes found animal origin foods are getting increasingly resistant to antibiotic drugs. • Food safety management bodies have been regularly monitoring the complete aspects of irradiated foods and use of growth hormones in the foods. This is also applicable to screening of contaminants and radionuclides which can bio-accumulate in the food chain.	• European Union regulations takes necessary actions through the risk assessment and management bodies which work based on scientific information. • EU also directs food handlers at all stages from production to consumption to report about safety issues. In the case of genetically modified organisms, tracing, and labeling are also dealt to ensure food safety and quality. • Certification and clearance are essential in case of animal origin foods. Product recalling by producers is observed. Food control laboratories have also been set up. • Monitoring food products for contaminant level is done in every market. Proper care is taken about hygiene from production to consumption stage with reference to all kinds of challenges to food safety.

TABLE 1.6 Food Safety Along with Policies and Plans of Action Followed in Central and South American Region

Region	Countries in the Region	Food Safety Concerns	Policies and Plans of Action
• Central and South American Region • The Central and South American region comprises of the greatest biodiversity on the planet. The diversified food quality is a result of this difference in climatic conditions, which in response raises food safety concerns. There is an increase in the prevalence of pathogens owing to increased travel for tourism and trading by the population.	• Argentina, Bahamas, Turks and Caicos Islands, Belize, Bolivia, Brazil, Chile, Colombia, Costa Rica, CPC Barbados, Cuba, Dominican Republic, Ecuador, El Salvador, Guatemala, Guyana, Haiti, Honduras, Jamaica, Nicaragua, Panama, Paraguay, Peru, Puerto Rico, Suriname, Trinidad and Tobago, Uruguay, and Venezuela.	• Large number of foodborne illnesses in this region arises due to variations in geographic and cultural factors. Outbreaks are attributed to handling practices and population issues such as growth of vulnerable population groups, lack of personal hygiene, long-term storage of food, inappropriate storage facilities, cross-contamination, and improper handling. Mycotoxins are also believed to be the causative agents of foodborne illness. • Many of the countries in this region adopted pre-inspection programs like Integrated Pest Management (IPM) and post-harvesting management programs. Those programs help ensure that food that reaches consumers does not carry excessive pesticide residues. • Due to lack of education among producers and consumers, unnatural ways of trading like barter system came into expansion. This resulted in extensive development of foodborne pathogenesis due to poor handling of food commodities.	• Central and South American Region has been identified internationally for their increasing foodborne diseases that has led to many innovative programs for promoting collaborative activities to protect the population from such diseases. • Epidemiological surveillance would be improved if physicians and other public health officials were trained and then linked to laboratories to ensure that illnesses were being fully reported. • The Central and South American Region has developed an integrated system for educating consumers about food safety. In 2002, the Pan American Sanitary Bureau (PASB) developed a consumer Internet site to provide food safety information in English, Portuguese, and Spanish.

TABLE 1.7 Food Safety Along with Policies and Plans of Action Followed in North American Region

Region	Countries in the Region	Food Safety Concerns	Policies and Plans of Action
• North American Region • The North American Region is renowned for its advanced food safety programs, yet this region is not spared of food security issues. Factors such as advancing technologies, sophisticated distribution system, increased demand that corresponded to an increase in the production and supply of food substances including agriculture and livestock farming. These have contributed tremendously to increased susceptibility of foodborne pathogens and other additional challenges. The three countries within North America should collaborate in an integrated surveillance program to counter this issue.	• Canada, Mexico, and the United States of America	• Despite having many programs and resources devoted to fighting foodborne disease in this region, the incidence of foodborne illnesses in North America is still quite high. • Outbreaks are expected in case of imports. • Usage of antibiotics on farm animals results in antibiotic resistance. This should be effectively monitored. Monitoring on genetically engineered foods is also essential. • Contamination in case of animal and food products. Special attention is given to the reports of bioterrorism, where pandemics are expected. Irradiated foods are preferred to reduce foodborne illness. • Consumer education and establishment of awareness centers is also done by the governments.	• Both US and Canada have established more than 55 agencies to effectively monitor food safety and security issues in the country. • Mexico has only recently developed an integrated food safety program, and legislation is currently being revised to improve food safety in the country.

1.5 ISSUES OF SAFETY AND QUALITY OF FOOD

A study has shown that food safety challenges may arise due to variations in the factors like region, differences in income level, local conditions, workers, and their knowledge in handling food processing (Tessema et al., 2014). The food has a long way to reach the consumer from the site of production to its consumption. Most of the markets trade fresh foods as well as a few packaged or processed foods. Home-cooked foods are also supplied by the street vendors. Yet, such prepared foods have low shelf-life and needs to be consumed immediately (Almanza et al., 2014). Therefore, food can be contaminated in any of the stages of its production. There are different reasons that affect the food safety in a negative way. These factors are listed below:

1.5.1 MICROBIOLOGICAL SAFETY

The biological nature of food supports the growth of microbes. Some of these microbes may act as pathogens hence are considered as potential sources of foodborne diseases. Although viruses are responsible for the majority of foodborne illnesses, bacterial agents cause the greatest number of hospitalizations and deaths. The toxins produced from pathogens cause illnesses ranging from mild to severe gastroenteritis, renal, hepatic, and neuronal syndromes. Bacterial agents can cause severe and fatal foodborne illnesses among foodborne pathogens (FBP). A majority of foodborne illnesses were due to the contamination of foods with species of *Salmonella*, *Staphylococcus*, *Campylobacter*, *Clostridium*, *Bacillus*, *Vibrio*, *Listeria*, and *E. coli* (Vaillant et al., 2005; Bintsis, 2017).

1.5.2 CHEMICAL SAFETY

Non-food grade chemical additives like colorants and preservatives may act as contaminants. It also includes pesticide residues from primary crops. Food samples may also contain metals such as lead, arsenic, cadmium, mercury, and copper at greater concentrations than normal samples, suggesting possible leaching from utensils and inadequate food hygiene (Rather et al., 2017).

1.5.3 PERSONAL HYGIENE

Personal hygiene is an important factor in the food production and packaging industry, failing which can lead to several health concerns. From production to consumption stage, food commodities undergo exchange between numerous handlers. This leads to an increased threat of contamination by microbes. Simple activities like thorough hand washing and sanitizing production area can prevent foodborne illnesses (Ismail et al., 2016).

1.5.4 LIFESTYLE CHANGES

The changes in lifestyle of people have led to practice abnormal eating habits and eating outside homes. Foods prepared outside are more likely to get contaminated as they are handled by many persons. Along with this, food items prepared many hours in advance may get contaminated, if not stored properly (Dudeja and Singh, 2017).

1.5.5 PUBLIC HYGIENE

Public hygiene plays a significant role in reducing further effects of outbreaks. Maintaining safe practices like sanitizing living places can reduce contamination of food commodities. Poor living conditions, production of food in unhygienic places, unhygienic conditions of workers are responsible for food contamination (Kibret and Abera, 2012).

1.5.6 ENVIRONMENTAL HYGIENE

Inadequate recycling and disposal of waste materials results in a long-term accumulation of food causing microbial growth and food spoilage. Poor sanitary conditions in food storage units, transport, and selling places also leads to food contamination (Rather et al., 2017; Thompson and Darvish, 2019).

1.5.7 ANIMAL HUSBANDRY

A great deal of progression has been witnessed in the field of animal husbandry. But it has also resulted in negative effect on food safety. Contamination may occur from animal carcasses, fecal matter, and leftover foods. Pathogens like *Salmonella* and *Campylobacter* have evolved into more potent and resistant microbes that can affect the animal health, which in turn affects food quality (Attrey, 2017; Ibarra et al., 2018). Crowded animals tend to increase the usage of antibiotics to prevent infections. Known as "factory farms" usage of such antibiotics has led to the development of resistance in bacteria for these antibiotics (Martin et al., 2015).

1.5.8 INTERNATIONAL TRADE AND TRAVEL

International trade may promote the positive effect of all four dimensions of food safety and security but also increase the risk of migration of foodborne pathogens between the two trading countries. In addition, the lapse between the food packaging to that of its consumption creates additional opportunities for contamination. International trading is also associated with the introduction of certain novel pathogenic types that are not previously exposed to the population of that country and therefore not immune. Meanwhile, people exposed to foodborne illness in one country are expected to infect the healthy population of another country (Keiichiro et al., 2015).

1.5.9 BIOTERRORISM

Following rising terror incidents worldwide, concerns about international food adulteration by terror groups, criminals, or any other antisocial associations have elevated and exposed the lacunae in the presently available preparations by the country to face such a situation. According to WHO, it is essential that to maintain strict food safety laws and national policies to support the infrastructure for the betterment of food legislation, monitoring, surveillance, inspection, education, and training are adequate and up to date (Rasco and Bledsoe, 2009; Solodoukhina, 2011). The challenges to food safety and the chances of occurrence of contamination are depicted in Figure 1.1.

1.6 DIAGNOSTIC ADVANCEMENTS ENSURING FOOD QUALITY AND SAFETY

Since food becomes an easy pathway for microorganisms to enter the host body through the gastrointestinal (GI) tract, it becomes a pavement for

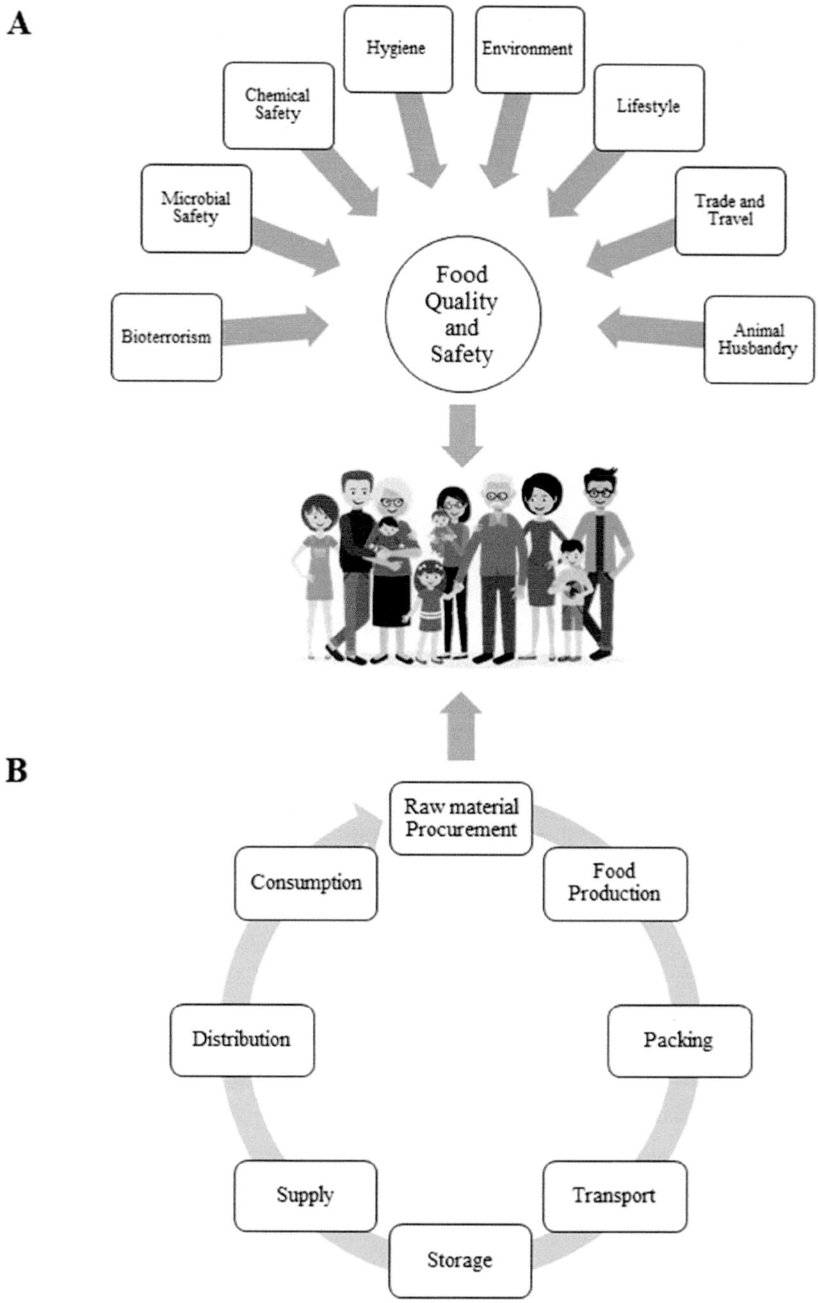

FIGURE 1.1 (A) Challenges to food safety; and (B) chances of occurrence of food contamination.

pathogens to reside inside the host (Godfrey et al., 2010). Timely and precise detection of the presence of pathogenic contaminations in food sources is required to ensure public safety. Bio-surveillance of pathogenic bacteria results in complete monitoring of microbials from its source to target species. The methods of detecting these microbials have improved over time (Mangal et al., 2016; Priyanka et al., 2016). Tedious cell culture-based tests are being replaced by faster, accurate, and more sensitive culture-independent assays. These normally include molecular assays like polymerase chain reaction (PCR) panels and antigen-antibody assays (Huang et al., 2016). However, these tests are performed only in public health and not available for practitioners and people in non-clinical and non-research fields.

Non-culture-based molecular techniques have an edge over conventional assays due to their relatively quick results. Animal food sources are primarily contaminated with *Salmonella* and *Vibrio* spp. that are known to cause several foodborne diseases. Generally, they can be detected using simple assays and instruments leading to an improvement in the sample efficiency by the use of simple techniques. This in turn helps in quick processing of a large amount of sample with increased accuracy (FDA, 2015). Normally these samples include soil, animal tissues, fruits, feces, water, and blood. As a result of faster methodologies followed, time, and overhead expenses can be saved. However, point need test (PONT) devices for the diagnosis of field samples from agricultural farming, animal husbandry as well as aquaculture are not available for all the pathogens and therefore its efficacy is questionable. Currently, nuclear magnetic resonance (NMR), nanotechnology, and PCR-based techniques are the most used approaches to detect the microbial origin (Yang et al., 2017, 2018).

1.7 FOOD STANDARDS

Different methodologies described above have a soul reason to be performed to maintain food standards. Food standards and control systems are essential to integrate quality in every stage of food production. These standards aid in not only supplying of wholesome and hygienic food to the public but to facilitate trade within and between nations. There are four levels of standards which are well coordinated (FAO, 2010):

1. **Company Standards**: Prepared by copying of the National Standards, these are the guidelines for the food production units.
2. **National Standards**: These are issued by government bodies.

3. **Regional Standards**: Regional groups with similar geographical, climate, etc., have legislation standardization bodies.
4. **International Standards**: The International Organization for Standardization (ISO) and CAC publish international standards.

1.8 ROLE OF GOVERNING BODIES IN ENSURING FOOD QUALITY AND SAFETY

Food quality and safety depend entirely upon thorough research along with stringent policies for its maintenance. Supply of food products should be carefully regulated but enforcing specific and updated laws to keep a check on the production companies for the wellness of the people. In most of the countries, an agency like food and drug administration (FDA) has been established which is responsible for assenting laws to keep a check on the public health and safety:

- To edify consumers of nutritive and other components of food products;
- To assent existing laws and regulations to ensure supply of safe food products; and
- To analyze and remove potential toxic elements and prosecute economic fraud via monitoring the chain of food supply.

Regulations and policies should be in place for food production, packaging, labeling, transportation, storage, distribution, and sales. The governing bodies are provided with resources and authority to form the etiquette, assemble experts to achieve the three-prong aim. Along with the law enforcement, risk management should be dealt to get the accurate estimate of the upcoming risk using current information. But to assure safety, these agencies generally go beyond scientific risk. Using toxicology and available human reports, they bring out maximum effort to reduce the risk caused. Some of these agencies work independently, some under government (Fung et al., 2017). A brief introduction about these working bodies is given below.

1.8.1 INTERNATIONAL GOVERNING BODIES

1. **Codex Alimentarius Commission (CAC)**: It is an international governing body established with the objective of assenting international standards to ensure the health of the consumers and facilitate food and agricultural trade. In 2017, it has become the single most

international reference point including 187 member countries with one Member Organization (European Community). It is a comprehension of guidelines, codes for practice and recommendations collectively called 'The Food Code.' Many countries use these guidelines to protect consumer laws and evaluate the standards of food trade (FAO, 2005).

2. **International Organization for Standardization (ISO)**: It is an NGO of national standards bodies (ISO member bodies) working across the world. It aids the standardization of etiquette and related activities across the world in order to promote the exchange of international goods and services. It also helps to enhance cooperation in intellectual, scientific, technological, and economic activity relating to food safety. The work done by ISO are tabulated as International Standards (ISO, 2012).
3. **World Trade Organization (WTO)**: Established in the year 1995, this organization assists in a smooth, free, and regulated monitoring of trading, by monitoring trade agreements, resolving trade disputes, assisting countries in policies of trading issues. The agreement spans various aspects such as goods, services, and intellectual property. A strong food control system is essential to enforce the adoption and implementation of standards. WTO also focuses on both food inspection and analytical capability of food management systems (WTO, 2004).
4. **Food and Agriculture Organization (FAO)**: It is one of the oldest specialized agencies of the UN that was established in 1945. It works across the globe, over 130 countries. FAO aids in the coordination of governments and other agencies towards improvising and development of agriculture, aquaculture, forestry, resources such as land and water. It also collects data on food production, safety, and agricultural development through its research programs. Meanwhile, it also provides technical assistance to projects and operates educational and training programs worldwide (FAO, 2020b, c).
5. **World Health Organization (WHO)**: It is an international agency of the UN responsible for maintenance of public health, established in 1948. In part, this agency thrives to protect the population from various diseases, including foodborne illnesses; the agency entirely works towards establishing a better environment for a healthy living of the entire population of the world. The WHO Food Safety Program aims at strengthening of food safety systems and educating the industries that are involved in food processing and packaging. In

addition, consumers and retailers also need to be educated on the safe handling of the food products in order to prevent them from being contaminated (WHO, 2020).

1.8.2 FOOD SAFETY MANAGEMENT SYSTEMS (FSMS)

1. **Good Manufacturing Practices (GMP)**: These are the part of quality assurance (QA) to ensure food safety with reference to protective steps taken from manufacturers. It facilitates minimalization of food contamination and false labeling, thus protecting consumer being misled and helping in purchasing products that are not harmful. GMP is referred to be good business tool that refine compliance and performance by the manufacturers (De Oliviera et al., 2016).
2. **Good Handling Practices (GHP)**: They indicate comprehensive approach from all the food handlers, starting from farm manufactures to consumer at the selling shop to identify potential sources of risk and indicates steps and procedures need to be adopted in order to lower the risk factors associated with contamination (FAO, 2017).
3. **Hazard Analysis Critical Control Point (HACCP)**: It refers to an approach to evaluate every stage of manufacturing food and storage in detail, which may contribute to foodborne illness in case of mishandling. HACCP plays an important role in the identification of hazards and their chance of occurrence in each step of the food chain, that is procurement of raw material, manufacturing, packing, distribution, and consumption (FAO, 2017). Figure 1.2 depicts all the working mechanisms of these governing bodies.

1.9 ROLE OF PUBLIC AND EDUCATION IN FOOD QUALITY AND SAFETY

Foodborne illnesses can be lethal in case of lack of bio-surveillance and effective implementation of food safety measures. Some of these illnesses can spread tremendously in healthy populations. Apart from the measures taken by governing bodies, the public must take it as an initiative to stop further expansion of pandemic. Public shares an equal responsibility alongside researchers and non-clinical staff working on food quality and safety. Following are some of the practices that public can implement on themselves for the betterment of society:

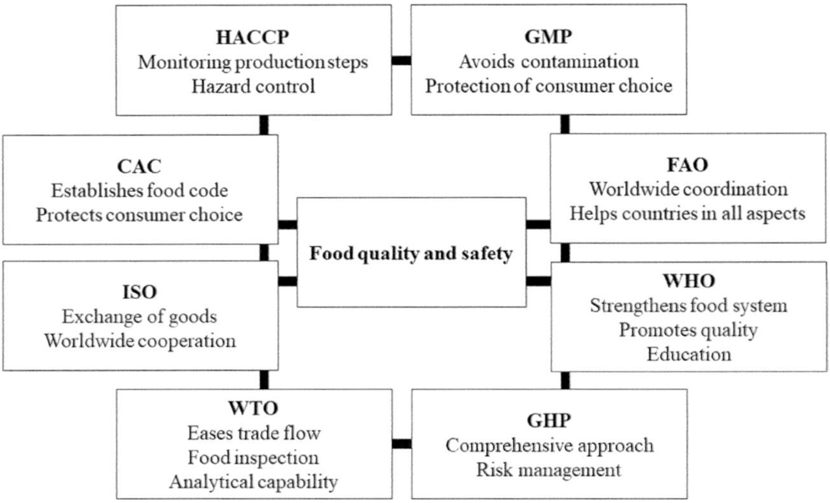

FIGURE 1.2 Working mechanisms of governing bodies.

1. **Buying Certified Foods**: One should always look for the certified foods from reliable food-management agencies. Buying unhealthy food commodities with improper labeling, poor packaging and unhygienic handling leads to the entry of food pathogens resulting in foodborne illness.
2. **Reporting Errors in Food Handling**: Consumers must be aware about misleading related to food safety issues. In case of erroneous handling of any food commodity irrespective of its stage and site of production, it should be reported to food inspectors. This can stop further infection from adulterated/contaminated foods.
3. **Suggesting Good Initiatives to Bodies**: Valuable suggestions related to effective maintenance of food commodities from the public are appreciated. These laws not only help governing bodies to identify the problem, but also help them to create or improvise the set of regulations.
4. **Maintaining Hygienic Practices**: Either personal or public, maintenance of hygiene plays a significant role in food safety. From the site of food production to the food selling, pathogens can likely enter the food sources at any point causing either food spoilage or foodborne diseases. It is advised that food handlers follow the standard operating protocols for handling the food from the field to the shop delivery.

5. **Avoiding Food Adulteration and Contamination**: Adulteration of food commodities with food additives can also result in outbreaks and other foodborne illnesses. It is essential that food handlers avoid the food adulteration and contamination at every step of food production.

Harmonizing the food regulations across the globe enables improvisation of nutritional security and increased the accessibility of ready-to-eat food products easily for the entire population. The successful harmonization would result in effective reduction of foodborne illness, particularly among women and children. This approach would sure benefit the consumers. Along with harmonization, one health approaches to problem-solving are the best described as one of the best approaches to solve issues related to food safety issues. One health focuses on integrated working of experts than isolated working. It believes in the collective approach to solve the issues related with food quality and safety. In the modern-day outbreaks resulting from unknown sources, one health promises effective resolving societal health problems. As the modern-day food safety issues are complex, it is advisable to go for one health approach which not only considers the problem but also the interconnected web of upstream factors to a problem (Angelos et al., 2016).

Education has been proven as one of the best solutions to avoid the causes as well as consequences related with food safety. Area of food security has traditionally comprised of disciplines like microbiology, sanitation, hygiene, food science and public health as well as good agricultural, manufacturing, processing, and packaging practices. It also deals with the formation and implementation of principles of risk management through hazard analysis and critical control points (HACCP). It should also shed enough light on food management systems and laws enforced by them. These subjects have been essential in the provisioning of both plant and animal origin food sources to modern societies. Thus, it becomes important to frame curriculum that involves food safety as an integral part (Angelos et al., 2016).

1.10 CONCLUSION

Being a basic human necessity, food collectively supports the health as well as the economy of a nation. Supply of healthy and nutritious food substances constitute an important aspect of sustainable development of a nation. From the observations made throughout the history, it becomes clear that improving food safety is a prerequisite in every part of the world. The positive and negative situations encountered in different countries with well-developed food safety monitors provide a reference to improve the systems across the world. Globalization has created a wider niche for a large number of foods at the global market, which results in longer and complex global food chain. The increasing demand for food has led to the use of fast and non-sustainable products for improvement of products from both agriculture and animal husbandry. This may result in the creation of new and unfamiliar challenges for food safety. Primary responsibility of food safety should be taken from food handlers of all stages, producers, distributors, handlers, and vendors. Consumers must be vigilant and literate on food safety issues. They must enforce the laws equitably and with fairness.

Legal professionals, medical and healthcare personnel, and researchers should work coordinately to resolve the issues with food safety in the global market. A healthy population is recognized by the supply of nutritionally rich food sources. A close collaboration between all stakeholders irrespective of taking the lead, is essential to achieve a meaningful food safety. Approaches like one health and educating the mankind with management of food substances create a niche that is free from foodborne illnesses as well as hunger and malnutrition.

In summary, food safety and nutrition are closely related. Unhealthy food creates a vicious cycle of disease and malnutrition affecting different age groups, including vulnerable groups like children and women. A comprehensive approach like one health will ultimately ensure food safety for all.

KEYWORDS

- **diagnostic tools**
- **education**
- **food quality**
- **food safety**
- **food standards**
- **governing bodies**
- **hygiene**

REFERENCES

Alders, R., De Bruyn, J., Wingett, K., & Wong, J., (2017). One health, veterinarians and the nexus between disease and food security. *Australian Veterinary Journal*, *95*(12), 451–453.

Almanza, B., Ghiselli, R., & Khan, M. A., (2014). *Food Safety: Researching the Hazard in Hazardous Foods*. CRC Press, Florida.

Angelos, J., Arens, A., Johnson, H., Cadriel, J., & Osburn, B., (2016). One health in food safety and security education: A curricular framework. *Comparative Immunology, Microbiology and Infectious Diseases*, *44*, 29–33.

Attrey, D. P., (2017). Food safety issues in production of foods of animal origin and from farm to plate. In: Dudeja, P., Gupta, R. G., & Minhas, A. S., (eds.), *Food Safety in the 21st Century* (pp. 217–228). New York: Academic Press.

Ayala, A., & Meier, B. M., (2017). A human rights approach to the health implications of food and nutrition insecurity. *Public Health Reviews*, *38*(1), 10–22.

Bintsis, T., (2017). Foodborne pathogens. *AIMS Microbiology*, *3*(3), 529–563.

Damayanti, F. N., & Wahyati, E., (2019). Food safety in the protection of the right to health. In: *IOP Conference Series: Earth and Environmental Science*, *292*(1).

De Oliveira, C. A. F., Da Cruz, A. G., Tavolaro, P., & Corassin, C. H., (2016). Food safety: Good manufacturing practices (GMP), sanitation standard operating procedures (SSOP), hazard analysis and critical control point (HACCP). In: Barros-Velázquez, J., (ed.), *Antimicrobial Food Packaging* (pp. 129–139). Academic Press, New York.

DeWaal, C. S., & Robert, N., (2005). *Global and Local: Food Safety Around the World*. Center for Science in the Public Interest. https://www.cspinet.org/sites/default/files/attachment/global.pdf (accessed on 18 February 2022) (accessed on 11 February 2022).

Di Renzo, L., Colica, C., Carraro, A., et al., (2015). Food safety and nutritional quality for the prevention of non-communicable diseases: The nutrient, hazard analysis and critical control point process (NACCP). *Journal of Translational Medicine*, *13*(1), 1–3.

Dudeja, P., & Singh, A., (2017). Food safety in modern society – changing trends of food production and consumption. In: Dudeja, P., Gupta, R. G., & Minhas, A. S., (eds.), *Food safety in the 21st Century* (pp. 217–228). New York: Academic Press.

FAO, (2005). *Understanding the Codex Alimentarius*. http://www.fao.org/3/w9114e/w9114e00.htm (accessed on 11 February 2022).

FAO, (2006). *Food Security: Policy Brief*. http://www.fao.org/fileadmin/templates/faoitaly/documents/pdf/pdf_Food_Security_Cocept_Note.pdf (accessed on 11 February 2022).

FAO, (2008). *An Introduction to the Basic Concepts of Food Security*. http://www.fao.org/3/a-al936e.pdf (accessed on 11 February 2022).

FAO, (2010). *Private Food Safety Standards: Their Role in Food Safety Regulation and Their Impact*. http://www.fao.org/3/ap236e/ap236e.pdf (accessed on 11 February 2022).

FAO, (2012). *Food Security: Definition, Four Dimensions, History*. http://www.fao.org/fileadmin/templates/ERP/uni/F4D.pdf (accessed on 11 February 2022).

FAO, (2012). IFAD (Food and Agriculture Organization of the United Nations, world food program and International Fund for Agricultural Development). *The State of Food Insecurity in the World*. http://www.fao.org/3/i0876e/i0876e.pdf (accessed on 11 February 2022).

FAO, (2014). Food and Agriculture Organization of the United Nations. Global agriculture towards 2050. *High Level Expert Forum: How to Feed the World in 2050*. http://www.fao.org/wsfs/forum2050/wsfs-forum/en/ (accessed on 11 February 2022).

FAO, (2017). *Good Hygiene Practices and HACCP*. http://www.fao.org/food/food-safety-quality/capacity-development/haccp/en/ (accessed on 11 February 2022).
FAO, (2020a). *Food Security: Concepts and Measurements*. http://www.fao.org/3/y4671e/y4671e06.htm#fn30 (accessed on 11 February 2022).
FAO, (2020b). *About FAO*. http://www.fao.org/about/en/ (accessed on 11 February 2022).
FAO, (2020c). *Building a Common Vision for Sustainable Food and Agriculture: Principles and Approaches*. http://www.fao.org/3/i3941e/i3941e.pdf (accessed on 11 February 2022).
FDA, (2015). *Acceptance Criteria for Confirmation of Identity of Chemical Residues Using Exact Mass Data within the Office of Foods and Veterinary Medicine*. https://www.fda.gov/media/96499/download (accessed on 11 February 2022).
Fryer, P. J., & Versteeg, C., (2008). Processing technology innovation in the food industry. *Innovation, 10*(1), 74–90.
Fukuda, K., (2015). Food safety in a globalized world. *Bulletin of the World Health Organization, 93*(4), 212.
Fung, F., Wang, H. S., & Menon, S., (2018). Food safety in the 21st century. *Biomedical Journal, 41*(2), 88–95.
Godfray, H. C. J., Beddington, J. R., Crute, I. R., et al., (2010). Food security: The challenge of feeding 9 billion people. *Science, 327*(5967), 812–818.
Gorris, L. G., (2005). Food safety objective: An integral part of food chain management. *Food Control, 16*(9), 801–809.
Grace, D., (2015). Food safety in low and middle income countries. *International Journal of Environmental Research and Public Health, 12*(9), 10490–10507.
Gross, R., Schoeneberger, H., Pfeifer, H., & Preuss, H. J., (2000). *The Four Dimensions of Food and Nutrition Security: Definitions and Concepts* (Vol. 20, No. 20, pp. 20–25). SCN News.
Havelaar, A. H., Kirk, M. D., Torgerson, P. R., et al., (2015). World health organization global estimates and regional comparisons of the burden of foodborne disease in 2010. *PLoS Medicine, 12*(12), e1001923.
Hsieh, P. Y. H., & Ofori, J. A., (2007). Innovations in food technology for health. *Asia Pacific Journal of Clinical Nutrition, 16*(S1), 65–73.
Huang, J. Y., Henao, O. L., Griffin, P. M., et al., (2016). Infection with pathogens transmitted commonly through food and the effect of increasing use of culture-independent diagnostic tests on surveillance–foodborne diseases active surveillance network, 10 US sites, 2012–2015. *Morbidity and Mortality Weekly Report, 65*(14), 368–371.
Hussain, M. A., & Dawson, C. O., (2013). Economic impact of food safety outbreaks on food businesses. *Foods, 2*(4), 585–589.
Ismail, F. H., Chik, C. T., Muhammad, R., et al., (2016). Food safety knowledge and personal hygiene practices amongst mobile food handlers in Shah Alam, Selangor. *Procedia-Social and Behavioral Sciences, 222*(2016), 290–298.
ISO (2018). *Food Safety Management Systems: Requirements for any Organization in the Food Chain*. https://www.iso.org/standard/65464.html (accessed on 11 February 2022).
ISO. (2012). *About ISO*. https://web.archive.org/web/20120919212143/http://www.iso.org/iso/home/about.htm#2012_aboutiso_iso_name-text-Anchor (accessed on 11 February 2022).
Kearney, J., (2010). Food consumption trends and drivers. *Philosophical Transactions of the Royal Society B: Biological Sciences, 365*(1554), 2793–2807.
Keiichiro, H., Otsuki, T., & Wilson, J. S., (2015). Food safety standards and international trade: The impact on developing countries' export performance. In: Hammoudi, A., Grazia,

C., Surry, Y., & Traversac, J. B., (eds.), *Food Safety, Market Organization, Trade and Development* (pp. 151–166). Switzerland: Springer, Cham.

Kibret, M., & Abera, B., (2012). The sanitary conditions of food service establishments and food safety knowledge and practices of food handlers in Bahir Dar town. *Ethiopian Journal of Health Sciences*, *22*(1), 27–35.

Lem, A., Bjørndal, T., & Lappo, A., (2014). *Economic Analysis of Supply and Demand for Food up to 2030: Special Focus on Fish and Fishery Products*. http://www.fao.org/3/i3822e/i3822e.pdf (accessed on 11 February 2022).

Lund, B. M., & O'Brien, S. J., (2011). The occurrence and prevention of foodborne disease in vulnerable people. *Foodborne Pathogens and Disease*, *8*(9), 961–973.

Mangal, M., Bansal, S., Sharma, S. K., et al., (2016). Molecular detection of foodborne pathogens: A rapid and accurate answer to food safety. *Critical Reviews in Food Science and Nutrition*, *56*(9), 1568–1584.

Martin, M. J., Thottathil, S. E., & Newman, T. B., (2015). Antibiotics overuse in animal agriculture: A call to action for health care providers. *American Journal of Public Health*, *105*(12), 2409–2410.

Mathur, P., & Pillai, R., (2019). Overnutrition: Current scenario and combat strategies. *The Indian Journal of Medical Research*, *149*(6), 695–705.

McFarland, P., Checinska, S. A., Rasco, B., et al., (2019). Efficacy of food safety training in commercial foodservice. *Journal of Food Science*, *84*(6), 1239–1246.

Napoli, M., De Muro, P., & Mazziotta, M., (2011). Towards a food insecurity multidimensional index (FIMI). *Master in Human Development and Food Security*. http://www.fao.org/fileadmin/templates/ERP/uni/FIMI.pdf (accessed on 11 February 2022).

Odeyemi, O. A., (2016). Public health implications of microbial food safety and foodborne diseases in developing countries. *Food and Nutrition Research*, *60*, 29819.

Oldewage-Theron, W. H., & Egal, A. A., (2016). Food quality and food safety. In: Temple, N., & Steyn, N. P., (eds.), *Community Nutrition for Developing Countries* (pp. 430–448). Edmonton, AB: AU Press, Athabasca University.

Pal, M., Gerbaba, T., Abera, F., et al., (2015). Impact of food safety on one health. *Beverage and Food World*, *42*(7), 21–25.

Panghal, A., Chhikara, N., Sindhu, N., et al., (2018). Role of food safety management systems in safe food production: A review. *Journal of Food Safety*, *38*(4), e12464.

Priyanka, B., Patil, R. K., & Dwarakanath, S., (2016). A review on detection methods used for foodborne pathogens. *The Indian Journal of Medical Research*, *144*(3), 327–338.

Radovanovic, R., (2011). Food safety: The global problem as a challenge for future initiatives and activities. In: Hefnawy, M., (ed.), *Advances in Food Protection* (pp. 27–48). Springer, Dordrecht.

Rasco, B. A., & Bledsoe, G. E., (2009). *Bioterrorism and Food Safety* (pp. 573–600). John Wiley and Sons, Inc., Hoboken, NJ.

Rather, I. A., Koh, W. Y., Paek, W. K., et al., (2017). The sources of chemical contaminants in food and their health implications. *Frontiers in Pharmacology*, *8*, 830.

Scharff, R. L., (2012). Economic burden from health losses due to foodborne illness in the United States. *Journal of Food Protection*, *75*(1), 123–131.

Shaw, D. J., (1945). *World Food Security, a History Since 1945* (pp. 3–461). Springer, Switzerland.

Solodoukhina, D., (2011). Food safety and bioterrorism from public health perspective. In: Hefnawy, M., (ed.), *Advances in Food Protection* (pp. 27–48). Springer, Dordrecht.

Tessema, A. G., Gelaye, K. A., & Chercos, D. H., (2014). Factors affecting food handling Practices among food handlers of Dangila town food and drink establishments, North West Ethiopia. *BMC Public Health*, *14*(1), 571–576.

Uçar, A., Yilmaz, M. V., & Cakiroglu, F. P., (2016). Food safety – problems and solutions. *Significance, Prevention and Control of Food Related Diseases* (pp. 3–15). IntechOpen, United Kingdom.

Uyttendaele, M., Franz, E., & Schlüter, O., (2016). Food safety, a global challenge. *International Journal of Environmental Research and Public Health*, *13*(1), 67.

WHO, (2018). *Delivering Quality Health Services: A Global Imperative for Universal Health Coverage*. https://apps.who.int/iris/handle/10665/272465 (accessed on 11 February 2022).

WHO, (2020). *About WHO*. https://www.who.int/about (accessed on 11 February 2022).

WTO, (2004). *Understanding the WTO*. https://www.wto.org/english/thewto_e/whatis_e/tif_e/tif_e.htm (accessed on 11 February 2022).

Yang, P., Hash, S., Park, K., et al., (2017). Application of nuclear magnetic resonance to detect toxigenic *Clostridium difficile* from stool specimens: A proof of concept. *The Journal of Molecular Diagnostics*, *19*(2), 230–235.

Yang, P., Wong, C., Hash, S., et al., (2018). Rapid detection of *Salmonella* spp. using magnetic resonance. *Journal of Food Safety*, *38*(4), e12473.

CHAPTER 2

Assessment of Food Quality and Management of Food Safety: International Regulations and Consumer Preferences

SOUMYA SINGH, ANURAG SINGH, ANKUR OJHA, and
MEENATAI KAMBLE

ABSTRACT

The safety of food at every single step of food handling till consumption is important to accomplish the high standard of food safety and to ensure the buyer's wellbeing. The quality of food includes the quality attributes that may impact the product value, such as biochemical, physical, visual, textural, and microscopic attributes. There are various methods to estimate the food quality parameters. The safety of the consumer needs to be ensured by the regular check of various food quality parameters. Therefore, to keep up the integrity and security in food supply chain, it is required to enforce the food laws which protect the consumer against unsafe, impure, and fraudulently presented food by prohibiting the sale of food not of the nature, substance or quality demanded by the purchaser. There have been different national, international organizations and agreements that upgrade the food safety, quality, and security. This chapter covers various aspects of food quality analysis and food safety as well as various international organizations and agreements related to food that control the quality and safety of food globally.

Food Microbial and Molecular Biology: From Fundamentals to Applications. Saher Islam, Devarajan Thangadurai, Jeyabalan Sangeetha, Zaira Zaman Chowdhury (Eds.)

2.1 INTRODUCTION

Food what we eat affects the health, nutritional status as well as our productivity. It is therefore very much essential that the food we consume should be of good quality, wholesome, nutritious, and most importantly safe. Inferior quality food consumption can cause the problems like malnourishment. Whereas consumption of unsafe food can cause various types of foodborne diseases. There is several news related to food poisoning or health problems due to the consumption of unsafe foods. Foodborne illness is a global issue, especially in under-developed countries where insufficient food availability itself is an issue. Foodborne illness not only can lead to deaths but can also affect the economic growth of the nations. Therefore, food safety and quality (FSQ) have gained worldwide significance.

Assuring food quality and food safety are the foundation pillars of wholesome food production. There are many properties associated with food quality which can help in the proper assessment of food. Moreover, scientific knowledge provides the liberty to analyze food quality through various food attributes and certain outlined procedures. As far as food safety is concerned, there is an efficient food safety management system (FSMS) in place which helps to achieve this goal. The FSMS primarily comprises Hazard Analysis Critical Control Point (HACCP) and its prerequisites (Good Manufacturing practices (GMPs) and Good Hygienic practices (GHPs)). To ensure that the food is safe and is of good quality, there also exists a very effective regulatory mechanism at the global level which mainly includes Codex Alimentarius Commission (CAC), Sanitary and Phytosanitary (SPS), and Technical Barriers to Trade (TBT) agreements. Various aspects of food quality and safety are discussed in this chapter emphasizing the need in a global context.

2.2 FOOD QUALITY

Quality is a term that denotes a degree of excellence, a high standard or value. Quality of foods may be defined as the composite of those characteristics that differentiate individual units of a product and have significance in determining the degree of acceptability of that unit to the user (Angel et al., 1965). Food quality can be defined in various ways as "Degree of acceptability by the consumer" or "fitness for use." Food quality is a subjective concept with many facets. These facets can be primarily categorized into the following five groups:

1. **Quantifiable Properties**: Freshness, viscosity, water content, chemical nature, physical attributes, bacteriological traits, shelf life.

2. **Organoleptic Inclinations**: Choice, smell, taste, color, size, shape, texture, eating characteristics, after taste.
3. **Environmental Circumstances**: Related to absence or presence of aspired environmental conditions.
4. **Wildlife Well-Being Traits**: Related to absence or presence of aspired wildlife wellbeing results.
5. **Societal Traits**: Related to absence or presence of aspired societal consequences.

The above diversification of facets takes into account the final food product only. After we apply these five facets to the process implied for food production, many more aspects have to be added. In today's scenario where processed foods/ready-to-eat meals have become so common, numerous quality aspects have to be included. For instance, one has to ensure the good quality of additional ingredients (coloring agents, flavor enhancers, emulsifiers, etc.), as well during processed food preparation. From a buyer's perspective, food packaging and marketing also holds a great deal of significance. So, these factors also form a part of the quality dimensions list. Sometimes, if we look deeper, consumer's quality perception is formed by a particular processing step character. Food quality and food safety are two different concepts. Unlike food quality, food safety cannot be measured in terms of "intensity" while framing healthcare policies. In such cases, food is either "safe" or "not safe." There is no midway (Figure 2.1).

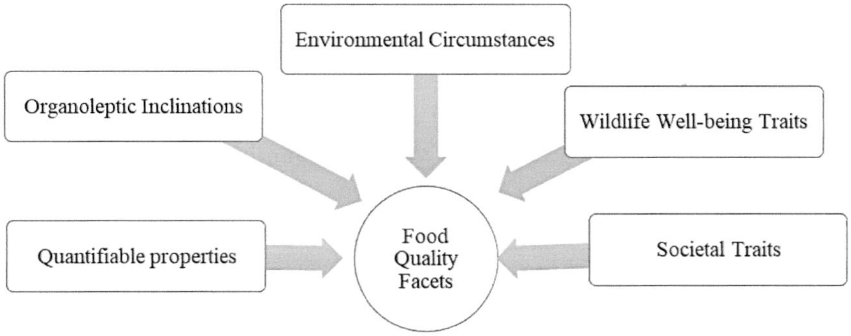

FIGURE 2.1 Food quality facets.

2.3 FOOD QUALITY ATTRIBUTES

There are numerous elements which ascertain food quality. The consumer criterion in selecting a particular food commodity becomes even more profound when there is a wide range of choices available. If we look deeper into the picture from a consumer's perspective, food quality can be explained as the amalgamation of all the traits which separate independent entities of a food item and also play an important role in regulating the extent of acceptability for that entity. In other words, quality regulation is the preservation of quality at stages of stoicism allowable to the consumer. This in turn leads to striking a good balance between consumer demands and the product price of the seller.

The whole idea of "food quality and its attributes" becomes very inclusive when we look at the whole picture from a research-based viewpoint and not merely from the consumer's frame of reference. The research-based study includes mechanical, physical, synthetic, microbe-based, nutritive, and organoleptic variables to attain the target of healthy food. These variables are monitored by a certain class of attributes like food color, tastebuds, smell, amount of nutrients (macro and micro), enzyme percentage, etc. One thing which needs to be kept in mind is that a single food item does not require the analysis of all the existing quality attributes. We need to be selective depending upon the type of product:

1. **Physical Attributes**: Those are the ones that humans recognize through their sense organs. Hence, these attributes are also called sensory attributes, e.g., color, texture, size, shape, physical deformations, mouthfeel, etc.
2. **Hidden Attributes**: Those are the ones which can be evaluated only through set chemical or microbiology-based processes (Shabir et al., 2008). These can be further classified into positive or negative. If the attribute is positive such as the nutritive value of food, required efforts should be taken to perpetuate it. On the other hand, if it is negative, there are very high chances that food will be unsuitable for eating. Therefore, preventive measures should be taken accordingly.

2.4 ASSESSMENT OF FOOD QUALITY

Various quality parameters of the food product can be analyzed by different instrumental methods. However, for the consumer, sensory analysis is the best option for analysis of the quality. The procedures that can be followed for food quality evaluation are discussed as follows:

2.4.1 SYNTHETIC PROCEDURES

Decaying of foodstuffs provides an opportunity for chemical determination. For instance, starch determination in milk, metanil yellow detection in turmeric powder, and approximation of mineral content loss at the time of cooking.

2.4.2 BIOCHEMICAL PROCEDURES

2.4.2.1 QUANTIFICATION OF HYDROGEN ION ACCUMULATION

It is done through a pH meter. Glass indicating electrode and a reference electrode is used to build the electric loop.

2.4.2.2 SUGAR PERCENTAGE

Sugar is measured using a refractometer in terms of °Brix. The instrument is constructed on the theory that the refractive index increases with the rise in the density of the substance. For instance, as the sugar concentration in water increases, the refractive index also increases accordingly.

2.4.3 MICROSCOPIC STUDY

The design of certain foods plays a very vital role in determining their characteristics. Microscopic methods can help in studying the structure of such foods so that important details can be derived. Few examples related to this are as follows:

- Kind of microbes in fermented foodstuffs like idli batter
- Spotting starch cells through microscopic vision
- Discovering food decay microbes by microscope
- Knowing smoothness of the item through size study of sugar crystal
- Quantity and magnitude of the air cells in doughs and froths

2.4.4 PHYSICAL PROCEDURES

2.4.4.1 WEIGHT

It helps to examine food quality, e.g., identification of egg or apple quality based on weight.

2.4.4.2 VOLUME

In the case of liquids, measuring cups are used, whereas in the case of solids, the displacement method is used. This method involves the use of two receptacles. The two receptacles will be treated as follows:

Receptacle 1 = Seeds (mustard seeds) + Baked product

Receptacle 2 = Seeds (mustard seeds)

Food volume will be calculated using the following formula:

$$\text{Food volume} = \text{Volume of seed in receptacle 2} - \text{Volume of seed in receptacle 1}$$

2.4.4.3 SPECIFIC VOLUME

Precision is required in specific volume measurements. It is always advised to take multiple readings of the same sample so that mean can be taken. This will avoid experimental errors, which is a big problem in specific volume measurements. The most tedious task is the bulk volume measurement of permeable products like idli. Idli has small pores which hinder correct bulk volume measurement. To deal with this problem, idli is immersed in molten wax solution so that pores can be closed. Then, a rise in volume is noted as the idli's bulk volume.

2.4.4.4 INDEX TO VOLUME

A planimeter is used for this purpose. First of all, a portion of food which can serve as the best representative of the whole item should be chosen. It is generally the center portion of the dish. Then, the area of the portion needs to be calculated. Index to volume measurements cannot be made without a well-developed sketch of the food cross-section. The cross-section can be generated using a pencil, pen, or ink.

2.4.4.5 SPECIFIC GRAVITY

$$\text{Specific gravity} = \frac{\text{Density of substance}}{\text{Density of water}}$$

It is a measure of the relative density of a substance in relation to that of water (Yadav, 2011). Some food items cannot be measured volumetrically as their physical form does not allow doing so. In such cases, specific gravity calculations are applied to know the comparative airiness of the products, e.g., Egg white foams. The specific gravity also has an impact on the culinary qualities of a product. For instance, cooking traits of low specific gravity potatoes (Potatoes coated with wax) varies from high specific gravity potatoes.

2.4.4.6 MOISTURE

1. **Press Fluids**: The weight of the original food specimen is taken. Weight is taken for the second time after pressing the sample by applying the required amount of pressure for a particular time period. The pressed sample's weight is subtracted from the initial weight. This is how the quantity of juice present in the native specimen is known, e.g., Juiciness of meats, poultry, and fish.
2. **Drying**: The weight of the original specimen is noted. After drying the product until the weight remains unchanged, the final weight is taken.

2.4.4.7 WETTABILITY

There exists a very simple experiment to know the wettability of products, especially over-cooked items. Take the product weight. Immerse it in water containing vessel for 5 seconds. Take it out after 5 seconds. Take the weight of the wet sample. Good moisture-holding capacity is synonymous with good wettability.

2.4.4.8 CELL CONSTRUCTION

Cell conformity, dimensions, bulkiness cannot be known without the proper study of cell structure. To know the cell structure, xerox of cut food portion needs to be taken. Such xerox helps to generate a 3-D picture which in turn gives a very clear idea of the cell dimensions and uniformity. Grain snapshots, ink images or sand holding capacity are methods used to know grain size. A classic example of the ink image method is that of idli where it is divided into two halves. Then, one half is taken and pushed down on the stamp pad and the ink image is taken on a white page. However, ink images might not be the best option for grain size study because of their poor image quality at times. As far as the sand retention method is concerned, food can hold more sand if its grains are rough.

2.4.4.9 COLOR QUANTIFICATION

The first and the foremost feature which the buyer pays attention to after buying the food product is "color." As color modifications can be very much related to taste changes, it is an important factor to study. Following is the detailed outline of the instruments used for color quantification:

2.4.4.9.1 Color Lexicon

The most popular source is Maerz and Paul wordbook. Wordbook has 56 charts. There are seven primary color tone categories according to their scales. Each category comprises 8 platters. Other sources apart from color lexicons are painted test boards, rings, discs, or platter replicas.

A mask of colorless non-descript nature comprising of two orifices is utilized. Both the orifices should have the same size as that of the color mark in the leaflet. One of the orifices is put on the top of the specimen and the other one is put one by one on the different color marks on the leaflet. As soon as a suitable color mark is found such that the specimen color matches with the color mark, the process is completed.

2.4.4.9.2 Disc Colorimeter

As the name suggests, this instrument comprises a number of discs. The discs are designed in such a manner that they possess spiral cuts. These spiral cuts help to accommodate the next disc which results in an arrangement where all discs can be seen but in different portions. The entire disc set-up is fitted on a pivot and rotated at 2,700 rpm. As the discs rotate, colors integrate to produce a solitary shade without any flutter. At this point in time, the specimen is positioned beside the rotating disc and a concurrent look is taken under proper light.

2.4.4.9.3 Colored Chips

If chip color goes well with the food color, it is chosen. However, this technique is not very reliable because it is a tedious job to find that one-minute colored chip which matches the color of food. Moreover, the information produced is complicated to organize and examine.

2.4.4.9.4 *Spectrophotometer*

There are many flaws attached to the whole process of color matching through the naked eye. To overcome these flaws, a spectrophotometer has been designed. A spectrophotometer works on the principle that solutions of different concentrations absorb light differently. Cuvette containing sample is put into the space provided in the spectrophotometer and light of a particular wavelength is thrown against the sample. As light passes through the sample, it is absorbed based on the intensity and color of the food sample. For instance, if we consider two different samples, i.e., juice, and (juice + water), the second one will exhibit lesser absorption of light. In other words, the second sample will emit a comparatively more amount of light passing through it. As a result, more deviation of the spectrophotometer detecting needle will be noted in the case of the second sample. In addition to this, the spectrophotometer also provides results related to sample haziness based upon the amount of light passing through it.

2.4.4.10 *TEXTURE ASSESSMENT*

There are numerous equipment available to assess the texture of all three forms of food. Rheology is defined as the science of deformation and flow of matter (Yadav, 2011). It is based on primarily three features: elasticity, viscosity, and plasticity. The concept of rheology is used to assess numerous mechanical characteristics of foodstuffs. Following are the two main causes behind the rheological study of food products:

- To analyze liquid flow behavior; and
- To evaluate solid mechanical characteristics at the time of consumption and manufacture.

2.4.4.10.1 *Texture Assessment Equipment's in Case of Liquids and Semisolids*

A very significant attribute in texture analysis of liquids and semisolids is food viscosity. It plays a very vital role in classifying a wide range of food items based on quality. Some of the examples of such food items are tomato-based items, different kinds of jams and jellies, mayonnaise, creamed corn, salad cream, syrup, and fruit flesh. The consumer acceptance of these food articles is very much controlled by their viscosity. Moreover, there are a number of advantages attached to the viscosity assessment of products. First of all, analyzing the viscosity of base material at different levels provides an idea to the manufacturer regarding finished item viscosity. Secondly, it helps to evaluate the quantity of a particular food element to be added while preparing a dish. Thirdly, heat perforation capacity and viscosity of food items are intensely related. Therefore, one can get to know the time and quantity of heat to be put in a particular food preparation process.

1. **Percent Sag**: With the help of a probe, the deepness of a food specimen like jelly is computed in the vessel. Then, the food specimen is demolded on a leveled platform. The following formula is used to calculate Percent Sag:

$$\text{Percent Sag} = \frac{(\text{Specimen deepness in vessel} - \text{Specimen deepness in the platform})}{\text{Specimen deepness in the vessel}} \times 100$$

2. **Stromer Viscometer**: The viscosity of some particular food items is calculated using a Stromer viscometer. This instrument uses a very simple method to measure viscosity. The method is based on the amount of time needed by the rotor to complete a 100 revolutions.
3. **Brookfield Syncrolectric Viscometer**: Hindrance offered to the spinning of pivot submerged in food specimen is calculated using this equipment. Some examples of food products whose viscosity can be very suitably measured using this equipment are custards, pie fillings, tomato-based items, creamed corn, mayonnaise, salad dips, and milk-based products.
4. **Efflux–Tube Viscometer**: It calculates the amount of time required for a particular volume of liquid to exit through an opening under certain pressure. For instance, tomato puree.
5. **Adams Consist Meter**: Viscosity analysis of creamed corn was the main reason behind designing this equipment. Presently, it is used for analyzing the viscosity of some other food items as well, e.g., tomato puree, apple dip, and fruit flesh.
   i. **Equipment Design**: It comprises a sizeable metal round slab which has 20 concentric circles carved on it. A cone with a circumference same as that of the smallest circle sits firmly on the slab and can be picked upright.
   ii. **Operating Procedure**: Specimen (liquid or semisolid) is loaded into the cone. Cone is then uplifted rapidly. As soon as the time

duration becomes 30 seconds, the viscosity of the sample is calculated by measuring the degree of product flow at four points situated at an equal distance from each other on the slab. Finally, viscosity is found by taking the mean of the four numbers.

6. **Penetrometer**: The softness of foods is judged using a penetrometer. There is a force cup fitted with a needle or cone which is pierced through the food specimen using gravity for a particular time duration. There is a direct correlation between instrument reading, depth traveled by needle/cone, and product softness. Higher the reading means more is the depth traveled and softer is the food item. The most common food items subjected to penetrometer measurement are jellies and oven-based products.

   One particular kind of penetrometer discussed here is the bloom gelometer. The operating principle is based on the fact that as the lead balls are put into the force cup, the force cup goes inside the food specimen because of its increasing weight. As soon as the desired depth has been traveled by force cup, the experiment is ended. The quantity of lead balls needed ascertains the experiment results.
7. **Barbender Farinograph**: The pliability of wheat dough for bread preparation is analyzed using a Brabender farinograph. As the blades of the instrument mix the dough, the force required to do this job are noted. This gives an idea of the physical characteristics of dough. The trend is that value of force needed to mix dough keeps on increasing until a point comes when it starts decreasing due to excess mixing. This is because excess mixing leads to the breaking of dough molecules which in turn leads to a decrease in the force applied.

2.4.4.10.2 Texture Assessment Equipment in Case of Solids

Food texture can also be explained in terms of a kind of force:

a. **Compression**: When the food remains as one intact entity even after a force is applied, e.g., bread.
b. **Shearing**: When one food segment slides over another after a force is applied, e.g., chewing gum.
c. **Cutting**: When the food is split up into two parts after a force is applied, e.g., apple slicing.
d. **Tensile Strength Calculation**: When the food gets stretched and finally breaks as the force is applied in the direction away from the material, e.g., chapathi.

1. **Magness-Taylor Pressure Tester (Compression)**: There is a plunger and a sprint. The plunger is pushed into the food specimen till a certain depth has been traveled. As the plunger is pushed, the sprint dwindles and calculates the compression force. For instance, a food commodity's acceptability for reaping or the time needed for fruit/vegetable to fully mature can be known using this tester.
2. **Succulometer (Compression)**: It gives results related to corn ripening and apple storage characteristics. To make the calculations, it makes use of the amount of juice drawn out at a particular pressure and time.
3. **Tenderometer (Compression and Shearing)**: First, shearing force is applied. Then, compression force is used, e.g., aptness of peas for conservation can be checked using this instrument.
4. **Fibrometer (Cutting)**: This instrument is used to separate ripened food from the whole lot. It works on the cutting mechanism.
5. **Shortometer (Cutting)**: Product softness is noted using this equipment based on the amount of force needed to fracture the food specimen.
6. **Christal Texturometer (Cutting)**: This equipment comprises a collection of rods which are pressed into the food specimen (usually meat). It operates on the principle that the extent of rod penetration into the flesh depends on how hard/soft the flesh is.
7. **Volodkevich Bite Tenderometer (Cutting and Shearing)**: It tries to emulate the operating mechanism of human teeth on food. The instrument takes note of the force required to chew a food item. As the food is distorted during chewing, the entire energy needed to cause this distortion is recorded, for instance, meat and meat-based products.
8. **Kramer Shear Press (Multiuse Equipment)**: This instrument is very popular as it serves a variety of functions. It has a number of test cell assemblages with the same power unit.

2.5 FOOD SAFETY

Food safety refers to "human health" (Buckwell, 2003). Here, human health does not include wildlife wellbeing and the environment. Health-care policy on food safety should be a notion based on "discrete scientific knowledge" wherein individual viewpoints hold no significance. After all, the prime focus of policymakers is to acknowledge whether the product is safe to consume or not, irrespective of what chitter-chatter media talks about. Allan Buckwell in his publication titled "Food Safety, Food Quality and the CAP" beautifully explains this by stating that "Science itself is fallible but the scientific process and criteria provide ideals to be constantly striven for" (Buckwell, 2003). In simpler terms, non-attendance/negligible presence of any kind of contaminant (innate or synthetic) in a food item is food safety.

One cannot form a definition of food safety based on absolutes. This is because there are numerous parameters involved which determine whether a particular food component can be detrimental to a particular person's health or not. Such parameters comprise age group, sex, body mass index, metabolism, physiology, hereditary characters, steps involved in food preparation, eating habits, etc. Moreover, policymakers have to accept the truth that once food reaches people's abode, it is not feasible to regulate food hygiene, storage, and preparation.

As there is a high degree of variation and uncertainty involved in the process, policymakers shall rely on three steps, namely, risk evaluation, communication, and risk management to make sure that the food is safe. In a nutshell, the food safety domain is rigid enough to leave no room for opinions and at the same time, flexible enough to include absolute's dependency. Hence, it is a concept which involves precautionary safety measures interspersed with broad edges of fallacy.

2.6 FOOD SAFETY MANAGEMENT SYSTEM (FSMS): HACCP AND ITS PREREQUISITES

Food safety cannot be ensured merely by 'final product evaluation' (FSSAI, 2016–2017). Each unit operation associated with food production needs to be monitored. FSMS is one such tool which helps in achieving this target. It is a "compilation of procedures and practices to be observed by food companies to prevent the occurrence of foodborne illnesses by actively controlling hazards throughout the food chain starting from food production till final consumption" (FSSAI, 2016–2017).

The very foundation of FSMS in food businesses comprises three elements: (i) GMPs; (ii) GHPs/SSOPs; and (iii) HACCP (De Oliveira et al., 2016). GMP and GHP/SSOP are prerequisite programs (PRP) for HACCP execution. The three elements of PRPs are discussed below in detail. These programs, i.e., GMP, and GHP/SSOP deal with the "good housekeeping" issues in the facility and may prevent a hazard from occurring. Whereas, HACCP manages specific hazards within the process (De Oliveira et al., 2016). The operating food business must be well-equipped in the documentation. They must properly maintain all the data related to PRPs that effectively reinforce their HACCP process. A very crucial factor at this stage is the commitment of the FBOs towards attaining the HACCP process goals. Moreover, the HACCP crew should time and again keep on checking that these PRPs are in alignment with the hazard analysis step of HACCP (Vilar et al., 2012).

2.6.1 GOOD MANUFACTURING PRACTICES (GMPs)

GMPs are measures taken during the preparation of food, drug, or medical instruments. The conceptualization of GMPs happened way back in the 1970s, but it was in the mid-1990s that official definitions of GMPs occurred in different parts of the world (Bennet et al., 1999). These practices are established around four principles: (i) elimination; (ii) erasing out any unpleasant alienated substance; (iii) impediment; and (iv) killing of unwanted bacteria. The elements that makeup GMPs are the facility and its surroundings, the staff, cleaning and sanitization processes, equipment and utensils, process and controls, and storage and distribution (De Oliveira et al., 2016). The prime motto of these GMP parameters is to produce good quality food that is acceptable in the eyes of the consumer and is safe to eat.

GMP execution is a recurring procedure that depends on PDCA series governance. According to the PDCA series, the application of the GMPs can be classified into four segments: (i) conducting early identification; (ii) sharing the detailed plan; (iii) marking nonconformities; and (iv) re-examining remedial steps. The first and last segments are performed using an inspection checklist formed in accordance with the GMP rules. Detailed plans might be chalked out after inspection. As far as taking remedial action is concerned, this calls for decision-making on certain specifically significant steps labeled as "priority areas" (Dias et al., 2012).

2.6.2 GOOD HYGIENE PRACTICES (GHPs)/SANITATION STANDARD OPERATING PROCEDURES (SSOPs)

GHP/SSOP is a documented methodology designed and applied in an organization to avert direct spoilage of food items. There might be instances where the company is answerable to higher authorities, so it is mandatory for the company to maintain proper records to substantiate its claims as and when needed. SSOP includes a complete description of the specific activities required to maintain utensils and equipment free of pathogenic microorganisms and minimal deteriorating microbiota, preventing the contamination of foodstuffs that get in contact with these utensils and equipment (Da Cruz et al., 2006). The duty of following the steps outlined in the SSOP catalog keeping documentation updated relies entirely on the organization itself. In addition to this, the written data must be capable enough to demonstrate the right execution, administration, and needed rectification of practices.

The most troublesome issue in SSOP is dealing with the food handlers. This matter has a multifaceted structure and comprises various psychic features. Quality participation of food handlers can be ensured through food safety training (e.g., FOSTAC, Preventive Control for Human Food, etc.), as this helps to lower stress and increase the knowledge base of employees, thus boosting their self-confidence (Da Cunha et al., 2015).

2.6.3 HAZARD ANALYSIS AND CRITICAL CONTROL POINT (HACCP)

HACCP is an approach applied in food businesses to make sure that food safety is restored by all means. It is a structured plan utilized to monitor food production systems so that no unwanted alterations occur in food items. The credit of HACCP evolution and its early utilization goes to the combined hard work of three companies: Pillsbury Company, National Aeronautics and Space Administration (NASA), and Natick U.S. Army Laboratories (Gangahar et al., 2018). There was a need for a system that could guarantee that space foods are safe to consume. This is how HACCP came into being.

Pillsbury Company realized that the prevailing quality control systems were not effective enough to prevent adulteration while food production. Also, if they relied only on finished product examination, it would be so large-scale that very little food will be left for space travelers. In such circumstances, the company concluded that "process change" is required which will ward off hazards during production. This new process called HACCP became the preferred approach for food hazard prevention after this event. The USFDA used HACCP in 1973 for the first time. The principles were employed to prevent food poisoning through *Clostridium botulinum* in the case of canned foods. One drawback of the HACCP approach is that although it lowers the possibility of food safety hazards but does not eliminate them (Table 2.1).

TABLE 2.1 Evolution of HACCP

Year	Happening
1959	HACCP was formed by the joint efforts of Pillsbury, NASA, and Natick US Army Laboratories.
1971	HACCP approach was publicized and record-keeping was done in America.
1985	The system was nominated as the preferred choice for hazard prevention by the National Academy of Sciences (NAS). As the system became widely accepted globally, the Food and Agriculture Organization (FAO)/World Health Organization (WHO) quoted it in Codex Compilation.
1993	The European regulation 93/43 from 14.7.93 provides the use of the system for the production of food.
1997	The Codex Code on general principles of food hygiene was redrafted to incorporate guidelines based on Codex HACCP implementation.
2007	HACCP was combined with ISO 9001 and Food Safety and Standards Act (ISO 22000) was designed.

2.6.3.1 HACCP PRINCIPLES

To understand the HACCP principles in a better way, we should go through the associated terminology. Table 2.2 provides an account of the HACCP associated terminology. The seven principles of HACCP are as follows (De Oliveira et al., 2016; Gangahar et al., 2018):

- **Principle 1**: Conduction of hazard analysis
- **Principle 2**: Determination of critical control point(s)
- **Principle 3**: Establishment of critical limit(s)
- **Principle 4**: Establishment of control systems to monitor critical control point(s)
- **Principle 5**: Establishment of remedial measures
- **Principle 6**: Establishment of validation systems
- **Principle 7**: Establishment of data maintenance

Each of the principles listed above is discussed in detail in Table 2.3.

TABLE 2.2 HACCP Associated Terminology

Terminology	Definition
Hazard	Unacceptable biological, physical, or chemical contamination renders food inadequate for consumption.
Risk	Estimated probability of the occurrence of a hazard.
Critical control point (CCP)	Production step where preventive measures are applied to maintain the given product under control and to eliminate, prevent or reduce risks to the health of the consumer. There are different types of CCPs according to the level of control of the hazard: CCPe, when hazards are eliminated; CCPp, when hazards are prevented; and CPP, when hazards are reduced, minimized, or delayed to significant levels.
Critical limit	Value or attribute determined for each variable related to a critical point. Noncompliance leads to risks to consumer health. Critical limits are determined by guidelines or legal standards, specialized literature, practical expertise, previous surveys, internal company regulations, and other sources.
Corrective Action	Immediate and specific actions to be put into place when non-compliance with critical limits occur.
Validation	Use of supplementary tests or review of monitoring records to determine if the HACCP system is functioning according to the plan.
Decision tree	The logical sequence is used to determine if a raw material, ingredient, or process step is a CPP for a given hazard.

2.7 PRODUCT TRACEABILITY: FOOD QUALITY AND FOOD SAFETY DIMENSIONS

In the present times, the usual range of miles through which the food product moves to reach the end-user has gone up because of food trade globalization (Aung et al., 2014). As a result, ensuring food quality and safety through the supply chain has become a much more tedious task than before. Furthermore, experiences of food-related emergencies in the past few years have led to the loss of dependability on the food sector. For instance, the buyer lost trust in the food industry when epidemics occurred due to sickness caused by *Campylobacteria*, *E. coli*, and *Salmonella*. These kinds of incidents made consumers skeptical regarding the quality and safety of food, and thus, they now demand food that comes with high quality and safety promises (Beulens et al., 2005; Bertolini et al., 2006; Regattieri et al., 2007; Trienekens et al., 2008). All this paved way for robust food product traceability (Food Standards Agency, 2002). Product traceability related to food quality and safety should be understood from two aspects: The regulatory aspect and the Food traceability paraphernalia (Hong et al., 2011).

TABLE 2.3 HACCP Principles with Their Corresponding Goals

Principle	Aim
Conduction of hazard analysis	Clarity of the kind of hazard present/plan of hazard analysis conduction/recognition of serious hazard/pointing out of preventive actions.
Determination of critical control point(s)	Finding out that which CCP will be best to deal with a particular hazard through the decision tree method or Q&A sessions.
Establishment of control systems to monitor CCPs	Deciding on what control systems will be used and who will monitor.
Establishment of remedial measures	Identification of corrective actions to be employed/develops an outline of corrective action process to be followed/maintaining files related to corrective actions.
Establishment of validation systems	Listing out of validation methods to be utilized and how to implement them.
Establishment of record-keeping	Identifying essential points to be noted for HACCP execution/developing schedule to timely keep track of the recorded facts/methods for data utilization.

2.7.1 REGULATORY ASPECT

There is a separate section dedicated in this chapter for discussing regulations regarding food quality and safety in general. Here, we are focused on discussing those rules and regulations which primarily deal with food traceability. This will help us to build a clear picture regarding "product traceability" specific regulations which at the same time contribute to food quality and safety.

A new set of guidelines are being developed which can establish a much more effective link between product traceability, safety, and quality (Petersen et al., 2004). Two countries that are at the forefront of filling the gaps of food traceability through policy amendments are the EU and the USA. Let us first discuss the traceability rules which exist in the EU and then we must move on to the USA. January 1, 2005, was the day when EU order 178/2002 was first implemented according to which each food, as well as fodder item traded inside European boundaries to be essential, has traced (Folinas et al., 2006). On the other hand, the United States of America released the

Bioterrorism Act (2002) which states that the one who is involved in the business of producing, operating, loading, delivering, collecting food must keep a proper account of the related data. Moreover, FDA has the authority to look through this data in case there is a doubt that a particular food product can have grave adverse health effects (Levinson et al., 2009).

The apex body internationally in formulating food product rules and regulations is CAC. CAC, through the implementation of HACCP and its prerequisites (GHPs and GMPs) in the food business, has ensured food quality and food safety in a very successful manner. CAC and its role in assuring FSQ have been discussed in much more detail in the later section titled "international regulations for food quality and safety: Codex, SPS, and TBT."

The ISO, the biggest body for formulating and printing global standards, has taken significant initiatives to lay down food traceability rules by making the necessary improvements in the present regulations and also by adding some new objectives. For instance, required additions have been made in ISO 9001 (2000) related to the safe food and traceability concept with special attention on traceability in particular. Also, another standard ISO 22005 (2007) elucidated the fundamentals and purpose of traceability and outlined the elementary needs regarding planning and execution of food and fodder traceability equipment. The best part is that planning and execution are done in such a way that any food firm working at any stage in the chain can implement the traceability system.

The common thread between EU order, ISO 22005, and USA Bioterrorism Act 2002 is the "one step up/one step down" traceability ideology. As per this approach, one can recognize the last distributor and the end-user of the product in the whole chain. This approach serves as the fundamental need for the planning and execution of food and fodder traceability apparatus indicated in all the three systems of regulations (Ruiz-Garcia et al., 2010).

2.7.2 FOOD TRACEABILITY PARAPHERNALIA

It is very important to have a robust product traceability system in place to ensure full safety and quality check. The efficiency of the traceability system depends on how fast the problem is spotted by the business operator and solved (Golan et al., 2004). In no way traceability system itself has the potential to make good and safe food items. It also does not hold any accountability for the product. It only serves as a part of the supply chain or quality/safety check by confirming whether the entire chain is operating in an intended manner or not. It provides us with the data which is interpreted in the correct way, can help us to take timely action to prevent any hazard (Aung et al., 2014).

Although both food quality and food safety are very much connected to traceability but the most common association can be found between "food safety and traceability" rather than "food quality and traceability." This conception among the public needs improvement. Product recalls and food purity testing procedures have made the public relate traceability to food safety primarily, "food quality and traceability" is the second big thing that comes to their mind. This thinking needs to be changed.

From the buyer's perspective, consumer credence is uplifted when traceability systems attest to food quality and safety processes (Van Rijswijk et al., 2006). In the article titled "Perspectives on traceability in food manufacture" by Moe (1998), the author has stated the fact that traceability is a self-contained system within a larger system called "quality management" (Moe et al., 1998). This necessitates the need for a potent intramural traceability system for quality management. Information gathering, product manufacture flow check, and good quality attestation will be greatly enhanced after having efficient traceability paraphernalia as a part of quality management. Figure 2.2 highlights the relationship between food quality, food safety, and traceability from a management point of view.

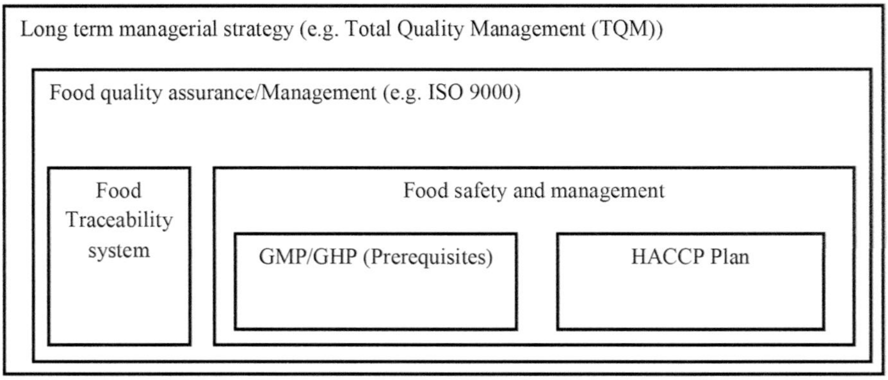

FIGURE 2.2 Food safety, food quality and traceability relationship.

Source: Adapted from: Huss (2007); and Aung et al. (2014).

2.8 INTERNATIONAL REGULATIONS FOR FOOD QUALITY AND FOOD SAFETY: CODEX, SPS, AND TBT

In the international domain, CAC, SPS, and TBT agreements play the whole game when it comes to assuring FSQ. A lot of amendments have been noticed in the last few years which have in turn resulted in reforms in FSQ regulations. The major driving force behind these changes is the "requirement for increased productivity." Assets are limited but consumers want the best from the government. There is an increased pressure on the government to fulfill consumer demands without compromising consumer health in any way (Randell et al., 1997). As a result, the government is investing in reframing FSQ policies so that these issues can be addressed. A few of the strategies include getting rid of all the facilities which are not directly associated with ensuring food safety. The second biggest reform is the "redefined outlook" of the industry as well as the government officials regarding the duties related to FSQ. In response to this, it has been realized that the food industry must play a bigger role in promising FSQ to consumers. In order to corroborate this idea, the industry is eager to participate much more actively in drafting food regulations. The most appreciable thing is the fact that the industry is ready to take control of the product's safety and quality. In the end, all that matters are the persisting faith of the buyer in the food item quality and safety. So, an industry with the help of the government must consistently work towards achieving this goal. In relation to this, government officials are giving inputs in the food standards sector which will ultimately enhance the FSQ domain.

The formulation of food regulations is based on the needs of the international import-export market. CAC, SPS, and TBT agreements play a very crucial role in facilitating smooth food trade on the international front. Also, as the economy is growing; food trade is experiencing flexibility; the food science and technology sector is evolving; transport and communication are developing, import-export in the food domain is reaching newer heights (World Health Organization, 2003) which will demand greater involvement of Codex, SPS, and TBT.

2.8.1 CODEX ALIMENTARIUS COMMISSION (CAC)

CAC is an inter-governmental body established by Food and Agriculture Organization (FAO) in 1961 (FAO, 1962; Randell, 1997). It strives to execute the Joint FAO/WHO Food Standards Program, which consists primarily of three goals: end-user wellbeing, equitable food business practices, and synchronization of food regulations-related assignments. Any country that is a part of FAO and/or WHO can be a member of the Commission.

2.8.1.1 CODEX AND FOOD TRADE

As stated earlier, one of the roles of Codex is to ensure equitable food trade. Here, "insurance of equitable food trade" means the elimination of fake or misleading business operations and monitoring of wrong assertions. These points are included in the WTO Agreement on TBT and address topics like labeling and dietary regulations, discretionary quality points mentioned in regulations, or some supplementary books. As per the TBT Agreement, WTO participants are motivated to apply international policies such as Codex regulations, laws, recommendations, and associated works. There must be instances where the country's government might apply excerpts of international regulations. On similar lines, any nation's government might use extracts of international laws when non-mandatory regulations are being framed.

The Codex 'General Guidelines on Claims' and the 'Guidelines on Nutrition Labeling' are advisory texts supporting the Codex 'General Standard for the Labeling of Prepackaged Foods' (Randell et al., 1997). As the laws related to claims and nutrition labeling from governing bodies are majorly influenced by technical recommendations, these works should be utilized by government officials while developing domestic technical laws in these sectors.

2.8.1.2 CODEX AND FOOD QUALITY

In the area of food quality, the scope of codex lies within purchaser health, finances, and fair trade only. Codex does not deal with conservation of nature, wildlife wellbeing, or preservation of threatened varieties, until, and unless these factors have an immediate impact on food safety. Codex regulations provide the needed degree of pliability so that there is enough room for technological revolutions. Moreover, it also takes into account the difference which exists in various locations, nations, and businesses. All these features help in achieving the goal of consumer protection in a better way.

Earlier, Codex regulations focused only on final stage examination and dealt with a limited number of methods to meet a particular aim. This factor has transformed. Now, the aimed target can be achieved through a much more number of methods which has, in turn, lead to 'equivalence.' An integral part of the formulation of Codex regulations is the "Risk Analysis" method. This method is especially very important when it comes to safeguarding consumer health. It consists of two parts: risk assessment and risk management. Lately, the Codex Executive Committee has instructed each and every Codex Committee to explain the reason behind the chosen risk analysis method. This explanation will subsequently form a significant segment of all upcoming regulations.

Commitments related to providing better food quality are the main feature where Codex regulations are undergoing more additions. For instance, rules, and regulations for HACCP implementation have been recently incorporated into Codex. HACCP implementation has made CAC realize that mere final product evaluation is not enough and one has to take precautionary steps to control hazards at each stage of food production. The most powerful characteristic of HACCP is that the concept can be utilized at each stage of the food chain, thus strengthening overall food safety. Furthermore, HACCP implementation helps in much more effective utilization of the available facilities and complications are now being resolved in a much more punctual manner due to the use of HACCP. The usual examination and attestation of food processes through certification courses are also very much benefitted because of HACCP implementation. Codex very rigorously supports the whole idea of HACCP implementation along with the use of GMPs to achieve food safety.

2.8.2 SPS AND TBT AGREEMENTS

Earlier, international trade in the food sector was not free-flowing. This led to the signing of AoA with the support and approval of WTO. AoA helped in achieving the free-flowing nature of international trade through the removal of tariffs and subsidies. But this was not enough. This is because even after the removal of tariffs and subsidies, there were some aspects that remained unaddressed and this hindered trouble-free trade among countries. There were problems of non-uniformity of food products, safety, and quality regulations in different parts of the world. Moreover, the consumer cannot judge the quality of any food item only based on physical examination. These unsolved challenges led to the signing of two more WTO agreements: SPS and TBT (Figure 2.4).

2.8.2.1 THE BASIS FOR SPS AND TBT (EXPERIENCE AND CREDENCE GOODS)

If a marketplace is characterized by the free flow of knowledge, there will be a variety of food items produced and buyer will purchase products which will be best suited for them. Figure 2.3 helps to explain the idea in a simpler manner.

Consumer C1 purchases lower quality products whereas consumer C2 purchases the higher quality product. These choices are primarily influenced by product price and buyer contentment. In this case, the calculations are sorted and henceforth, food authorities also need not bother to interfere. But if we look deeper, this case has a major disadvantage which can be described by discussing the difference in three types of goods: search goods, experience goods, and credence goods (Nelson et al., 1970, 1974; Darby et al., 1973; Deodhar et al., 2001). These goods have been explained in detail in the following paragraphs:

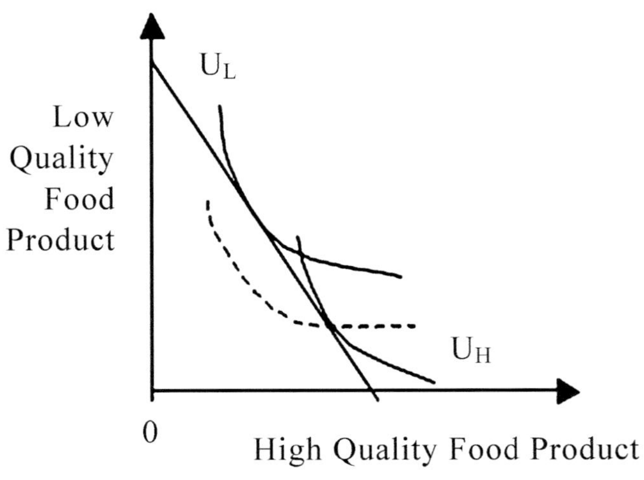

FIGURE 2.3 Food quality alternatives under complete awareness.

Source: Adapted from: Deodhar et al. (2001).

2.8.2.1.1 Search Goods

Those goods whose qualities can be assessed by the buyer even without purchasing them are called search goods. The physical examination of the product will be enough to judge the product quality. For example, when bananas are being exported to other countries, they are physically checked before being loaded for transportation.

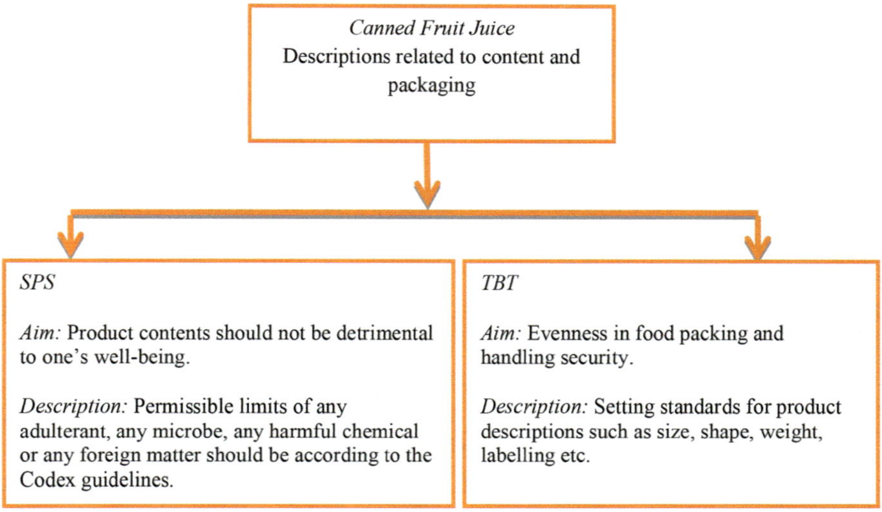

FIGURE 2.4 SPS and TBT juxtaposition in a food item.

2.8.2.1.2 Experience Goods

As the name suggests, the customer must purchase and utilize the item to understand its quality. Depending on the response of the customer towards the product, there are two categories under experience goods: repeat purchase attribute goods and one-time purchase attribute goods (Deodhar et al., 2001).

Repeat purchase attribute goods help in establishing the company's name. This is because if an item is being bought, again, and again, this implies that it is of superior quality which in turn builds the company brand. Also, the company may charge a higher cost for the same item as the brand grows. These all factors contribute in eradicating market deformities by compensating through the company's good name and incessant product buying. In this case, meat products serve as a good example because if food poisoning takes place after the consumer eats meat, a brand name is at stake. As far as one-time purchase goods are concerned, the name itself is self-explanatory.

2.8.2.1.3 Credence Goods

Those goods where the product quality cannot be judged by the buyer even after buying it. From the producer's side, product know-how might be a dicey case but as far as the buyer is concerned, proper product judgment is not there at any level. There can be a number of reasons behind this. One of the major reasons can be that the consumer does not have a clear picture regarding the side-effects of the product on health. For instance, pesticide remnants and other harmful chemicals might be very detrimental to the health of the consumer if proper precautions are not taken to avoid exposure to such products.

Amongst the above three kinds of goods, search goods are the ones which can survive very well in the present market conditions. This was not the case with the other two types of products. There was a need for appropriate support in the form of rules and regulations. Appropriate rules and regulations are present at the national level for all countries but the same was not the scenario at the world level. The direct outcome of this was an unwanted outcropping of non-tariff trade barriers, which could not be dealt with in the present AoA. Finally, SPS, and TBT agreements threw light on these problems and solved them.

2.8.2.2 INFERENCES FROM SPS AND TBT

As mentioned earlier, SPS and TBT agreements were inscribed in addition to AoA with the support and approval of WTO. The following lines of AoA Article 14 very thoroughly indicate that AoA has always been in full support of SPS (Deodhar et al., 2001):

> *"Members agree to give effect to the Application of Sanitary and Phytosanitary Measures."*

No matter how the desired level of importance has neither been given by industries nor by researchers to the two agreements. An appreciable amount of uncertainty is involved in comprehending the distinction between SPS and TBT agreements. The difference is that the SPS agreement deals only with foodstuffs and farming whereas the TBT agreement encompasses all other items in addition to food items. The primary motive of the SPS agreement is human, flora, and fauna security. SPS tries its best to provide health security from insects and illnesses which result due to the import of food items. Now, if we consider the TBT agreement. It takes care of item descriptions such as size, shape, weight, labeling, etc.

> *"Members shall base their sanitary and phytosanitary measures on international standards, guidelines, and recommendations. The sanitary and phytosanitary measures that conform to the international standards, guidelines and recommendations will be deemed necessary to protect human, animal or plant life or health."*

Rules and regulations outlined by CAC serve as the referral point when any international food law has to be formulated. Furthermore, SPS, and TBT agreements provide strong support to the CAC rules which makes CAC implementation a compulsory affair.

2.8.2.3 SPS AND TBT: DEVELOPED AND DEVELOPING NATIONS

Both developed and developing nations follow SPS and TBT agreements, but things are much more twisted when it comes to developing nations. As indicated from Article 3.3 excerpt of SPS, developed nations much more benefit and privilege. In this context, Article 3.3 excerpt (Nelson et al., 1974) is as follows:

> *"Members may introduce or maintain sanitary or phytosanitary measures which result in a higher level of sanitary or phytosanitary protection than would be achieved by measures based on the relevant international standards, guidelines or recommendations if there is a scientific justification..."*

Above Article 3.3 excerpt has been introduced primarily for developed countries like the USA. This article excerpt however is demeaning towards CAC and the entire process of WTO recommendation of CAC regulations. This is because once CAC rules have been laid down after much technical research and SPS and TBT support it in an effective manner, there seems to be no need to introduce more stringent laws.

2.8.2.3.1 India in Particular

India in particular abides by SPS and TBT rules to avoid non-tariff hindrances to commerce. Some of the classic examples where India has faced non-tariff hindrances to food commerce have been discussed below:

1. **Peanut Aflatoxin Amount**: CAC allows 15 parts per billion (ppb). India allows 30 ppb. European Commission (EC) allows just 4 ppb. The direct barrier to food commerce here is that India cannot export peanuts to European countries because even if it sets its limits

according to the standard CAC rules, EC limits will still be left to be achieved.

2. **Lead Amount in Milk**: India allows 0.2 parts per million (ppm). Global requirements are 0.02 ppm.
3. **Sulfur Amount in Sugar**: 20 ppm was the limit decided by CAC. After further contemplations, Indian researchers concluded that the permissible limit of 75 ppm is also very much fine.

Another aspect that needs to be addressed is the low engagement of India in CAC guideline formulation gatherings. India must take the help of developed nations to increase their participation in these CAC meetings. For instance, many developed nations are ready to help countries like India to build the required facilities in this area. Such help can be done by providing suggestions, finances, or at times, sharing scientific knowledge (according to rules and regulations) may be enough.

2.9 CONSUMER PERCEPTION ABOUT FOOD QUALITY AND FOOD SAFETY

Studies have been conducted across nations to explore numerous aspects of consumer perception about food quality and safety. Some of the prominent articles have been discussed in the following paragraphs. In an article by Röhr et al. (2005), it has been concluded that the public is unable to draw a clear line between the concept of "food quality" and "food safety" (Van Rijswijk and Frewer, 2008). Since consumers accept the fact that they cannot comprehend food safety by merely reading the product label while product purchase, they are more interested in verifying the product quality as a buyer. Moreover, the public strongly believes that regulatory authorities are competent enough to assure food safety, hence they focus more on food quality when purchasing any food item.

A research group led by Röhr et al. (2005) studied the results of a consumer survey by comparing them with previous survey conclusions. It was found that two kinds of consumers exist in the market when it comes to food safety. One category is of those people who are more worried about the rising food costs than anything else. Collected information corroborates this fact because these cost-dominated consumers do not intend to consider safety aspects such as the product buying location, manufacture process, etc. The primary thing for this class is the product price. Another category consists of the people exactly opposite to the first ones. This category is safety-dominated.

Another paper on consumer perception emphasized the role of media in shaping consumer viewpoints regarding food safety (Ergönül, 2013). The study highlights the importance of consumer awareness about customs that encourage food safety at the household level. Data particulars proved that television and wireless are the two most popular means of communication for educating the purchaser regarding food safety. Something which works even better than TV and radio in spreading consumer awareness is "Ministry notification" (Rimal et al., 2001). People have full confidence in these ministry circulars which in turn can serve as a powerful medium to empower consumers by providing them the right information at the right time.

The eating pattern in an individual's life should be influenced by the food safety levels prevailing in the surroundings. Necessary changes in the diet should be introduced accordingly. But this is not the case as per the studies (Rimal et al., 2001). The most vital component which influences consumer mind regarding the relationship between food safety and the dietary pattern is "education." It is their degree of knowledge and awareness which makes them different from each other. More consumer education, better is the understanding about required dietary changes.

2.10 NOVEL TECHNIQUES FOR IMPROVING FOOD QUALITY AND FOOD SAFETY

High pressure processing (HPP) is a recent technology that has immense capability to assure food safety through minimum processing (Considine et al., 2008). Hurdle technology, another innovation, has gained a lot of attention for improving microbial safety (Khan et al., 2017). PEF, Cold Plasma, and light therapy are some more innovations that contribute towards enhancing food quality and food safety (Morales-de la Peña et al., 2019).

2.11 CONCLUSION

Proper assurance of food quality and food safety has immense significance because not only does it affect the health of the people but also impacts the economic sector. Once the health will deteriorate, people will be inflicted with all sorts of diseases. As a result, this will have a drastic effect on the financial progress of the nation. Therefore, one cannot even think of consuming a food item that has not met quality and safety requirements as this may lead to certain irreplaceable adverse outcomes. As food quality and food safety have been dealt with as separate topics, one can gain the required clearance between the two concepts. Special importance should be given to food product traceability. It is through good product traceability in place that one can time and again check the exact location as well as the condition of the food article. Item can be monitored for any kind of spoilage very easily if there is an effective product traceability system available. Two global agreements, i.e., SPS, and TBT to foster food quality and food safety have been signed by major countries. An international organization, CAC constantly strives to ensure food quality and food safety through its principles based on food hygiene. The prime focus of the whole subject of food quality and food safety should be the consumer and that is why consumer perception is also very important. Consumer perception can be formed due to a number of factors, but according to studies, consumer education plays a very significant role. Last but not the least, many new technologies have come into the market to propel us to make sure that food is good and hygienic.

KEYWORDS

- **chemical hazards**
- **foodborne illness**
- **food quality**
- **food safety and management system**
- **hazard analysis critical control point**
- **physical hazards**
- **quality assessment**

REFERENCES

Angel, S., Kramer, A., & Yeatman, J. N., (1965). Physical methods of measuring the quality of canned peas. *Food Technology*, *19*(8), 1278.

Aung, M. M., & Chang, Y. S., (2014). Traceability in a food supply chain: Safety and quality perspectives. *Food Control*, *39*, 172–184.

Bennet, W. L., & Steed, L. L., (1999). An integrated approach to food safety. *Quality Progress*, *32*, 37–46.

Bertolini, M., Bevilacqua, M., & Massini, R., (2006). FMECA approach to product traceability in the food industry. *Food Control*, *17*(2), 137–145.

Beulens, A. J., Broens, D. F., Folstar, P., & Hofstede, G. J., (2005). Food safety and transparency in food chains and networks relationships and challenges. *Food Control*, *16*(6), 481–486.

Buckwell, A., (2003). Food safety, food quality and the CAP: Studies in spatial development. In: *Policy Vision for Sustainable Rural Economies in an Enlarged Europe* (Vol. 4, pp. 153–161). Akademie für Raumforschung und Landesplanung, Hannover.

Considine, K. M., Kelly, A. L., Fitzgerald, G. F., Hill, C., & Sleator, R. D., (2008). High-pressure processing–effects on microbial food safety and food quality. *FEMS Microbiology Letters*, *281*(1), 1–9.

Da Cruz, A. G., Cenci, S. A., & Maia, M. C., (2006). Quality assurance requirements in produce processing. *Trends in Food Science and Technology*, *17*(8), 406–411.

Da Cunha, D. T., Cipullo, M. A. T., Stedefeldt, E., & De Rosso, V. V., (2015). Food safety knowledge and training participation are associated with lower stress and anxiety levels of Brazilian food handlers. *Food Control*, *50*, 684–689.

Darby, M. R., & Karni, E., (1973). Free competition and the optimal amount of fraud. *The Journal of Law and Economics*, *16*(1), 67–88.

De Oliveira, C. A. F., Da Cruz, A. G., Tavolaro, P., & Corassin, C. H., (2016). Food safety: Good manufacturing practices (GMP), sanitation standard operating procedures (SSOP), hazard analysis, and critical control point (HACCP). In: Barros-Velázquez, J., (ed.), *Antimicrobial Food Packaging* (pp. 129–139). Academic Press.

Deodhar, S., (2001). *WTO Agreements on SPS and TBT: Implications for Food Quality Issues*. https://core.ac.uk/download/pdf/6443608.pdf (accessed on 11 February 2022).

Dias, M. A. C., Santana, A. S., Cruz, A. G., José De, A. F. F., De Oliveira, C. A. F., & Bona, E., (2012). On the implementation of good manufacturing practices in a small processing unity of mozzarella cheese in Brazil. *Food Control*, *24*(1, 2), 199–205.

Ergönül, B., (2013). Consumer awareness and perception to food safety: A consumer analysis. *Food Control*, *32*(2), 461–471.

FAO. (1962). Resolution No. 12/61. *Report of the Eleventh Session of the Conference*. Rome.

Folinas, D., Manikas, I., & Manos, B., (2006). Traceability data management for food chains. *British Food Journal*, *108*, 622–633.

Food Standards Agency, (2002). *Traceability in the Food Chain: A preliminary study*. Food Chain Strategy Division, Food Standards Agency. http://www.adiveter.com/ftp_public/articulo361.pdf (accessed on 11 February 2022).

FSSAI, (2016–17). *Transforming the Food Safety and Nutrition Landscape in India*. https://archive.fssai.gov.in/dam/jcr:e790e0c6-b1c6-46b7-b320-067d7eb42c56/Fssai_Report_2017_28_06_2017.pdf (accessed on 11 February 2022).

Gangahar, P., & Tomar, K. P. S., (2018). *Block-1 Food Safety and Quality Management Systems*. IGNOU, New Delhi. https://egyankosh.ac.in//handle/123456789/42841 (accessed on 18 February 2022).

Golan, E. H., Krissoff, B., Kuchler, F., Calvin, L., Nelson, K. E., & Price, G. K., (2004). *Traceability in the US Food Supply: Economic Theory and Industry Studies (No. 1473-2016-120760)*. Traceability in the U.S. Food Supply: Economic Theory and Industry Studies (usda.gov).

Hong, I. H., Dang, J. F., Tsai, Y. H., Liu, C. S., Lee, W. T., Wang, M. L., & Chen, P. C., (2011). An RFID application in the food supply chain: A case study of convenience stores in Taiwan. *Journal of Food Engineering*, *106*(2), 119–126.

Huss, H. H., (2007). *Assessment and Management of Seafood Safety and Quality (No. 444)*. Assessment and management of seafood safety and quality: Current practices and emerging issues. FAO Fisheries and Aquaculture Technical Paper No. 574.

Khan, I., Tango, C. N., Miskeen, S., Lee, B. H., & Oh, D. H., (2017). Hurdle technology: A novel approach for enhanced food quality and safety – A review. *Food Control*, *73*, 1426–1444.

Levinson, D. R., & General, I., (2009). *Traceability in the Food Supply Chain*. Washington, DC: US Department of Health and Human Services.

Moe, T., (1998). Perspectives on traceability in food manufacture. *Trends in Food Science and Technology*, *9*(5), 211–214.

Morales-De La, P. M., Welti-Chanes, J., & Martín-Belloso, O., (2019). Novel technologies to improve food safety and quality. *Current Opinion in Food Science*, *30*, 1–7.

Nelson, P., (1970). Information and consumer behavior. *Journal of Political Economy*, *78*(2), 311–329.

Nelson, P., (1974). Advertising as information. *Journal of Political Economy*, *82*(4), 729–754.

Petersen, A., (2004). *Status of Food Traceability in the European Union (EU) and United States of America (US), with Special Emphasis on Seafood and Fishery Products*. MS Assignment. Danish Technical University, Copenhagen, Denmark.

Randell, A. W., & Whitehead, A. J., (1997). Codex Alimentarius: Food quality and safety standards for international trade. *Revue Scientifiqueet Technique–Office International des Epizooties*, *16*, 313–318.

Regattieri, A., Gamberi, M., & Manzini, R., (2007). Traceability of food products: General framework and experimental evidence. *Journal of Food Engineering*, *81*(2), 347–356.

Rimal, A., Fletcher, S. M., McWatters, K. H., Misra, S. K., & Deodhar, S., (2001). Perception of food safety and changes in food consumption habits: A consumer analysis. *International Journal of Consumer Studies*, *25*(1), 43–52.

Röhr, A., Lüddecke, K., Drusch, S., Müller, M. J., & Alvensleben, R. V., (2005). Food quality and safety – consumer perception and public health concern. *Food Control*, *16*(8), 649–655.

Ruiz-Garcia, L., Steinberger, G., & Rothmund, M., (2010). A model and prototype implementation for tracking and tracing agricultural batch products along the food chain. *Food Control*, *21*(2), 112–121.

Shabir, S., (2008). *Value Addition and Quality Attributes in the Marketing of Pulses and Spices in India*. http://hdl.handle.net/10603/53584 (accessed on 11 February 2022).

Trienekens, J., & Zuurbier, P., (2008). Quality and safety standards in the food industry, developments, and challenges. *International Journal of Production Economics*, *113*(1), 107–122.

Van, R. W., & Frewer, L. J., (2006). How consumers link traceability to food quality and safety: An international investigation. *European Association of Agricultural Economists 98th Seminar*. Chania, Crete, Greece.

Van, R. W., & Frewer, L. J., (2008). Consumer perceptions of food quality and safety and their relation to traceability. *British Food Journal*, *110*(10), 1034–1046. https://doi.org/10.1108/00070700810906642 (accessed on 11 February 2022).

Vilar, M. J., Rodriguez-Otero, J. L., Sanjuán, M. L., Diéguez, F. J., Varela, M., & Yus, E., (2012). Implementation of HACCP to control the influence of milking equipment and cooling tank on the milk quality. *Trends in Food Science and Technology*, *23*(1), 4–12.

WHO and FAO, (2003). *Assuring Food Safety and Quality: Guidelines for Strengthening National Food Control Systems*. Food and Agriculture Organization of the United Nations, Rome and World Health Organization, Geneva.

Yadav, V., (2011). *Food Analysis and Quality Control*. https://gpadampur.files.wordpress.com/2011/11/6-2-faqc-class-notes-08022014.pdf (accessed on 11 February 2022).

CHAPTER 3

Recent Approaches for Assuring Food Safety, Quality Systems, and Other Considerations

MUHSIN JAMAL, LILOMA SHAH,
SAYED MUHAMMAD ATA ULLAH SHAH BUKHARI,
MUHAMMAD ASIF NAWAZ, SAADIA ANDLEEB, SANA RAZA, and
REDAINA

ABSTRACT

Food could never be completely safe, and several types of microbes cause numerous foodborne diseases. Numerous foodborne diseases are triggered by microbes originating from animals or animal products that include bacteria, viruses, helminths, and protozoa. At any step-in food processing chain, food may be contaminated with microbes, hence, there is need to implement hazard analysis critical control point (HACCP) system. Food safety system focuses on identifications and prevention of hazards which might cause the product deterioration and also assures food safety and quality (FSQ). System of HACCP, uses the approach of critical control points (CCP) during the food handling to avoid food safety problems. Food spoilage is a major issue for food manufacturers. Hence, it is crucial to assure the safety of food and to avoid food spoilage. Ocins (e.g., enterocins), lactocins, and bacteriocins are anti-bacterial proteins produced by bacteria that suppress or destroy the growth of other bacterial strains. Ocins signify a hopeful strategy to ensure food safety. This chapter assesses the factors which affect the safety and quality of food, and how FSQ are regulated.

Food Microbial and Molecular Biology: From Fundamentals to Applications. Saher Islam, Devarajan Thangadurai, Jeyabalan Sangeetha, Zaira Zaman Chowdhury (Eds.)

3.1 INTRODUCTION

Food is basically any product and/or substance which could be drunk or eaten by people and/or animals. It increases and maintains living standards or generally used as medical or nutritional supplement. Hence, food safety and quality (FSQ) are critically important for both developing and underdeveloped countries (Adams and Moss, 1995). "Food and Agriculture Organization of the United Nations (FAO) defined that: quality of food is the basic characteristic of the food which helps to determine its acceptability or value to consumers" (Oldewage et al., 2016). Safe food is that type of food which does not contain any substance that may pose a threat to the health of a person (Oldewage et al., 2016). Generally, food safety is the prevention of diseases which results from the ingestion of contaminated food. In 2000, the foodborne illnesses occurred which declined the confidence of the public and consumers crosswise the Europe, and the European Union (EU) began to change all approaches related to food safety. A unique legislation was decided by the EU for the entire members concerning food safety in which it was suggested that there should be scientific risk assessment with strict measures (Terreri, 2009). Globally as a result of the severity and rising number of outbreaks of food-related diseases, the public awareness is increased regarding the food safety (Forsythe, 2000). The diseases caused by *Salmonella* and *Campylobacter*, the main agents responsible for foodborne infections, are dramatically increasing in some countries, including Denmark, Finland, Iceland, the Netherlands, Norway, Sweden, Switzerland, and the United Kingdom. Recently, toxoplasmosis is recognized as an emerging foodborne parasitic disease. This protozoan zoonosis is estimated to be the third leading cause of food-related deaths in the USA after salmonellosis and listeriosis (Krause and Hendrick, 2011). Foodborne illnesses occur when contaminated products and foods are consumed. Food contamination at any step, from food production-consumption, yields parasites, viruses, bacteria, toxins, and chemical agents which finally trigger the foodborne illnesses (WHO, 2014). To make sure that individuals are provided with a safe food product, it is extremely important to improve management systems related to food safety (Gorris, 2005). Management systems which ensure safety of food are the following; Good Hygiene Practice (GHP), good manufacturing practice (GMP) and hazard analysis critical control points (HACCP). In current years these systems are applied in numerous countries (Nguz, 2007). Ocins are proteins or anti-microbial peptides which are produced by bacterial species (Jack et al., 1995). Now, ocins are utilized for controlling the growth of microbes in an extensive range of beverages and foods. For instance, nisin, which is produced by *Lactococcus lactis* by fermentation of milk-based substrates and pediocin PA-1 (pediocin PA-1/AcH) is produced by *Pediococcus acidilactici* are used for food protection (Rodríguez et al., 2002). Phages are used as natural anti-microbials within food for inhibiting unwanted bacterial species making the food acceptable for consumers. Hence phages could also be used for ensuring food safety (Strauch et al., 2007).

3.2 SAFE FOOD

Food safety is the guarantee that food would not be hazardous for consumers (WHO, 1998). Food safety is grouped for two purposes: (1) safe for soul and (2) safe for body. Food safe for the soul is basically that food which is determined as *halal* by the government and religion, and a food which is safe for the body is actually that food which does not contain hazardous microbes. Regarding the food safety, certain regulations were implemented by the government so that to prevent the biological and chemical contamination of different food products which threatens human. Besides the implementation of regulations which is discussed above, the government also decides strategies for managing the storage of food and also ensuring the nutrition of food. As per the food safety government regulation, it was decided that food safety distribution should discover quality standard. Standards for food safety might be categorized in numerous kinds like: (a) usual process which defines the necessities of the management system, (b) typical process that defines that how food would be formed with certifiable performance objectives which might generate numerical, and (c) numerical standards that define the characteristic needed for the products for instance the maximum residue limits or contaminant limits (Sjafrina et al., 2013).

3.3 FOOD QUALITY

Food quality does not relate to physical characteristics of different foods, but it also relates to how the final products are being perceived through consumer (Grunert, 2005). This comprises flavor, microbial aspects and texture, process of distribution, auditing, and certification costs. It moreover suggests responses such as technological modernization for creating higher effectiveness and reducing costs. Novel technological developments like "time-temperature indicators" might be applied so that to bring improvement in monitoring of temperature all over distribution system (Giannakourou and Taoukis, 2003). This helps in improving the shelf life of different food products, for example, fish, poultry, and pork (Raab et al., 2008). International Organization for Standardization (ISO), safe quality food (SQF), global food safety initiative (GFSI), British retail consortium (BRC) and International Food Standard (IFS) are the quality monitoring systems within different food industries. The purpose of such standards is to support the sellers in ensuring security of different products and to monitor quality of food.

3.4 FOODBORNE ILLNESSES DUE TO UNSAFE AND LOW-QUALITY FOOD CONSUMPTION

Foodborne diseases threaten the health of human. Foodborne diseases are responsible for higher mortality and morbidity rates within the general population. Similarly, it is also responsible for higher mortality and morbidity rates for at-risk groups people such as children, elder age people, young, infants, and immunocompromised individuals. Within the USA, about 500 deaths, 300,000 hospitalization's and 76 million diseases are caused by foodborne infections. Many life-long and acute infections are triggered due to the consumption of unsafe food, which ranges from diarrheal illnesses to numerous kinds of cancer. Each year in developing countries, about 2.2 million deaths are caused by the foodborne illness. Foodborne outbreaks are linked with the improper processing and handling of food, and also it is linked with variations in food eating habits (Tajkarimi et al., 2013). Numerous of zoonotic illnesses which transmits naturally between humans and animals are spread by means of various kinds of food (Pal, 2007). Different types of biological agents that comprise protozoa, viruses, helminths, and bacteria are responsible for causing foodborne illnesses and such illnesses have variable degrees of severity (Olaimat and Holley, 2012). Viruses which are commonly linked with foodborne disease outbreaks comprise *Sapoviruses*, *Adenoviruses*, *hepatitis A*, *Astroviruses*, *Norovirus* (NoV), and *Rotaviruses* (Cliver, 2007). Because of *rotavirus* about 870,000 Childs death occur each year in developing countries due to gastroenteritis (Pal, 2005). Numerous bacteria which are linked with foodborne diseases such as *Clostridium* spp., *Bacillus cereus*, *Aeromonas hydrophila*, *Listeria monocytogenes*, *Brucella* spp., *E. coli* O157:H7, *Salmonella* spp., *Shigella* spp., *Vibrio cholerae*, and *Campylobacter* spp. (Pal, 2013). In US about 500 deaths and 10,000–20,000 cases are caused each year because of *E. coli* 0157:H7 (Read and Kaplan, 1996).

Vibrio parahaemolyticus is the important reason for causing gastroenteritis in Japan which causes about 70% of entire food poisoning cases (Pal, 2005). In the USA, annually, 5 million cases of salmonellosis occur (Pal, 2005). Numerous protozoa, for example, *Cryptosporidium parvum*, *Toxoplasma gondii*, and *Cyclospora cayetanensis*, helminths like *Diphyllobthrium latum*, *Fasciolopsis buski*, *Fasciola hepatica*, and *F. gigantica* are moreover considered as foodborne microbes (Pal and Boru, 2010). Human-related fasciolosis is considered to be an important food-related parasitic disease with approximately 2.4 million cases (Dorny et al., 2009). Moreover, 20 million people are infected with *Diphylobothrium latum* infection worldwide. It is reported that about 35 million persons were infected with *Clonorchis sinensis* globally, of whom approximately 15 million were in China (Dorny et al., 2009).

3.5 FACTORS AFFECTING FOOD SAFETY AND QUALITY (FSQ)

Foods serve as fundamental building blocks of living beings; however, they might cause harm and threaten human health in numerous conditions (Yiğit and Duran, 1997). Several individuals worldwide get ill due to food which they consume. Such illnesses which are linked with consumption of food are termed as foodborne illnesses, and these might outcome from hazardous microbes (WHO, 2012). Foods could get hazardous or lethal for human health when it combines with viruses, parasites, bacteria, chemical toxins, and mold (Yiğit and Duran, 1997). Hence, it is essential that a safe food should be provided to consumers. Our food may be threatening in several ways like absence of preventive controls during preparation and processing of the food, poor hygiene practices during food processing, unsuitable agricultural practices, contaminated raw materials, improper usage of the chemical materials, and improper storage of food (WHO, 2003). Such issues were categorized into four groups: food hygiene, personal hygiene of food handlers, kitchen sanitation and equipment hygiene.

3.5.1 FOOD HYGIENE

Food hygiene is undermined by numerous factors. Foods hygienic quality is undesirably affected by purchasing low quality foods, food storage in unsuitable conditions, combine storage of cooked and raw foods, cooking, and preparing food by using improper means (Yiğit and Duran, 1997). The food hygiene is compromised if food get contaminated at any point, from stage of production to stage of consumption, dependent on the pH values, humidity, temperature of environment in which it is stored so the food gets possibly hazardous for health of human. Foodborne intoxication or infection produced through the ingestion of a contaminated drink or food is termed as food poisoning (Bas, 2006). Food poisoning might be caused by pathogens, metabolic disorders, naturally created food toxins, parasites, radioactive substances, allergic reactions, and naturally created fish toxins (Uçar, 2016). *Campylobacter*, Enterohemorrhagic *E. coli* (EHEC) and *Salmonella* are foodborne microbes affecting millions of individuals annually. Food poisoning caused by these microbes leading to symptoms such as upset stomach, diarrhea, abdominal pain, headache, and vomiting. Mostly bacteria are involved in food poisoning but numerous viruses and parasites (e.g., *T. gondii* and *T. spiralis*) could also be involved. Viruses such as Hepatitis A could act like parasites and it affects individuals along with the whole food chain (Duyff, 2012). Several bacteria such as *Campylobacter*, *C. perfringens*, *C. botulinum*, *S. aureus*, *Streptococcus*, *Bacillus cereus*, *Salmonella*, *E. coli* 0157:H7 and *Shigella* are mostly involved in food poisoning (Adams et al., 1995).

3.5.2 PERSONAL HYGIENE OF FOOD HANDLERS

Within the food chain, the most noteworthy stage is the food processing stage, and those who are responsible for performing their duties involved in such a stage take key responsibilities for preventing cases of food poisoning (Walker et al., 2003). The staff involved in the processing of food must include healthy persons who do not have any illnesses, and their medical check-ups should be done regularly. Along with their health, it is noteworthy also that the employees involved in food processing should take care of their own hygiene and implement appropriate handling of food. It is particularly important as handlers could bring about cross-contamination among cooked and raw foodstuffs, and they might threaten the hygiene of food through inappropriate cooking, storage, and preparation of foodstuffs (Walker et al., 2003). "Food and Drug Administration (FDA)" confirmed that about 81 food-related illnesses were produced through contaminated foods by workers during food processing (Ross and Guzewich, 1999). Public health is remarkably influenced by food handlers. In order to reduce food poisoning or foodborne illnesses, it is essential that food handlers should take care of their personal hygiene practices (Green et al., 2006).

3.5.3 KITCHEN SANITATION

Another critical concern in food safety provision is sanitation of kitchen. In South Carolina and North Carolina states, an investigation was carried out in child care centers, and it was confirmed that most of the kitchens were not meeting the requirements of "FDA's 2009 Food Code" (Wohlgenent et al., 2014). So as to reduce the hazard of foodborne diseases during the food processing and production, it is essential to reduce contamination hazard in arrangement, decking, and placement of kitchen utensils, all working area is cleaned and disinfected, and also to guarantee that materials and surfaces within the kitchen are anti-toxic. Besides this, it is essential that the kitchen should have mechanisms for controlling humidity and temperature. Effective measures should be taken in order to control pests (Codex Alimentarius, 2009).

3.5.4 EQUIPMENT HYGIENE

All the equipment which come in daily contact with foods must be made from such material which is capable of disinfection and cleaning. It should be non-toxic and resistant to corrosion. Arrangement of equipment must be in such a way so that it should be easily cleaned and all the area surrounding it should be easily cleaned. According to instructions, the chemicals should be used for cleaning the equipment. There must be regular and recorded calibration checks of all the equipment (Uçar, 2016).

3.6 OTHER MAJOR FACTORS AFFECTING FOOD SAFETY STANDARDS

3.6.1 CHEMICALS

Chemical threats are commonly produced by the existence of chemicals which could reason for intoxication. Heavy metals like mercury (Hg) and lead (Pb) are the poisoning triggering chemicals. Hormones, heavy metals, pesticides, antibiotics are utilized in the production of foods are examples of chemical contamination and are frequently foodstuff, particularly fruits, vegetables, and certain animal-related food products. Chemicals used in agriculture for beneficial and positive effects can also cause health hazards, i.e., when pesticides residues are left. So, it is very important to minimize chemical agents and pesticides and also to implement good agriculture practice (GAP) (Sjafrina et al., 2013).

3.6.2 PHYSICAL HAZARDS

Physical hazards include wood pieces, metal, stone, nails, and hairs. These may be arose from raw materials which are being contaminated, from damaged equipment, or it may also come from workers who process the food. Though physical hazards may not always reason for health issues or disease, but this hazard could act as a carrier of pathogenic bacteria and definitely interferes with food esthetic value. Further materials which are likewise unsafe for health are additives, preservatives, and different coloring agents. The usage of additives unsuitable are: (1) Borax for making of meatballs, crackers, and rice cake, (2) Formalin for preservation of noodles and tofu, (3) Dietary artificial sweeteners (for example saccharin and cyclamate) are applied for snack foods, and (4) Hazardous color (amaranth, methanyl yellow and rhodamine B) are mostly found on lemonade, crackers, product syrup, jellies, bread, snack food (fried chicken, fried bananas, cendol, and tofu) (Sjafrina et al., 2013).

3.7 FOOD SAFETY MANAGEMENT SYSTEM (FSMS)

For producing nutritious, good quality and a safe food, there is a requirement of effective management for the entire system. "Food Safety Management" is coordinated activity to control and direct a company so that to produce safe and quality foods (Griffith, 2000). Health hazards are generally caused by food safety hazards, that are categorized as physical, chemical, or biological. "Food safety management system (FSMS)" carries a set of procedures and practices which are intended to manage all the hazards related to the safety of food (Pozo et al., 2018). Operative food control systems are required for controlling the safety of food (McMeekin et al., 2006). Presently, PAS 220, ISO 22000 and HACCP are commonly applied and globally accepted FSMS.

3.7.1 HACCP

Hazards Analysis and Critical Control Points (HACCP) is basically a systematic approach which analyzes all risks and hazards regarding the food safety. HACCP is basically scientific and organized system for ensuring the safety of food. In 1960s Pillsbury established this system for the NASA and U.S. army in order to accomplish "zero defects" and to guarantee complete food safety. Basically, this is management, implementation, and development of efficient processes to guarantee the safety of food (Buchanan, 1990). HAACP could be utilized for controlling any step of the food supply chain (like distribution, processing, and supply) (Ropkins and Beck, 2000). It is recognized as one of the finest systems to guarantee the safety of the product, and it is getting internationally accepted (Unnevehr and Jensen, 1999) and has its key aims to control, assess, and identify risks for safety of food (Stevenson and Bernard, 1999). HACCP is utilized appropriately to avoid foodborne disease outbreaks (Bhuiyan and Alam, 2005). The systems which are based on HACCP are thought out to be the most efficient means to manage the safety of food (Griffith, 2000).

3.7.2 ISO 22000

In 2005, a standard was issued by "The International Organization for Standardization (ISO)" for managing the safety of food which is recognized as ISO-22000. Actually, system of ISO-22000 is mixture of HACCP principles. Afterwards its introduction, it started to be applied within about 50 countries in two years (Surak, 2007). ISO-22000 standard used a fundamental approach of implementing a defensive system which helps in securing consumers from all food-related illnesses. ISO-22000 standard controls and regulates entire processes within the food chain, comprising equipment, staff members and infrastructure. Within business organizations, implementation of "FSMS" comprise of equipment control, general hygiene practices, product control, maintenance, production control, hygiene of visitor and staff, transportation, storage, training, product information, communication, evaluation, and selection of suppliers (Arvanitoyannis et al., 2009). The key objective of ISO 22000 is to have such a system which determines all undesirable hazards and also its main aim is the protection of consumer health and product safety. Food safety is not only the important role of ISO 22000 but this system also helps in improving the nutritional and sensory qualities of food. Moreover, it performs a key part in assurance of quality in industrial production. Finally, ISO 22000 aids to decrease operational losses through introducing a more operative usage of resources to rise productivity, and thus, it establishes such a system which assure quality (Surak, 2007).

3.7.3 PUBLICLY AVAILABLE SPECIFICATION 220 (PAS 220)

Worldwide different food manufacturers in collaboration with the CIAA (i.e., Confederation of Food and Drink Industries) created this standard. PAS 220 objective was to eliminate the flaws of the ISO 22000. Unilever, Nestle, Kraft, and Danone (the sector leaders commonly recognized as "G4"), jointly issued this standard. In 2008, PAS-220 standard was accessible, and it is appropriate for all kinds of companies. It was proposed that globally recognized ISO 22000 standard and PAS 220 must be used in combination (Boisrobert et al., 2009). The topics and content of PAS 220 elaborate on the 10 sub-titles in the ISO-22000 standard and includes five of its own, and hence it results in following 15 items (Boisrobert et al., 2009).

- Management of purchased materials
- Adequacy of the equipment, cleaning, and preventive care
- Supporting plants (air, water, energy)
- Placement of work site, buildings, and their wings
- Product recall procedures
- Supporting facilities, including sewage and wastes
- Informing consumers about products
- Measures against cross-contamination
- Re-processing
- Storage
- Staff hygiene and workers' lodgings
- Cleaning and sanitation
- Food defense, biodefense, and bioterrorism
- Structure and placement of buildings
- Pest control

3.8 GMP AND GHP

A management system of processes and products is necessary for every food producer (Mårdén, 1995). GMPs could be termed as good house-keeping within the food production (Bernhardt and Raschke, 1998). GMPs guarantees safe food production which having consistent quality. Additional profits of GMP are enhanced profits and reduced waste. Primarily the recommendations for GMP were established just for drug industries, but soon it was improved to suit food production too. GMPs encompass routines for managing production premises, raw material, equipment, and hygiene (Mårdén, 1995). Proper GMPs are essential when working according to the systems like ISO 22000 or HACCP standards. System of HACCP goes deeper as compared to GMP into the examination of possible risks and then how to control these risks. According to Bernhardt and Raschke (1998) it is: essential that managements are convinced of the need to introduce GMP. They must understand not only the benefits of GMP but also appreciate the resources required to make it work.

This can suppose for all types of quality management systems (QMS) which are established for the production of food. Whereas GMP mainly discusses all the technical features of the process of production. The focus of good hygiene practice (GHP) is basically on the hygiene aspect in the production of food (Buncic, 2006). Both the terms like GMP and GHP are frequently applied instantaneously because within processing of food it is difficult to consider hygiene when technical perspective is not considered (Nordenskjöld, 2012). In brief, GHP and GMP offer fundamental principles for the production of food in a hygienic manner. Those industries which implemented the GMPs have the following results as compared to others:

- Greater productivity and employee motivation and better psychological conditions
- Safer products, better quality and decrease in incidence of consumer complaints (da Cruz et al., 2006)
- Better, cleaner more agreeable, and safer working environment

3.9 ASSURING SAFETY AND QUALITY OF FOOD IN PROCESSING

Some factors are considered to assure FSQ in processing of food which are given in subsections.

3.9.1 SANITARY DESIGN AND MAINTENANCE OF PLANTS AND EQUIPMENT

By "Federal Food Drug and Cosmetic Act, Chapter IV, Section 402, it was defined that food as contaminated; if it has been prepared, packed, or held under unsanitary conditions whereby it may have become contaminated with filth" keeping equipment and facilities in such a way which avoids these conditions is just not a best idea, but its regulatory and legal prerequisite for foods production. Numerous checklists are there (counting in the regulations) which provide particular requirements for grounds and equipment conditions. This information is included in the audit scheme of GFSI. Generally, there should be no standing water in the surrounding grounds, discarded equipment or accumulation of waste that might captivate or harbor pests like insects, birds or rodents. Windows, doors, and exterior walls should be sealed in order to stop the access of pests into the working area (FDA, 2007). Area of food processing should be physically detached from worker welfare zones, for example, locker rooms, lunchrooms, and restrooms so that to avoid contamination where food is processed. Walls, ceilings, floors within the food processing should be built in such a manner that avoid water or food debris accumulation and could be effortlessly sanitized and cleaned. Within production area porous materials and wood are not allowed. Proper sinks should be available for handwashing. Sinks for washing the utensils should be separate from the sinks for handwashing. Merely drinkable water should come into interaction with equipment. Dripping water, condensation, and improper drains could be a main contamination source, and it should be adjusted (FDA, 2010c). Equipment should be built in such a manner that it is easily sanitized and cleaned. There should be no rough welds, dead ends, or further areas within the equipment which could permit water and food debris accumulation. Furthermore, the equipment must be able to resist cleaning chemicals. Biofilms formation (sticky bacterial layer that adheres to a surface) could cause food product cross-contamination. Moreover, rusted equipment could cause a physical pollutant risk. There should be routine maintenance schedule for all equipment. Proper standard operating procedures (SOPs) for maintaining equipment are important, and the equipment should be formally handed off to the sanitation group after maintenance and before production resumes (FDA, 2011b).

3.9.2 CLEANING AND SANITATION

Proper sanitation and cleaning are important in order to maintain a good quality and safe food. Appropriate sanitation and cleaning means that protective equipment is utilized and strict procedures are followed by entire workers at all times. Appropriate sanitation and cleaning would only occur if suitable sanitizers and cleansers are selected, and definite mechanical action, concentration, temperature, and dwell time are followed. Generally, the equipment and environment of processing must be maintained in sanitary and clean condition. Moreover, sanitation and cleaning actions must be carried out in such a manner that it does not contaminate or threaten the food, environment, or equipment. Acid cleaners are effective in removing inorganic soils, while alkaline cleansers help to remove organic soils. Verification and validation of sanitation and cleaning methods of food contact surfaces are noteworthy so that to assure that approaches are operative. Numerous methods are applied: protein swabs, visual inspection, microbiological swabbing, and ATP swabs. Use care when choosing which method to use (FDA, 2011b).

3.9.3 TRANSPORTATION AND STORAGE

An important role is played by the storage and transportation in maintaining quality and safety of foods. Appropriate control of temperature, allergen segregation and pest control help to maintain food safety throughout transportation and storage. In 2005, Congress passed the Sanitary Food Transportation Act, but rulemaking was not initiated by the FDA until 2010, and proposed regulations that have not yet been published. However, general guidance was issued in 2010 to assist food and transportation companies (FDA, 2010a). In spite of the lack of particular regulatory instructions, food-related companies are even now accountable for food security and safety throughout the whole supply chain involving transportation. Before loading a truck, which carries all the finished products or raw materials, it is essential that it must be checked carefully so that to detect any conditions which can influence FSQ. Workers must check holes within the trailer that might permit entry of rodent excreta, insects, dirt, debris (like pallet wood) and pests. Unusual odors must be checked by the workers. Inspection results must be recorded so that help should be taken from it in the future. For frozen and refrigerated loads, the definite temperature of food and trailer must be examined and recorded upon receiving and before distribution in order to guarantee that temperature is in that range which is required (FDA, 2011a). Foods which are transported in frozen state must yet be in frozen state when it is received (US Public Health Service and FDA, 2013). Upon loading all the full truckload and bulk load shipments are sealed, and particular seal numbers are noted on documentation (for example, lading bill). Former to receiving the load, seal numbers must be confirmed against the numbers which are recorded in documentation so that to determine that during the shipment the seals was removed or not. Then one who inspects must reject the loads which seals are changed or missing. Rail cars must be inspected and sealed in the same way. During transportation in warm months of summer, commonly the rail cars (which contains grain products, for example, flour) are fumigated with insecticides in order to avoid insects infestation (FDA, 2011b).

3.9.4 PERSONNEL PRACTICES

The aim of personnel practices is to avoid environment or food contamination through workers. Those workers who are sick that might possibly contaminate food contact surfaces, food, or packaging should not be allowed to work within the food industry till their health problem is solved. It is important that injuries or minor must be covered via proper bandage. Those bandages which are applied within processing zones are characteristically blue colored, so that it is visible within food (Schoenfuss and Janet, 2014). Cleanliness must be maintained by the workers while working within food environment. Appropriate cleanliness is achieved by following steps like; correct washing of hands and suitable use of glove, good personal hygiene, wearing hair restraints, removing jewelry, protective clothing, refraining from drinking, eating, chewing gum or smoking within areas of food processing, storage of personal objects and clothing within an area which is isolated from the area of food processing and avoiding food contamination with fragrance or perfume, lotions, perspiration, medicine, and further cosmetics products. Hand washing must be followed and gloves should be changed after using the restroom. If non-food contact surfaces are touched, such as garbage, pallet jacks, cleaning utensils, pens, or objects that are on the floor so after this glove must be changed or hands should be washed. Furthermore, one's face or hair touched so after this handwashing should be done or gloves are changed. If gloves are damaged or dirty, so there is a need to replace such gloves. After removing the disposable gloves, it is important that these gloves must be discarded, and should never be used again. All the visitors and workers should be trained in appropriate personnel practices, and rules should be reinforced by the supervisors (Schoenfuss and Janet, 2014).

3.9.5 WATER QUALITY

During processing of food, water is utilized in numerous ways, for example:

- As an ingredient to rehydrate other ingredients or standardize a formula
- Sanitation, cleaning, and rinsing

- For irrigation of crops
- For postharvest cooling of fruits and vegetables
- For hand washing and other sanitary functions
- As a cooling or heating medium (either indirect contact with food or indirectly)

The foremost concern with regard to water is its role in food safety. The quality attributes of water (flavor, pH, mineral composition, etc.), its composition and amount of wastewater from a facility are also important to food processors. Drinking water from municipal sources that is generally considered safe has been a source of contamination and led to outbreaks such as the 1993 *Cryptosporidium* incident that affected approximately 400,000 people and led to 100 deaths (CDC, 1997). Microbes like hepatitis A virus (HAV) from fecal-contamination, together with environmental contaminants like arsenic and nitrates are a concern with well water. In the US within 2008, pathogens like *Legionella*, *Campylobacter*, *Shigella*, *Salmonella*, and *E. coli* O157:H7 caused 16 illness outbreaks which affected 1672 individuals and caused 3 deaths (CDC, 2013). New concerns are for "emerging contaminants" that are not removed by traditional municipal water treatments and are found in drinking water that, while not at toxic levels, present health concerns (Richardson et al., 2014). These comprise:

- Hormones (from the animal industry and urine, especially from women on birth control – estradiol and estrogen)
- Pesticides
- Phthalates (plasticizers in plastics, used in inks and cosmetics)
- Surfactants (from cleaning agents)
- Endocrine-disrupting compounds
- Perfluoro-octane sulfonate (PFOS), a chemical used in fabric protectors

Furthermore, food processing plants provide certain of such chemicals to the aquatic habitat in wastewater form. Local agencies and EPA help in its regulation. Relatively, plants need to be constructed to lessen the water release which is un-treated and required to carry out such activities which regulate pH and lessen the biological oxygen demand (BOD). Numerous of the endocrine active substances present in the drinking water are because of food processing and agricultural activities (Lundgren and Novak, 2009).

3.10 APPROACHES FOR ASSURING FOOD SAFETY AND QUALITY (FSQ)

Different factors affect FSQ. So, it is extremely important to protect our food. Following approaches are adopted for assurance of FSQ such as ocins, bacteriophages, and some other approaches are used.

3.10.1 OCINS FOR FOOD SAFETY

Ocins are basically proteins or anti-microbial peptides which are produced by the bacterial strains and species (Jack et al., 1995). Presently, ocins are applied for controlling growth of microbes within an extensive range of beverages and foods. Hence, there are commercial benefits of ocins. Nisin is formed when *Lactococcus lactis* ferments milk and *Pediococcus acidilactici* helps in formation of pediocin PA-1/AcH, and it is utilized as a protectant of food (Rodríguez et al., 2002). Although nisin is expansively applied within food industry, but there are several limitations of its use like, at higher temperature it has low stability (Holcapkova et al., 2017). It is susceptible to biological enzymes (e.g., chymotrypsin, trypsin, and proteinase Ks) and has effectual action on a very slight range of pH. Consequently, it is essential that new protein-based anti-microbials are developed for use within the food industries. Chikindas et al. (1995) reported the applications of bacteriocins, and Yang et al. (2014) discussed the antimicrobial activities of bacteriocins and their applications in foods and pharmaceuticals. The authors argued that several purified bacteriocins have been used in the food industry to extend food preservation time and treat certain diseases. Bacteriocins play significant roles in ecological homeostasis wherein they maintain the population dynamics within the species (Margaret and David, 1999).

3.10.1.1 USE OF OCINS IN DAIRY INDUSTRY FOR FOOD SAFETY

A huge range of foodstuffs are formed from milk, comprising dried, pasteurized, and raw milk, yogurt, paneer, cottage cheese, butter, yogurt-based drinks, cream, sour cream and further fermented dairy products, processed cheese, cream cheese, ripened cheeses, and fresh soft cheeses. Within dairy industries, the pathogenic bacterial species such as *E. coli*, *Salmonella* spp., *L. monocytogenes* and *S. aureus* are of main concern (De Buyser et al., 2001). The hard and semi-soft cheeses decomposition because of the formation of gas through *C. tyrobutericum* is accountable for considerable economical losses (Bergère and Lenoir, 2000). In different types of dairy products, nisin is widely tested. For instance, in cheese it is used for inhibiting the production of gas through *C. tyrobutyricum* (De Vuyst and Vandamme, 1994) and it is also used in thermally treated cream for inhibiting multiplication of *Bacillus cereus* throughout storage (Nissen et al., 2001). In packaged sliced cheese, *S. aureus*, *Listeria innocua*, and lactic acid bacterial (LAB) population is decreased by nisin-coated with polyethylene/polyamide packaging (Scannell et al., 2000). In preservation of dairy food, a potential application was shown by the Lacticin 3147 and Lactococcal ocin (Ross et al., 1999). Infant formulae which contain lacticin 3147 reduced cell counts of *S. aureus* and suppressed the *L. monocytogenes* growth (Morgan et al., 1999). When lacticin 3147 powder was used so within 2 hours, viable cell numbers of *L. monocytogenes* were reduced by 99% in cottage cheese and 85% in yogurt (Morgan et al., 2001). As concentrated enterocin CRL 35 was added to goat cheese so it decreases population of *L. monocytogenes* and does not affect the quality of cheese (Farías et al., 1999). Correspondingly, in numerous dairy products, counting cheese sauce, and dressed cottage cheese, *L. monocytogenes* was decreased by pediocin PA-1/AcH (Pucci et al., 1988). Propionicin PLG-1 and *Propionibacterium* P127 bacteriocin caused inhibition of numerous bacterial species (*Corynebacterium* sp., *Yersinia enterocolitica, Vibrio parahaemolyticus, P. fluorescens* and *L. monocytogenes*), which suggests that it act as an anti-microbial preservative of food (Lyon et al., 1993). Variacin is a type of ocin which is made by *Kocuria varians*, and when this was added to chocolate, vanilla desserts, and chilled milk products, it stopped proliferation and growth of *B. cereus* (O'Mahony et al., 2001). Several investigations (Fenelon et al., 1999) have shown that when lacticin 3147-producing starters were used so that it enhanced the quality of cheese through preventing non-starter LAB flora throughout ripening. Within cottage cheese, growth of *L. monocytogenes* was stopped via the trans-conjugants of lacticin 3147 (McAuliffe et al., 1999).

3.10.1.2 USES OF OCINS IN CEREALS, PULSES, AND LEGUMES

Bacteriocins made by LAB guarantee the quality and safety of legume-based fermented food products and cereals. Nisin is produced by *L. lactis* subsp. *lactis* IFO12007, and it is utilized as a starter culture in the fermentation of cooked rice. When the growth of such bacteriocin making strain occurred within cooked rice so it produced nisin which stop the *B. subtilis* growth (Kato et al., 2001).

3.10.1.3 USES OF OCINS IN FRUITS, SPROUTS, AND VEGETABLE INDUSTRY FOR FOOD SAFETY

This group of food comprises untreated vegetables and fruits, fermented vegetables, processed ready-to-eat vegetables, drinks, beverages, canned products, and fruit juices. For many centuries sprouts are being used in Asia and in other regions of the world (Rosas and Escartin, 2000). Sprouts is thought of as healthy food, and these are sources of food-related diseases produced by *B. cereus*, *E. coli* O157 and *Salmonella* (Beuchat, 2002). Fresh foods such as lettuce, tomatoes, and raw celery are involved in numerous outbreaks of listeriosis (Beuchat, 1996). *Enterococcus mundtii* were isolated from slightly processed vegetables which produced the mundticin and inhibited the growth of *L. monocytogenes* (Bennik et al., 1999). Mundticin helped in protection of mung bean sprouts (Bennik et al., 1999). *Erwinia carotovora* subsp. *carotovora* causes soft rot in fruits and vegetables and is involved in decay. No damaging impacts of ocins have been reported on vegetable tissues or cells. Therefore, ocins may be suitable for preventing spoilage in vegetables and fruits, when is applied either separately or along with sanitizers. Bacteriocin are produced by LAB and prevents decay of vegetables and fruits. When Pediocin is applied alone or in combination with organic acid so it is much effective as compared to nisin alone in decreasing populations of *L. monocytogenes* (Bari et al., 2005).

Antagonistic action was shown by enterocin AS-48 against *S. aureus* and in lettuce juice it entirely inactivated the multiplication of *B. cereus* and *L. monocytogenes* (Grande et al., 2005b). The growth of *L. monocytogenes* was inhibited in slices of honeydew melon using nisin and its lytic activity was increased when phage was used in combination with it (Leverentz et al., 2003). No effectiveness was shown by nisin in controlling fruit spoilage within yogurt (Penney et al., 2004). The growth of *C. tyrobutyricum* and production of gas was inhibited by bovicin HC5 and nisin in tinned mango pulp (de Carvalho et al., 2007). Endospore-formers bacterial species are key contamination source in cooked and canned vegetables. It was confirmed by numerous investigations that bacteriocins might eradicate the proliferation and growth of endospores, and improve the effectiveness of thermal treatments to eliminate endospores in cooked and canned vegetables. Nisin inhibited the canned vegetables spoilage by endospore-forming, aciduric

(*B. coagulans*, *B. macerans* and *C. pasteurianum*) and non-aciduric (*C. thermosaccharolyticum* and *B. stearothermophilus*) microbes (Thomas et al., 2005). Nisin inhibited the spoilage of beverages and fruit juices by *Alicyclobacillus* (Yamazaki et al., 2000). *Alicyclobacillus acidoterrestris* was entirely inactivated by enterocin AS-48 (for 3 months) subsequently to inoculation of numerous fruit juice samples (Grande et al., 2005a). Nisin and enterocin AS-48 inactivated the endospores which were made by *A. acidoterrestris* (Grande et al., 2005a).

3.10.2 BACTERIOPHAGE APPLICATION IN FOOD SAFETY

Bacteriophages are viruses that infect bacteria and do not cause any harm to plants, animals, and humans. Since the phages discovery in 1915, they are extensively applied not only in veterinary medicine and human but likewise in numerous agricultural surroundings. Viruses are obligate parasites and take control of host cell machinery upon multiplication. Phages either integrates their genetic material into chromosome of bacteria without causing death of host cell (lysogenic pathway) or cause lysis of cell so that to release the virus particles which are newly formed (lytic pathway). From a food safety perspective, lytic phages are significant. As a bio-control agents, the phages are noteworthy for numerous reasons: (i) phages have long shelf life; (ii) as bacterial species develop defense mechanisms against phage so that to ensure survival, so in result of this phages also undergo adaption in them to such modified host systems; (iii) phages are easy to propagate, isolate, and relatively cheap; (iv) bacteriophages have lower intrinsic toxicity, meanwhile they comprise nucleic acids and proteins; (v) phages could usually survive food processing environmental stresses (comprising food physio-chemical conditions); (vi) bacteriophages are specific to their host cell and leave remaining microbiota as unaffected; (vii) phages are found abundant in different foodstuffs and can be isolated from extensive range of different raw products (for example chicken and beef) (Atterbury et al., 2003), fermented products (e.g., yogurt, and cheese), processed foods (e.g., roast turkey, biscuit dough and pies), and seafoods (e.g., oysters, and mussels) (Suárez et al., 2002). Phages usage for promoting safety of food could be done on four different phases in the food chain. Reducing the colonization of microbes within animals throughout primary production (Rosenquist et al., 2003). Phages could be incorporated in food or drinking water or given orally for controlling *Campylobacter* and *Salmonella* within poultry, or sprayed so that the avian pathogenic *E. coli* is targeted within poultry, and administered rectally/orally for controlling *E. coli* in ruminants. Reduction in colonization on different foods (bio-control) throughout processing could be achieved using phages on food surfaces, for instance, in the case of processed foods, meats, and fresh products. It was shown by an experimental data that bacteriophages were much effective towards those bacteria which were actively growing and shown less activity against those bacteria which were not growing (Snyder and Champness, 2007). Biofilms are made on surfaces of different instruments which are used in the food processing industry, for instance, in food processing, storage, or food handling. Biofilms are particularly formed in spots which are not easily sanitized or cleaned. In a study *in-vitro*, when phages were used against the biofilm-forming pathogenic and spoilage bacteria under ideal conditions, viable cell counts were reduced significantly and hence, phage usage for bio-sanitation could be hopeful though it is a challenging task because of the range of bacterial species present in different surroundings. Phages also help in bio-preservation since they were reported to cause lysis of hosts at temperatures as lower as 1°C (Greer, 1988).

3.10.2.1 BACTERIOPHAGE FOR CONTROLLING ESCHERICHIA COLI O157:H7 CONTAMINATION

Pathogens emergence, like *E. coli* O157:H7, is public health-related hazard. Ruminants serves as main reservoirs for *E. coli* O157:H7 and different animals products are polluted during slaughtering or milking. Before the slaughtering, phage treatment helps in reducing the contamination caused by the pathogen. When phage CEV1 was orally given, so in 2 days, it caused a 2-log-unit reduction of *E. coli* O157:H7 which was present within intestines of sheep (Raya et al., 2006). Though, no efficiency was shown by the phage KH1 which was orally given. When a mixture of phages like SH1 and KHI were given rectally to cows and bacteriophages were correspondingly kept at (10^6 PFU/ml) within the drinking water, so a minor cell count was noticed (Sheng et al., 2006). Phages are also used on meat surfaces so that to prevent the growth of pathogens. A beef was contaminated with *E. coli* O157:H7 (about 103 CFU g^{-1}) so a combination of 3 phages was used on such contaminated beef. No viable cells were noticed within most samples after storing at 37°C (O'Flynn et al., 2004). Similarly, in another study, phages, which were specific for *E. coli* O157, were isolated and collected from feces of cattle and applied on spinach blade, so afterwards 2 hours of phage application, it reduced *E. coli* by 4.5-log (Patel et al., 2011).

3.10.2.2 BACTERIOPHAGE FOR CONTROLLING CONTAMINATION BY SALMONELLA

Salmonella could be found in several species of animal and food poisoning is mostly caused by it. The average fecal counts of *S. enteritidis* was reduced by 0.3–1.3 log units when a cocktail of phage was used, but microbe was not entirely eliminated (Sklar and Joerger, 2001). In a study the broilers were infected artificially and were orally treated with higher amounts of phages (about 10^{11} PFU) (Fiorentin et al., 2005). Even though no bacterial eradication was noticed from birds, both investigations revealed that bacteriophage treatment might reduce levels of bacteria which enters the production line of poultry. *Salmonella* phage SJ2 activity was assessed in manufacturing of cheddar cheese. Phage SJ2 when applied on cheddar cheese, it reduced levels of *Salmonella* by 1.0–2.0 log in pasteurized and raw cheese (which was made by using milk that was phage treated) (Modi et al., 2001). Phage activity was also assisted on fruits using phage cocktails against *Salmonella*. In melon the numbers of phage remained comparatively stable and target bacteria was significantly reduced (Leverentz et al., 2001). In another study, *Salmonella*-phage "Felix-O1" was assessed in bio-control trials. A chicken hotdog was contaminated with *S. typhimurium* (300 CFU) and phage was applied. The phage resulted in a reduction of 1.8 and 2.1 log units in bacterial count (Whichard et al., 2003). Similarly, when phage F01-E2 was added to chocolate milk and Turkey deli meats, so it resulted in 5 log reductions of *Salmonella* (Guenther et al., 2012).

3.10.2.3 BACTERIOPHAGE FOR CONTROLLING CAMPYLOBACTER CONTAMINATION

Enteric microbe such as *C. jejuni* colonized in broiler chickens. In industrialized nations, oral infection by such microbe is the most common reason of foodborne illness. The presence of phages negatively correlates with *Campylobacter* levels within cecal contents and this indicates that some level of bio-control might happen (Atterbury et al., 2005). Good efficacy was shown by the phages in decreasing the counts of *Campylobacter* in cecal contents, thus it lowers the cross-contamination risk throughout slaughtering (Wagenaar et al., 2005). Phage F2 was shown to be very effective when applied on artificially contaminated chicken skin (Atterbury et al., 2003). The phage anti-bacterial action was noticed at –20 and 4°C using a higher MOI (about MOI 10^5). At room temperature, efficiency was shown by the minor MOIs of about 100, and more than 95% of the targeted cells killing was noticed (Goode et al., 2003). Similarly, when phage NCTC 12673 was applied on chicken skin, so it caused a 1.0-log reduction in counts of *Campylobacter* in the treated group (Atterbury et al., 2003).

3.10.2.4 BACTERIOPHAGE FOR CONTROLLING STAPHYLOCOCCUS AUREUS CONTAMINATION

Food poisoning is frequently caused by *S. aureus* and it is an etiological agent which causes nosocomial and opportunistic infections. Staphylococcal food poisoning occurs when staphylococcal enterotoxins (SEs) are absorbed which are already formed in the matrices of food (Le Loir et al., 2003). Mastitis caused by *S. aureus* is a main issue for the dairy industry and this pathogen is the chief source for contaminating the milk. The capability of lytic *S. aureus* phage-K for eliminating the intramammary infection caused by bovine *S. aureus* during lactation was assessed, but in the gland the infused phage was inactivated or degraded (Gill et al., 2006a). Also, in the raw milk the inactivation of phage K was noticed, this was because of whey proteins adsorption to surface of cell which interferes with attachment of phage (Gill et al., 2006b). Though, when a combination of 2 lytic bacteriophages (which were of dairy origin) were applied at MOI 10^2 so it effectively causes inhibition of *S. aureus* during processes of curd manufacturing (Garcia et al., 2007). In another study phage cocktail (Φ35 and Φ88) was added to hard cheese and fresh cheese. Phage cocktail when added to pasteurized milk that contain *S. aureus*, so after 6 hours it reduced *S. aureus* to nondeductible levels within fresh cheese and also caused a steady reduction of *S. aureus* in hard cheese (Bueno et al., 2012) (Table 3.1).

3.11 CONCLUSION

Food could never be completely safe and several types of microbes cause numerous foodborne diseases. A large number of foodborne diseases are caused by microbes originating from animals or animal products that include bacteria, viruses, helminths, and protozoa. At any stage in the food processing chain, food may be contaminated with microbes and therefore there is a need to implement hazard analysis critical control point (HACCP) system. Production of a safe and good quality food products do not occur by coincidence, but it relies on an obligation from the entire workers in the food industry

TABLE 3.1 Application of Phages for Ensuring Different Foods Safety

Kind of Food	Phage	Target Pathogen	Results of the Study	References
Chicken skin	NCTC 12673	*Campylobacter*	1.0-log reduction seen in treated group compared to untreated group	Atterbury et al. (2003)
Spinach blade	Bacteriophages specific for *E. coli* O157 collected and isolated from feedlot cattle feces	*Escherichia coli* O157:H7	4.5-log reduction of *E. coli* after 2 hours of phage addition	Patel et al. (2011)
Cheddar cheese	SJ2	*Salmonella*	*Salmonella* levels were reduced by 1.0–2.0 log in raw and pasteurized cheese created using milk that was treated with phage.	Modi et al. (2001)
Fresh cheese and hard cheese	Phage cocktail (Φ88 and Φ35)	*Staph. aureus*	Phage cocktail added to pasteurized milk challenged with *S. aureus*, led to reduction of *S. aureus* to undetectable levels after 6 h in fresh cheese and continuous reductions in hard cheese.	Bueno et al. (2012)
Turkey deli meats and chocolate milk	Phage F01-E2	*Salmonella*	Phage addition resulted in 5 log reduction of CFU and a 3-log reduction when it was applied on hot dogs	Guenther et al. (2012)
Chicken skin	Phage F2 or numerous lytic bacteriophages	*Campylobacter*	More than 95% of the targeted cells killing was noticed	Atterbury et al. (2003); Goode et al. (2003)

to develop and implement methods for ensuring that the food is made in a safer environment. As regulatory requirements develop to be much more challenging as science explores innovative hazards, hence the industry should develop innovative approaches for guaranteeing safer and high-quality food as consumers become much choosier about what they consume. It is important for food manufacturers to constantly bring improvement in their food quality and safety management systems. HACCP and other FSQ management systems search for risks or everything that can go inappropriate regarding the safety of product and implements controls so that to guarantee that the food products would not be harmful for the consumer. FSQ is necessary in most countries, particularly in developing countries. Foodborne diseases due to unsafe and low-quality food have a substantial impact not just on health but also have impact on development. Therefore, novel approaches are needed in order to bring improvement in FSQ.

KEYWORDS

- **food poisoning**
- **food safety**
- **foodborne diseases**
- **food intoxication**
- ***Salmonella***
- ***Shigella***

REFERENCES

Adams, M. R., Moss, M. O., & McClure, P., (1995). *Food Microbiology* (Vol. 17, pp. 121, 122). The Royal Society of Chemistry, Science Park, Cambridge.

Alimentarius, C., (2009). *Foods Derived from Modern Biotechnology*. Codex Alimentarius Commission, Joint FAO/WHO Food Standards Program, Rome, Italy.

Arvanitoyannis, I. S., (2009). *HACCP and ISO 22000: Application to Foods of Animal Origin*. John Wiley & Sons.

Atterbury, R. J., Connerton, P. L., Dodd, C. E., Rees, C. E., & Connerton, I. F., (2003). Application of host-specific bacteriophages to the surface of chicken skin leads to a reduction in recovery of *Campylobacter jejuni*. *Applied and Environmental Microbiology, 69*(10), 6302–6306.

Atterbury, R. J., Dillon, E., Swift, C., Connerton, P. L., Frost, J. A., Dodd, C. E., et al., (2005). Correlation of campylobacter bacteriophage with reduced presence of hosts in broiler chicken ceca. *Applied and Environmental Microbiology*, *71*(8), 4885–4887.

Bari, M. L., Ukuku, D. O., Kawasaki, T., Inatsu, Y., Isshiki, K., & Kawamoto, S., (2005). Combined efficacy of nisin and pediocin with sodium lactate, citric acid, phytic acid, and potassium sorbate and EDTA in reducing the *Listeria monocytogenes* population of inoculated fresh-cut produce. *Journal of Food Protection*, *68*(7), 1381–1387.

Baş, M., Ersun, A. Ş., & Kıvanç, G., (2006). Implementation of HACCP and prerequisite programs in food businesses in Turkey. *Food Control*, *17*(2), 118–126.

Bennik, M. H. J., Van, O. W., Smid, E. J., & Gorris, L. G. M., (1999). Biopreservation in modified atmosphere stored mungbean sprouts: The use of vegetable-associated bacteriocinogenic lactic acid bacteria to control the growth of *Listeria monocytogenes*. *Letters in Applied Microbiology*, *28*(3), 226–232.

Bergere, J. L., & Lenoir, J., (2000). Cheese manufacturing accidents and cheese defects. *Cheesemaking: From Science to Quality Assurance*, pp. 477–508.

Bernhardt, H. W., & Raschke, A., (1998). Good manufacturing practice (GMP) in cane sugar factories. In: *Proc. S. Afr Sug. Technol. Ass.*, *72*, 215–218.

Beuchat, L. R., (1996). Pathogenic microorganisms associated with fresh produce. *Journal of Food Protection*, *59*(2), 204–216.

Beuchat, L. R., (2002). Ecological factors influencing survival and growth of human pathogens on raw fruits and vegetables. *Microbes and Infection*, *4*(4), 413–423.

Bhuiyan, N., & Alam, N., (2005). An investigation into issues related to the latest version of ISO 9000. *Total Quality Management and Business Excellence*, *16*(2), 199–213.

Boisrobert, C., Stjepanovic, A., Oh, S., & Lelieveld, H., (2009). *Ensuring Global Food Safety: Exploring Global Harmonization*. Academic Press.

Buchanan, R. L., (1990). HACCP: A re-emerging approach to food safety. *Trends in Food Science and Technology*, *1*, 104–106.

Bueno, E., García, P., Martínez, B., & Rodríguez, A., (2012). Phage inactivation of *Staphylococcus aureus* in fresh and hard-type cheeses. *International Journal of Food Microbiology*, *158*(1), 23–27.

Buncic, S., (2006). *Integrated Food Safety and Veterinary Public Health*. CABI.

Castro-Rosas, J., & Escartin, E. F., (2000). Survival and growth of *Vibrio cholerae* O1, *Salmonella typhi*, and *Escherichia coli* O157: H7 in alfalfa sprouts. *Journal of Food Science*, *65*(1), 162–165.

Centers for Disease Control and Prevention (CDC), (1997). *Cryptosporidium and Water: A Public Health Handbook*. Atlanta, CA: Centers for Disease Control and Prevention.

Centers for Disease Control and Prevention (CDC), (2013). *Waterborne Disease and Outbreak Surveillance System (WBDOSS)*. www.cdc.gov/healthywater/surveillance/index.html (accessed on 11 February 2022).

Chikindas, M. L., Venema, K., Ledeboer, A. M., Venema, G., & Kok, J., (1995). Expression of lactococcin A and pediocin PA-1 in heterologous hosts. *Letters in Applied Microbiology*, *21*(3), 183–189.

Cliver, D. O., (2000). Viruses and protozoan parasites in food. *Microbial Food Contamination* (pp. 77–86). CRC Press.

Da Cruz, A. G., Cenci, S. A., & Maia, M. C., (2006). Quality assurance requirements in produce processing. *Trends in Food Science and Technology*, *17*(8), 406–411.

De Buyser, M. L., Dufour, B., Maire, M., & Lafarge, V., (2001). Implication of milk and milk products in foodborne diseases in France and in different industrialized countries. *International Journal of Food Microbiology*, *67*(1, 2), 1–17.

De Carvalho, A. A. T., Mantovani, H. C., & Vanetti, M. C. D., (2007). Bactericidal effect of bovicin HC5 and nisin against *Clostridium tyrobutyricum* isolated from spoiled mango pulp. *Letters in Applied Microbiology*, *45*(1), 68–74.

De Vuyst, L., & Vandamme, E. J., (1994). Nisin, a lantibiotic produced *by Lactococcus lactis* subsp. *lactis:* Properties, biosynthesis, fermentation and applications. In: *Bacteriocins of Lactic acid Bacteria* (pp. 151–221). Springer, Boston, MA.

Dorny, P., Praet, N., Deckers, N., & Gabriël, S., (2009). Emerging foodborne parasites. *Veterinary Parasitology*, *163*(3), 196–206.

Duyff, R. L., (2012). *American Dietetic Association Complete Food and Nutrition Guide.* Houghton Mifflin Harcourt.

FAO/WHO, (2003). Assuring Food Safety and Quality: Guidelines for Strengthening National Food Control Systems. https://www.fao.org/3/y8705e/y8705e.pdf.

Farías, M. E., Nuñez De, K. M., Sesma, F., Palacios, J., De Ruiz, H. A. P., & Oliver, G., (1999). Inhibition of *Listeria monocytogenes* by the bacteriocin enterocin CRL35 during goat cheese making. *Milchwissenschaft*, *54*(1), 30–32.

Fenelon, M. A., Ryan, M. P., Rea, M. C., Guinee, T. P., Ross, R. P., Hill, C., et al., (1999). Elevated temperature ripening of reduced-fat Cheddar made with or without lacticin 3147-producing starter culture. *Journal of Dairy Science*, *82*(1), 10–22.

Fiorentin, L., Vieira, N. D., & Barioni, Jr. W., (2005). Oral treatment with bacteriophages reduces the concentration of *Salmonella enteritidis* PT4 in caecal contents of broilers. *Avian Pathology*, *34*(3), 258–263.

Food and Drug Administration (FDA), (2007). *Letter to Food Manufacturers Regarding Legal Responsibilities for the Safety of Food Ingredients.* Washington, DC: US Food and Drug Administration.

Food and Drug Administration (FDA), (2010a). *Guidance for Handbook – The Transportation of Food.* www.fda.gov/food/guidanceregulation/guidancedocumentsregulatoryinformation/sanitationtransportation/ucm208199.htm (accessed on 11 February 2022).

Food and Drug Administration (FDA), (2010c). *Draft Guidance for Industry, Questions and Answers Regarding the Reportable Food Registry as Established by the Food and Drug Administration Amendments Act of 2007 (Edition2).* www.gpo.gov/fdsys/pkg/FR-2009-06-16/pdf/E9-14048.pdf (accessed on 11 February 2022).

Food and Drug Administration (FDA), (2011b) *Defect Levels Handbook–The Food Defect Action Levels.* www.fda.gov/food/guidanceregulation/guidancedocumentsregulatoryinformation/sanitationtransportation/ucm056174.htm (accessed on 11 February 2022).

Food, and Drug Administration (FDA). (2011a). *Grade a Pasteurized Milk Ordinance Revision.* Washington, DC: FDA, 2011.

Forsythe, S. J., (2000). Food poisoning microorganisms. *The Microbiology of Safe Food*, *87*, 148.

Garcia, P., Madera, C., Martínez, B., & Rodríguez, A., (2007). Biocontrol of *Staphylococcus aureus* in curd manufacturing processes using bacteriophages. *International Dairy Journal*, *17*(10), 1232–1239.

Giannakourou, M. C., & Taoukis, P. S., (2003). Application of a TTI-based distribution management system for quality optimization of frozen vegetables at the consumer end. *Journal of Food Science*, *68*(1), 201–209.

Gill, J. J., Pacan, J. C., Carson, M. E., Leslie, K. E., Griffiths, M. W., & Sabour, P. M., (2006). Efficacy and pharmacokinetics of bacteriophage therapy in treatment of subclinical *Staphylococcus aureus* mastitis in lactating dairy cattle. *Antimicrobial Agents and Chemotherapy*, *50*(9), 2912–2918.

Gill, J. J., Sabour, P. M., Leslie, K. E., & Griffiths, M. W., (2006). Bovine whey proteins inhibit the interaction of *Staphylococcus aureus* and bacteriophage K. *Journal of Applied Microbiology, 101*(2), 377–386.

Goode, D., Allen, V. M., & Barrow, P. A., (2003). Reduction of experimental *Salmonella* and *Campylobacter* contamination of chicken skin by application of lytic bacteriophages. *Applied and Environmental Microbiology, 69*(8), 5032–5036.

Gorris, L. G., (2005). Food safety objective: An integral part of food chain management. *Food Control, 16*(9), 801–809.

Grande, M. J., Lucas, R., Abriouel, H., Omar, N. B., Maqueda, M., Martínez-Bueno, M., et al., (2005). Control of *Alicyclobacillus acidoterrestris* in fruit juices by enterocin AS-48. *International Journal of Food Microbiology, 104*(3), 289–297.

Grande, M. J., Lucas, R., Valdivia, E., Abriouel, H., Maqueda, M., Galvez, A., et al., (2005). Stability of enterocin AS-48 in fruit and vegetable juices. *Journal of Food Protection, 68*(10), 2085–2094.

Green, L. R., Selman, C. A., Radke, V., Ripley, D., Mack, J. C., Reimann, D. W., Bushnell, L., et al., (2006). Food worker hand washing practices: An observation study. *Journal of Food Protection, 69*(10), 2417–2423.

Greer, G. G., (1988). Effects of phage concentration, bacterial density, and temperature on phage control of beef spoilage. *Journal of Food Science, 53*(4), 1226–1227.

Griffith, C., (2000). Food safety in catering establishments. *Food Science and Technology*, 235–256.

Grunert, K. G., (2005). Food quality and safety: Consumer perception and demand. *European Review of Agricultural Economics, 32*(3), 369–391.

Guenther, S., Herzig, O., Fieseler, L., Klumpp, J., & Loessner, M. J., (2012). Biocontrol of *Salmonella typhimurium* in RTE foods with the virulent bacteriophage FO1-E2. *International Journal of Food Microbiology, 154*(1, 2), 66–72.

Holcapkova, P., Kolarova, R. Z., Hrabalikova, M., Salakova, A., Drbohlav, J., & Sedlarik, V., (2017). Isolation and thermal stabilization of bacteriocin nisin derived from whey for antimicrobial modifications of polymers. *International Journal of Polymer Science*. 3072582. https://doi.org/10.1155/2017/3072582.

Jack, R. W., Tagg, J. R., & Ray, B., (1995). Bacteriocins of gram-positive bacteria. *Microbiological Reviews, 59*(2), 171–200.

Kato, T., Inuzuka, L., Kondo, M., & Matsuda, T., (2001). Growth of nisin-producing lactococci in cooked rice supplemented with soybean extract and its application to inhibition of *Bacillus subtilis* in rice miso. *Bioscience, Biotechnology, and Biochemistry, 65*(2), 330–337.

Krause, D. O., & Hendrick, S., (2011). *Zoonotic Pathogens in the Food Chain*. CABI, Oxfordshire, UK.

Le Loir, Y., Baron, F., & Gautier, M., (2003). *Staphylococcus aureus* and food poisoning. *Genetics and Molecular Research, 2*(1), 63–76.

Leverentz, B., Conway, W. S., Alavidze, Z., Janisiewicz, W. J., Fuchs, Y., Camp, M. J., et al., (2001). Examination of bacteriophage as a biocontrol method for *Salmonella* on fresh-cut fruit: A model study. *Journal of Food Protection, 64*(8), 1116–1121.

Leverentz, B., Conway, W. S., Camp, M. J., Janisiewicz, W. J., Abuladze, T., Yang, M., et al., (2003). Biocontrol of *Listeria monocytogenes* on fresh-cut produce by treatment with lytic bacteriophages and a bacteriocin. *Applied and Environmental Microbiology, 69*(8), 4519–4526.

Lundgren, M. S., & Novak, P. J., (2009). Quantification of phytoestrogens in industrial waste streams. *Environmental Toxicology and Chemistry, 28*(11), 2318–2323.

Lyon, W. J., Sethi, J. K., & Glatz, B. A., (1993). Inhibition of psychrotrophic organisms by propionicin PLG-1, a bacteriocin produced by *Propionibacterium thoenii*. *Journal of Dairy Science, 76*(6), 1506–1513.

Mårdén, P., (1995). *Att Arbeta Med HACCP: Hur Och Varför*. SIK Institutet för livsmedelsforskning, Göteborg, Sverige.

McMeekin, T. A., Baranyi, J., Bowman, J., Dalgaard, P., Kirk, M., Ross, T., et al., (2006). Information systems in food safety management. *International Journal of Food Microbiology, 112*(3), 181–194.

Modi, R., Hirvi, Y., Hill, A., & Griffiths, M. W., (2001). Effect of phage on survival of *Salmonella enteritidis* during manufacture and storage of cheddar cheese made from raw and pasteurized milk. *Journal of Food Protection, 64*(7), 927–933.

Morgan, S. M., Galvin, M., Kelly, J., Ross, R. P., & Hill, C., (1999). Development of a lacticin 3147–enriched whey powder with inhibitory activity against foodborne pathogens. *Journal of Food Protection, 62*(9), 1011–1016.

Morgan, S. M., Galvin, M., Ross, R. P., & Hill, C., (2001). Evaluation of a spray-dried lacticin 3147 powder for the control of *Listeria monocytogenes* and *Bacillus cereus* in a range of food systems. *Letters in Applied Microbiology, 33*(5), 387–391.

Nguz, K., (2007). Assessing food safety system in sub-Saharan countries: An overview of key issues. *Food Control, 18*(2), 131–134.

Nissen, H., Holo, H., Axelsson, L., & Blom, H., (2001). Characterization and growth of *Bacillus* spp. in heat-treated cream with and without nisin. *Journal of Applied Microbiology, 90*(4), 530–534.

Nordenskjold, J., (2012). *Implementation of a Quality Management System in Food Production*. Master's Thesis, Swedish University of Agricultural Sciences, Uppsala, Sweden. https://stud.epsilon.slu.se/4676/7/nordenskjold_j_ 120815.pdf.

O'Flynn, G., Ross, R. P., Fitzgerald, G. F., & Coffey, A., (2004). Evaluation of a cocktail of three bacteriophages for biocontrol of *Escherichia coli* O157: H7. *Applied and Environmental Microbiology, 70*(6), 3417–3424.

O'Mahony, T., Rekhif, N., Cavadini, C., & Fitzgerald, G. F., (2001). The application of a fermented food ingredient containing 'variacin', a novel antimicrobial produced by *Kocuria varians*, to control the growth of *Bacillus cereus* in chilled dairy products. *Journal of Applied Microbiology, 90*(1), 106–114.

Olaimat, A. N., & Holley, R. A., (2012). Factors influencing the microbial safety of fresh produce: A review. *Food Microbiology, 32*(1), 1–19.

Oldewage-Theron, W. H., & Egal, A. A., (2016). Food quality and food safety. *Community Nutrition for Developing Countries*, 430.

Pal, M., (2005). Importance of zoonoses in public health. *Indian Journal of Animal Sciences, 75*(5), 586–591.

Pal, M., (2007). *Zoonoses*. Satyam Publishers, Jaipur, India.

Pal, M., (2013). *Food Safety is Becoming a Global Public Health Concern*. The Ethiopian.

Pal, M., & Boru, B. G., (2010). *Cyclospora cayetanensis*: An emerging foodborne pathogen. *Journal of Natural History, 6*(1), 9–11.

Patel, J., Sharma, M., Millner, P., Calaway, T., & Singh, M., (2011). Inactivation of *Escherichia coli* O157: H7 attached to spinach harvester blade using bacteriophage. *Foodborne Pathogens and Disease, 8*(4), 541–546.

Penney, V., Henderson, G., Blum, C., & Johnson-Green, P., (2004). The potential of phyto-preservatives and nisin to control microbial spoilage of minimally processed fruit yogurts. *Innovative Food Science and Emerging Technologies*, *5*(3), 369–375.

Pozo, H., Barcelos, A. F., & Akabane, G. K., (2018). Critical factors of success for quality and food safety management: Classification and prioritization. *Universal Journal of Industrial and Business Management*, *6*(2), 30–41.

Pucci, M. J., Vedamuthu, E. R., Kunka, B. S., & Vandenbergh, P. A., (1988). Inhibition of *Listeria monocytogenes* by using bacteriocin PA-1 produced by *Pediococcus acidilactici* PAC 1.0. *Applied and Environmental Microbiology*, *54*(10), 2349–2353.

Raab, V., Bruckner, S., Beierle, E., Kampmann, Y., Petersen, B., & Kreyenschmidt, J., (2008). Generic model for the prediction of remaining shelf life in support of cold chain management in pork and poultry supply chains. *Journal on Chain and Network Science*, *8*(1), 59–73.

Raya, R. R., Varey, P., Oot, R. A., Dyen, M. R., Callaway, T. R., Edrington, T. S., et al., (2006). Isolation and characterization of a new T-even bacteriophage, CEV1, and determination of its potential to reduce *Escherichia coli* O157: H7 levels in sheep. *Applied and Environmental Microbiology*, *72*(9), 6405–6410.

Read, C. A., & Kaplan, B., (1996). Help prevent *Escherichia coli* 0157: H7. *Journal of American Veterinary Medical Association, 209*, 1231.

Richardson, S. D., & Thomas, A. T., (2014). Water analysis: Emerging contaminants and current issues. *Analytical Chemistry*, *86*(6), 2813–2848.

Riley, M. A., & Gordon, D. M., (1999). The ecological role of bacteriocins in bacterial competition. *Trends in Microbiology*, *7*(3), 129–133.

Rodríguez, J. M., Martínez, M. I., & Kok, J., (2002). Pediocin PA-1, a wide-spectrum bacteriocin from lactic acid bacteria. *Critical Reviews in Food Science and Nutrition*, *42*(2), 91–121.

Ropkins, K., & Beck, A. J., (2000). Evaluation of worldwide approaches to the use of HACCP to control food safety. *Trends in Food Science and Technology*, *11*(1), 10–21.

Rosenquist, H., Nielsen, N. L., Sommer, H. M., Nørrung, B., & Christensen, B. B., (2003). Quantitative risk assessment of human campylobacteriosis associated with thermophilic *Campylobacter* species in chickens. *International Journal of Food Microbiology*, *83*(1), 87–103.

Ross, M., & Guzewich, J., (1999). *Evaluation of Risks Related to Microbiological Contamination of Ready-to-Eat Food by Food Preparation Workers and the Effectiveness of Interventions to Minimize Those Risks*. FDA White Paper, FDA, CFSAN.

Ross, R. P., Galvin, M., McAuliffe, O., Morgan, S. M., Ryan, M. P., Twomey, D. P., et al., (1999). Developing applications for lactococcal bacteriocins. *Lactic Acid Bacteria: Genetics, Metabolism and Applications*, 337–346.

Scannell, A. G., Hill, C., Ross, R. P., Marx, S., Hartmeier, W., & Arendt, E. K., (2000). Development of bioactive food packaging materials using immobilized bacteriocins lacticin 3147 and nisaplin®. *International Journal of Food Microbiology*, *60*(2, 3), 241–249.

Schoenfuss, T. C., & Lillemo, J. H., (2014). Food safety and quality assurance. *Food Processing: Principles and Applications*, 233–247.

Sheng, H., Knecht, H. J., Kudva, I. T., & Hovde, C. J., (2006). Application of bacteriophages to control intestinal *Escherichia coli* O157: H7 levels in ruminants. *Applied and Environmental Microbiology*, *72*(8), 5359–5366.

Sjafrina, N., & Yani, A., (2013). Food safety and the implementation of quality system in food. *International Journal on Advanced Science, Engineering and Information Technology, 3*(4), 309–313.

Sklar, I. B., & Joerger, R. D., (2001). Attempts to utilize bacteriophage to combat *Salmonella enterica* serovar enteritidis infection in chickens. *Journal of Food Safety, 21*(1), 15–29.

Snyder, L., & Champness, W., (2007). Lytic bacteriophages: Genetic analysis and use in transduction. *Molecular Genetics of Bacteria*, 293–305.

Stevenson, K. E., & Dane, T. B., (1999). *HACCP: A Systematic Approach to Food Safety*. Food Processors Institute, Washington, DC.

Strauch, E., Hammerl, J. A., & Hertwig, S., (2007). Bacteriophages: New tools for safer food?. *Journal für Verbraucherschutz und Lebensmittelsicherheit, 2*(2), 138–143.

Suárez, V. B., Quiberoni, A., Binetti, A. G., & Reinheimer, J. A., (2002). Thermophilic lactic acid bacteria phages isolated from Argentinian dairy industries. *Journal of Food Protection, 65*(10), 1597–1604.

Surak, J. G., (2007). A recipe for safe food: ISO 22000 and HACCP. *Quality Progress, 40*(10), 21.

Tajkarimi, M., Ibrahim, S. A., & Fraser, A. M., (2013). Food safety challenges associated with traditional foods in Arabic speaking countries of the Middle East. *Trends in Food Science and Technology, 29*(2), 116–123.

Terreri, A., (2009). Preventing the next product recall. *Food Logistics, 111*, 20–25.

Thomas, L. V., & Delves-Broughton, J., (2005). 7 nisin. *Antimicrobials in Food*, 237.

Uçar, A., Yilmaz, M. V., & Çakiroğlu, F. P., (2016). Food safety – problems and solutions. In: Significance, *Prevention and Control of Food Related Diseases*, Makun, H.A. (Ed.), IntechOpen Limited, London, http://dx.doi.org/10.5772/63176.

Unnevehr, L. J., & Jensen, H. H., (1999). The economic implications of using HACCP as a food safety regulatory standard. *Food Policy, 24*(6), 625–635.

US Public Health Service and Food and Drug Administration (FDA), (2013). *Food Code 2013*. www.fda.gov/Food/GuidanceRegulation/RetailFoodProtection/FoodCode/ucm374275.htm (accessed on 11 February 2022).

Wagenaar, J. A., Marcel, A. P., Van, B., Mark, A. M., Trudy, M. W., & Richard, M. C., (2005). Phage therapy reduces *Campylobacter jejuni* colonization in broilers. *Veterinary Microbiology, 109*(3, 4), 275–283.

Walker, E., Pritchard, C., & Forsythe, S., (2003). Food handlers' hygiene knowledge in small food businesses. *Food Control, 14*(5), 339–343.

Whichard, J. M., Sriranganathan, N., & Pierson, F. W., (2003). Suppression of *Salmonella* growth by wild-type and large-plaque variants of bacteriophage Felix O1 in liquid culture and on chicken frankfurters. *Journal of Food Protection, 66*(2), 220–225.

Wohlgenant, K. C., Cates, S. C., Fraser, A., Chapman, B., Jaykus, L. A., & Chen, X., (2014). Sanitation in classroom and food preparation areas in child-care facilities in North Carolina and South Carolina. *Journal of Environmental Health, 77*(4), 20–27.

World Health Organization, (1998). *Food Safety Programs in the South East Asia Region, Overview and Perspective*. WHO Regional Office South East Asia, New Delhi, India.

World Health Organization, (2003). Assuring food safety and quality: Guidelines for strengthening national food control systems. In: *Assuring Food Safety and Quality: Guidelines for Strengthening National Food Control Systems* (pp. 73–73).

World Health Organization, (2012). *Five Keys to Growing Safer Fruits and Vegetables: Promoting Health by Decreasing Microbial Contamination*.

World Health Organization, (2014). *WHO Initiative to Estimate the Global Burden of Foodborne Diseases: Fourth Formal Meeting of the Foodborne Disease Burden Epidemiology Reference Group (FERG): Sharing New Results, Making Future Plans, and Preparing Ground for the Countries*.

Yamazaki, K., Murakami, M., Kawai, Y., Inoue, N., & Matsuda, T., (2000). Use of nisin for inhibition of *Alicyclobacillus acidoterrestris* in acidic drinks. *Food Microbiology*, *17*(3), 315–320.

Yang, S. C., Lin, C. H., Sung, C. T., & Fang, J. Y., (2014). Antibacterial activities of bacteriocins: Application in foods and pharmaceuticals. *Frontiers in Microbiology*, *5*, 241.

Yiğit, V., & Duran, T., (1997). *Institutional Nutrition Technology I (in Turkish)*. Ekin Publishing, Istanbul, Turkey.

CHAPTER 4

Food Hazards and Their Risk Management

ADNAN KHALIQ, SAMREEN AHSAN,
MUHAMMAD FARHAN JAHANGIR CHUGHTAI, ATIF LIAQAT,
TARIQ MEHMOOD, NIMRA SAMEED, KANZA SAEED, and
SYED JUNAID UR RAHMAN

ABSTRACT

The priorities of the food-oriented sector have been diverted towards the production of high profiling food products with respect to food quality and safety standards. The confronted hazards associated with food handling are biological, physical, and chemical in nature. The alarming and most prevailing hazard among these are biological. Hence, the invasion of biological hazard prevails intimacy of food, which resultantly provokes inconsistency in food quality. The international policies and quality controlling standards are being implemented for assuring conformity and consistency in quality of food products with respect to international standards. However, quality-assuring standards, different food deteriorating hazards, situation coping recommendations and benefits are discussed in this chapter.

4.1 INTRODUCTION

Food safety and quality (FSQ) has been categorized as strong pillars of global food standards and these evolved with the antiquity of manhood. Food either in liquid or solid state constantly is susceptible by means of

Food Microbial and Molecular Biology: From Fundamentals to Applications. Saher Islam, Devarajan Thangadurai, Jeyabalan Sangeetha, Zaira Zaman Chowdhury (Eds.)

physical, chemical, and biological hazards (BIOHAZ). Physical hazards comparatively are easy to control however chemical (intrinsic and extrinsic) and biological (harmful bacteria, viruses, and foodborne pathogens) hazards always threatened to FSQ systems. Globally, many foods are being fortified or enriched due to processing losses, and these techniques may be unsafe in terms of quality and safety at the consumer end if did not monitor as per standards. Industrial wastes are similarly massive element that compromise food chain safety and quality directly or indirectly. To overcome these aforementioned issues of food quality and safety, various systems has been developed by international bodies like Food safety management system (FSMS) (ISO 22000), Hazard analysis and critical control point (HACCP), threat assessment critical control point (TACCP), Vulnerability assessment critical control point (VACCP) and food management system (ISO 9001). Compliance of these food quality and safety systems, wholesome, safe, and nutritious food will be assured at the end-user.

Food quality and safety are closely related acronyms in global food standards and these ensure all the standard operating procedures (SOPs) for producing safe and qualitative foods for the masses. Food either in solid or liquid is not only used for maintaining living standards but also to combat nutritional disorders and medical supplement these days. Therefore, food quality and safety are on top priority for both developed and underdeveloped economies. Additionally, safety of consumer or end-user must be the ultimate objective in food supply chain system (Savov et al., 2009).

Initially, consumers were not much cognizant about the safety and quality of foods, but after the rapid development in economies at the biosphere, the quality standards are being more focused by the consumers. The demand of safe and qualitative foods increased exponentially around the globe throughout the food supply chain. Food quality is one of the fundamental tools for any organization that is producing foodstuff or providing food related services. The food quality has diversified meaning and aspects among the different stake holders, but it can be interpreted generally as "excellency to conformance or superiority of a product or any system." Quality perception for food industry comprises subsequent significant features:

- Product or system conformance as per their intended use;
- Consumer's satisfaction level of expectations and perceptions;
- Excellence in distinct characteristics or traits.

Food safety is a very broad term that encompasses many international standards that ensure the consumers health throughout the food supply chain. Food safety is an ongoing approach that ensures the adequate set of procedures and protocols for safe food production are being executed, swotted, and sustained. Personal hygiene of supportive staff, utilities, and sanitary operations shall be monitored continuously to uphold an effective food safety environment. Additionally, many hazards are associated with the food supply chain which ultimately challenges the safety aspects of a particular product or a food system. The details of these are given below:

1. **Physical Hazards**: Human hairs, nails, glass piece, wooden piece, dust, dirt, metal piece, paper, etc.
2. **Chemical Hazards**: Pesticides residue, biocides, detergents, and cleaning agents.
3. **Biological Hazards**: Viruses, bacteria, yeast, molds, antibodies, pathogens, or toxins produced by them.

Currently, awareness regarding the food quality and safety among the individuals has been raised enormously, and that motivated all food stake holders to provide wholesome, safe, and nutritious foods/ingredients to the masses. This chapter will highlight some of the food quality and safety standards and their significant impact on safe and qualitative food supply chain. Additionally, it will comprise of future prospective for mankind to step up for the availability of safe foods globally.

4.2 HISTORY OF SAFE FOODS

Almost a century ago, folks were not so much conscious and concerned about the safety of foods. Moreover, no rules, regulations were available regarding the misbranded and filthy foods. The United States Department of Agriculture anticipated national law for food and drugs in 1880 to 1906 based on their research activities, but regrettably, that proposal was not accepted at that time. In 1906, the Pure Food and Drugs Act was unanimously accepted and adulterated/contaminated and misbranded foods were banned within the United States. Moreover, many food additives such as artificial dyes, colors, preservatives were also prohibited under this act with reference to consumer health and food safety aspects.

In the era of 1949 to 1973, a book entitled "Surveying Procedure for Toxicity of Synthesized Food" was published by Food and Drug Administration (FDA) for all the stake holders related to food industries. This book comprehensively explained about the foodborne diseases (FBDs) and raised the awareness level of the consumers. Subsequently, FDA established many other administrative programs for ensuring not only the safety status of food from plant and animal origin but also food processes and services. Many other regulations were also documented and executed regarding the processed foods to ensure the safety, particularly for low acid foods such as botulism outbreak by canned foods.

In the time period of 1993 to 2010, FDA published a food code for vendors and retailers regarding the food services and operations, and their code was adopted by different local bodies and agencies. The HACCP, a FSMS was implemented which ensured the phytosanitary regulation for seafood, eggs, and juices import and export mechanism. Authorities under the umbrella of public health security and bioterrorism outlined an act known as "Preparedness and Response Act" structured mechanism for record-keeping of trade and organizational confinement. Later on, in 2006, Food Allergen Labeling and Consumer Protection Act was executed that directed to pronounce all types of allergens from any food source should be mentioned on the label of food containing material. *Salmonella enteritidis* is the predominant cause of foodborne salmonellosis, particularly in chicken eggs, and in July 2010, Egg Safety Rule was implemented on producers and retailers (FDA, 2011).

4.3 BURDEN AND TRENDS IN FOODBORNE ILLNESSES

Globalization and development in the world have transformed dietetic patterns and customs altogether that ultimately created a new era in food export and import. This trend of food trading also played a huge role in transferring the foodborne illnesses from one region to another region. Many outbreaks have been reported by the Centers for Disease Control and Prevention (CDC) due to food safety issues in various developing and devolved economies. In the time spam of 2005 to 2010, there were 39 different epidemics reported in 15 different realms due to import and export of food items. However, some industrialized states reinforced food safety rule and curtailed foodborne illness cases. Moreover, abundant usage of pesticides in crop production and tradition of drugs and hormones in livestock are also responsible for foodborne illnesses in the world.

The capturing of every foodborne illness is not possible even in the developed countries due to shortcomings and missing pieces of information. However, the surveillance system of CDC reports that 48 million people get sick, 128,000 hospitalized and nearly 3,000 are dying annually in the United States due to foodborne illnesses. In Australia, at least 4.1 million cases are reported for foodborne illnesses with the impact of \$1.249 billion on the economy every year.

4.4 CONTEMPORARY FOOD SAFETY

In the last few decades, the awareness level of food safety has been raised among farmers, transporters, consumers, and the government officials as well. In fact, the FDA issued a regulation to carry out the FDA Food Safety Modernization Act (2011), and that was the most sweeping reform to reduce food safety issues in the past 70 years. The food stake holders are required to do more to identify and prevent the foodborne illness before they happen through research-based activities. Apart from these, industries are also concentrating on international training, accreditation, and certification process that has positive impact in the society and socioeconomic burden of foodborne illness also decreased enormously.

4.5 TYPES OF HAZARDS

4.5.1 PHYSICAL HAZARDS IN FOOD

Any object or matter unintentionally introduced to food is called physical hazard. It may cause any harm or illness to consumer but definitely compromises the quality of food. There are a number of physical hazards introduced by different sources during food processing, packing, storage, transportation or serving. There are generally two types of physical hazards: unnatural physical hazards (hairs, stone particles, metal fragment, insects, pieces of plastic, wood particles, blood, or piece of glass) and natural physical hazards (stems in barriers, dirt on different vegetables, microscopic air born debris and minute insect fragments). Lack of equipment maintenance techniques, improper training of workers and poor implementation of SOPs are the main reasons of physical hazards contamination in food. All food processors must develop and follow standard processing plans to avoid the risk of physical hazards.

4.5.1.1 PHYSICAL HAZARDS RISK IN FOOD

Physical hazards like sharp and hard objects can cause cut to the mouth, teeth, gums, throat, and gastrointestinal tract. Physical hazards can lead toward serious health issues and results in cuts, bleeding, infections, choking, trauma, illness, and sometimes needs surgery for the removal from the body. The presence of physical hazards can damage the food quality and more than this damages the brand name of a food processing company.

4.5.1.2 CONTROLLING MEASURES OF PHYSICAL HAZARDS

There are two ways to control contamination of physical hazards in food, prevention, and removal. Preventive measures are as follows:

- By following good manufacturing practices (GMPs)
- By implementation of hazard analysis critical control point (HACCP)
- By conducting GAP analysis
- Identification from raw materials, ingredients, packing materials, processors which are the major reasons of contamination
- Proper management and maintenance of buildings, facilities, grounds, processing plants and equipment
- Elimination of potential sources and proper training of processors and workers

Detection and removal are possible by:

- Physical removal of hazards through screens, filters, and sieves, etc.
- Proper equipment design, calibration, and maintenance along with online visual inspection
- Detection methods should apply for example metal detector, X-ray machine, optical system, and magnets to minimize the chances of contamination
- Screening of raw material and finished products
- Also focus on consumer feedback related to physical hazards

4.5.1.3 MAIN MATERIALS OF CONCERN AS PHYSICAL HAZARD

There are different types of materials concern as physical hazards contaminated food through different sources and also cause harm differently are described in Table 4.1.

4.6 CHEMICAL HAZARDS

4.6.1 CHEMICAL HAZARDS AND THEIR TYPES

Food safety is ensured by controlling all physical, chemical, and BIOHAZ. Out of all three hazards, chemical hazards impart long term effects on food consumer. Chemical food safety is related to prevent all the possible chemical risks associated with the food chain. Chemical present in food from endogenous or exogenous sources can affect our health. These chemicals may result in food intoxication and food allergic reactions in people (Bozoglu, 2012). There are many types of chemical hazards in food including:

- Plant-based hazards
- Animal-based hazards
- Additives induced hazards
- Hazards associated with processing operations
- Industrial chemicals
- Agricultural sources
- environmental contaminants
- Pesticide residues

TABLE 4.1 Physical Hazards with Contamination Sources and Harmful Outcomes

Physical Hazard	Contamination Source	Expected Harms
Glass	Bottles, jars, utensils, light fixtures	Cuts, bleeding; may require surgery for removal
Stones	Fields, raw material, building	Broken teeth, choking
Wood	Pallets, boxes, building, fields	Cuts, bleeding; may require surgery for removal
Bullet/needles	Animals shoot in field, hypodermic needles used for infections	Infection, cuts; may need surgery for removal
Bones	Fields, improper plant processing (mostly in meat industry)	Choking, trauma
Plastic	Packing material, pallets, employees	Choking, cuts, infection; may need surgery for removal
Jewelry	Pen/pencils, buttons, rings, nose pin (careless employees)	Cuts, infections, disturb gastrointestinal tract; may need surgery for removal
Personal effect	Employees or processors	Choking, cuts, broken teeth; may require surgery for removal
Metals	Machinery residues, wires, stapler pins (during packing)	Cuts, infections; may require surgery for removal
Insects/Filth	Fields, plant post-process entry	Trauma, choking, illness

4.6.1.1 PLANT-BASED CHEMICAL HAZARDS

Some edible plants may contain certain natural substances that can cause hazardous effects on human health. Coumarin is an example of a natural compound found in plant-based food sources that may act as a chemical hazard in food safety. Coumarin is commonly found in fruits, green tea, some vegetables such as chicory and in spices, i.e., cinnamon or cassia. Cassia finds its application in various recipes such as chocolates, desserts, biscuits, and cakes. On oral exposure Coumarin in rats and mice has shown carcinogenic effects, along with liver and kidney damage (Lungarini et al., 2008). Similar toxic effects of Coumarin can be shown in humans where they can result in health risks (WHO, 1995). Aflatoxins (AFs) are toxic chemical compounds produced by certain fungi *Aspergillus parasiticus* and *Aspergillus flavus* in peanuts, tree nuts and corn oil. Mycotoxins if accidentally ingested can result in abnormalities in central nervous system (CNS), pulmonary system, and heart functioning. Caffeine is present in many food products, including coffee, soda, tea, and energy boosters. Overconsumption of these products may result in intoxication leading to increased heartbeat, seizures, and hormonal imbalance. Solanine, a natural compound found in green potatoes in high concentration proves to be toxic. Due to the production of chlorophyll potatoes turn green in color. During sunlight exposure Solanine is produced in harvested potatoes. This toxic compound may result in gastric and pulmonary issues in human being. Solanine is converted to solanidine which have resulted in spontaneous abortions in laboratory animals (Schmidt and Rodrick, 2003). It is therefore suggested that pregnant ladies should consume green potatoes with caution. Bioactive natural chemical compounds, alkaloids are present in plant-based food including potato, tea, and coffee (Crews, 2014). They give bitter taste; overconsumption may lead to toxicity and ill health effects. Anti-nutritional or natural toxic chemicals present in some legumes pose adverse health effects. Phytic acid, saponin, and lectins are some anti-nutritionals, may result in nutrient unavailability in the body and act as growth inhibitors. These are linked to infertility, reduced blood sugar level, and reduced mineral absorption in the body.

4.6.1.2 ANIMAL-BASED CHEMICAL HAZARDS

Antibiotics are administrated in animals and humans to curb any bacterial illness. These are used in animals as therapeutic, growth promoters and prophylactic purpose. Antibiotic residues and antibiotic-resistant bacteria, including zoonotic pathogens originating in treated animals and their transfer to humans via foods of animal origin is a seriously growing chemical and biological hazard, i.e., vancomycin-resistant enterococci in animals fed avoparcin, a glycopeptide (Barton, 2000). Along antibiotic resistance it may lead to devastating health effects in human being in near future. The trend of producing seafood through aquaculture to meet the growing demand of high-quality protein is rising, and so the cultivation methods are changing as well. This is characterized by the use of antibiotics in feed, disinfectants, and pesticides to increase the productivity, which on the other hand rise potential health and chemical food safety risks. Organohalogens such as dioxin-contaminated seafood results in negative health impacts in human being. These toxic compounds are found in higher quantities in farm raised seafood animals as compared to wild ones (Antunes and Gil, 2004). GMO fish such as transgenic fish is developed by genetic engineering to carry transgene for expression of certain proteins such as Cecropin (anti-microbial peptide), growth hormones and lactoferrin for increased growth and enhanced nutritional quality. Transgenic fish are not approved for human consumption in China, the United States, and other aquaculture producing countries. The reason is the lack of available data regarding their health impacts on human being (Sapkota et al., 2008). Residues resulting from antibiotics administration in animals, i.e., Penicillin residues in milk can affect the starter cultures used in making fermented milk products (yogurt, cheese, sausages) and leads to technological problems during food processing. Penicillin contaminated milk can result in allergic reactions in human as well. Animal breeders and producers administrate growth promoting hormones and anti-microbial (ionophores) to animals in order to increase production, growth rate and quality of meat or milk. Natural or synthetic hormones such as estrogen, progesterone, and testosterone or trenbolone acetate, melengestrol acetate and zeranol, respectively, are used in animals (Al-Dobaib and Mousa, 2009). Hormonal residues in meat and milk obtained from treated animals if consumed by human being results in harmful health impacts including carcinogenic, neurological, developmental, and genotoxic implications.

4.6.1.3 ADDITIVES INDUCED HAZARDS

Monosodium glutamate (MSG) is a flavor enhancer commonly used in various product recipes in food industry and cuisine. The advantage of MSG is its readily availability, cost-effectiveness and consumer likeliness. It is suspected to have some negative health impacts in human beings including headache, chest pain, heart palpitations, tingling sensation, numbness, and nausea (Staples, 2016). But still it is one of the controversial food additives as FDA has approved its usage and recommended MSG as GRAS (Generally recognized as safe). Sulfiting agents are used as bleaching agent, preservative, and to prevent the spoilage and brewing reactions in foods and beverages. They can result in sulfite induced asthma; the symptoms may worse in asthmatic people (Walker, 1985). Other food additives that may cause potential health risk include nitriles (curing agents) results in the formation of N-nitrosamines, which are proven carcinogens. Aspartame can also result in health implications. Moreover, it is suspected that BHT, BHA, and tartrazine may also involve in producing undesirable products via associated reactions in food.

4.6.1.4 HAZARDS ASSOCIATED WITH PROCESSING OPERATIONS

Food manufacturing gives rise to many chemical reactions resultantly various desirable and undesirable compounds are produced. Some of these compounds show toxicity and act as chemical hazard in food safety. These include heterocyclic amine, polyaromatic hydrocarbons, sterols, triacylglycerol, nitrosamines, and lysino alanine. Acrylamides and heterocyclic amines are produced during smoking and grilling of food and in Millard reaction. These compounds have varying carcinogenic effects (Luning and Devlieghere, 2006). During milk processing, milk heating results in Maillard reactions of endogenous lactose and protein. Consequently, leads to the formation of new chemical contaminants fructosyl-lysine, carboxymethyl lysine, and pyrraline (Nguyen et al., 2014). A review showed the higher proportion of these carboxymethyl lysine compounds in infant formulas due to sterilization treatments. Process parameters control can reduce these contaminants in these sensitive products.

4.6.1.5 INDUSTRIAL CHEMICALS

In oil and ghee industries, deodorization of vegetable oil is done to remove light and heavy fractions by carbon treatment that may contaminate the oil. Heavy metals have deleterious effect on the human body; they can enter in human and animal body via consumption of contaminated food and water. Estimation showed that more than 70% of dietary intake of cadmium is caused due to food chain. In industrialized countries, heavy metal contamination of crops and vegetables due to field irrigation by industrial wastewater or sewage sludge is a serious concern. Many industries such as steel industries discard cadmium, lead, and iron scrap, textile, and leather tanneries, mining, and pigment industries release hazardous compounds. Owing to the lack of proper drainage facilities, wastewater containing chemical hazards is released in rivers and canals. In developing countries, it is a common practice to irrigate crops with this industrial wastewater containing toxins and heavy metals. The uptake of these heavy metals by plants thus their accumulation in crop or vegetable acts as a potential chemical hazard in the food chain. Food plant species of some families are highly tolerant to heavy metals without any damage to themselves they can accumulate high concentrations of heavy metals in their tissues. The plant families which can pose a serious health risk to consumers by high bioaccumulation of heavy metals include Leguminosae, Gramineae, Cruciferae, Asteraceae, and Caryophyllaceae. Moreover, leafy vegetables can accumulate more of these heavy metals. Vegetables having high tolerance for bioaccumulation of heavy metals without any visible symptoms of damage are a potential health risk for consumers, i.e., *Brassica pekinensis*, commonly known as Chinese cabbage (Xiong and Wang, 2005).

During dairy processing on-farm or industrial-scale wide range of chemicals are used for equipment, utensils, and plant cleaning and disinfection. These chemicals include caustic, hydrogen peroxide, acidic compounds, peracetic acid, aldehydes, ozone, isothiazolinones, phenolics, and surfactants (Simões et al., 2010). Sodium hypochlorite forms disinfection product chloroform increase the trichloromethane level in milk. Residues of trichloromethane and quaternary ammonium compounds remain in milk and ice-cream (Danaher and Jordan, 2013; Van Asselt et al., 2017). Chlorine dioxide used in industries for cleaning and disinfection generates harmful byproducts chlorite and perchlorate. Due to improper cleaning and water rinsing residues of these chemicals can contaminate milk and milk products. Certain compounds from equipment or packaging materials such as phthalates from plastic bottles can migrate to food product (i.e., dairy) results in health implications. 2-isopropyl thioxanthone and 2-ethylhexyl-4-dimethylaminobenzoate are chemicals found in printing inks. This ink has been used on cartons for food packing and is likely to pose a threat to the safety of food being packed. Aluminum present in aluminum foil used in packing of food products (cereal porridge, infant formulas, dairy, and beverage) may migrate to food and contaminate it. Melamine – a chemical used illegally for increasing protein content in dairy infant formulas and milk powders is highly toxic for kidneys. In 2008, melamine toxicity led to morbidity and mortality among children in China (Jun-Shi, 2009).

4.6.1.6 AGRICULTURAL SOURCES

Many plant species produce pyrrolizidine alkaloids and grow as weeds all over the world. In agricultural crops, these weeds contaminate various foods such as honey, grains, meat, milk, and eggs result in significant level of toxicants in food (Edgar et al., 2011). In central and south Asia, fatalities were caused due to the consumption of teas and herbal medicines containing these pyrrolizidine alkaloids. In Africa during 1920, the main reason of bread poisoning was dehydro pyrrolizidine alkaloids from *Senecio* species found to be growing in grains used to make bread flour. Consumption of honey contaminated with pyrrolizidine alkaloids can result in liver cirrhosis, cancer, and pulmonary hypertension. Pollens are said to be the main route of transferring these toxins to honey nectar via honey bees. The plants used for salads, condiments, and herbal teas, i.e., Symphytum, and Borago, due to their leaf similarity leaves of *Senecio vulgaris* common toxic weed containing pyrrolizidine alkaloids, were found. The presence of these alkaloids in condiments and spices resulted in liver failure and subsequently death of an infant (Rasenack et al., 2003).

4.6.1.7 ENVIRONMENTAL CONTAMINANTS

Smoke from industries and automobiles release many contaminants in the environment also leads to contamination of agricultural food products. Aromatic hydrocarbons, such as benzene, are present in urban air environment. Benzene is known to have carcinogenic effect. The most important chemical hazard in food chain from environment is dioxin or dioxin like compounds that are abundantly present in the environment in soil, air, and water. Although in low concentration, human beings are exposed to this contaminant through food supply by consuming fish, dairy products, and animal fat in meat. The half-life of dioxin in human body is 10 years (Luning and Devlieghere, 2006). Potential health effects include cancer, diabetes, damage to the endocrine system, neurological, and immune system damage in infants and skin damage (Islam and Hoque, 2013). Acrolein is an aldehyde generated as a result of heat treatment of amino acids, carbohydrates in vegetable and animal fat. It is formed in cooked food as a result of heating of food and in the environment by burning of fossil fuels (Stevens and Maier, 2008). A few sources of Acrolin include incomplete burning of wood, tobacco smoking, oil frying of food and lipid peroxidation (Esterbauer et al., 1991). More than 180°C cooking temperature in non-refined rapeseed oil results in Acrolin formation from 5 to 250 mg/Kg of oil. Further oil reheating increases the formation of toxicant. Acrolein has been found in cheese, potatoes, bread, doughnut, and fish in ranges 10 to 600 μg/kg (Henning et al., 2017). Increased exposure of Acrolein can cause cardiovascular disorders. The hazard of heavy metal toxicity has deleterious effects on physical health, physiological, and mental wellbeing. Arsenic reaches plants via soil hence in the food chain from arsenic contaminated soil. In marine plants and animals, arsenic is found in organic form such as arseniobetaine, arsenosugars in algae, arsenocholine, in fishes and mollusks as arseniolipid (Francesconi, 1994). Lead contamination in water can occur by the lead pipes in the water supply system. Mercury can bioaccumulate in food chain as methyl mercury, consumption of fish caught from contaminated waters can be a source of potential hazard. Human exposure in fetus leads to brain damage in unborn child (Davidson et al., 2004).

4.6.1.8 PESTICIDE RESIDUES

According to Codex Alimentarius, pesticides are chemical agents employed to control pests during food processing in agricultural commodities. These chemicals can gain entry to food chain via agricultural practices employed during crop cultivation such as application of pesticides, insecticides, and other phytochemicals. Pesticides can also be introduced in fruits and vegetables during harvesting or storage. Pyrethroids, organophosphates, organochlorine, and carbamates are used insecticide and pesticides in agriculture crops. Pyrethroids pesticide widely used against insect pests and as grain protectants. They have potential health impacts such as kidney and liver damage. Pesticide residues in food results in deleterious effects such as cancer, blindness, infertility, nervous system, and liver impairments (Gupta, 2006). Insecticides such as organochlorines are fat-soluble, when reaches the lakes, rivers, and sea. The residues adhere to organics consumed by fishes and invertebrates' inhabitants of the water body. Insecticides residues may accumulate in the fish muscles and become the part of the food chain. The accumulation of insecticide is 10 to 1,000 times higher than the actual concentration in water (Kannan et al., 1997). Human get exposed to these toxic chemicals when consume contaminated fish and seafood. The subsequent bioaccumulation of organochlorine insecticides is high in animals and humans. Dithiocarbamates are commonly used fungicides and dinitrophenol, an herbicide is suspected to be carcinogenic and teratogenic, respectively. Phenylcarbamates, phenoxyalkyl carbonic acid, diuron, triazines, and quaternary ammonium compounds (paraquats) are some herbicides that can contaminate the food chain (Luning and Devlieghere, 2006). Legislative control over administration of these chemicals in agriculture is required beforehand it is too late. Chemical hazards in food chain can be controlled by following good manufacturing practices (GMPs) and HACCP and food manufacturers' commitment toward ensuring chemical safety in food chain.

4.7 BIOLOGICAL HAZARDS (BIOHAZ) AND ITS TYPES

Food poisoning is taken into consideration when a severe illness including neurological or gastrointestinal (GI) symptoms influences 2 or more people who have consumed a common meal in the past 72 hours. The term food poisoning usually includes both food related intoxication and food related infection. Consumption of a broad range of foods and water containing pathogenic organisms such as bacteria, fungi, viruses, etc., or the toxins may result in food poisoning (Addis and Sisay, 2015). These BIOHAZ are characteristically different from each other.

4.7.1 BACTERIA

Bacteria are prokaryotic microorganisms. Morphologically bacteria exist as cocci (spherical), bacilli (rod-shaped), and comma-shaped (curved) (Ray and Bhunia, 2007). Major foodborne pathogenic bacteria include *Escherichia coli*, *Campylobacter* spp., *Staphylococcus aureus*, and *Listeria monocytogenes*, for which a great surveillance is provided by the food industry and government agencies (Schirone et al., 2017). However, there are a number of other bacteria, pathogenic for human beings, including *Bacillus* spp., *Clostridium* spp., *Shigella* spp., *Yersinia* spp., *Salmonella* spp., etc. (Bintis, 2017). These pathogenic bacteria are diverse regarding their physiological characteristics, epidemiology, virulence, and host association (Scallan et al., 2011).

Numerous species and serotypes of *Salmonella* produce diseases in humans and animals including salmonellosis. The foods linked with this disease include raw or improperly cooked meat, milk, eggs, and their products (Zahoor et al., 2018). Disease symptoms include headache, diarrhea, and chill. *Staphylococcus aureus* produces six different types of enterotoxins while growing in foods. After ingestion, these heat stable toxins result in vomiting (Addis and Sisay, 2015). *Brucella* species transmitted from animals to humans cause brucellosis and it is associated with milk.

Campylobacter species are associated with campylobacteriosis (Zahoor et al., 2018) for which diarrhea, fever, malaise, and severe abdominal pain are major indications of infection. The disease is associated with directly from human-to-human contact or animals. Food implicated in this disease include, meat, mushrooms, milk, shellfish, eggs, raw clams, etc. (Adams and Moss, 2008). The strains *Clostridium botulinum* cause botulism, characterized by some GI and neurological disorders. Association of this disease has been found with various foods such as fruits (peaches and figs), low acid vegetables (corn, spinach, green beans, asparagus, mushrooms, and pepper) fermented and smoked fish (Ray and Bhunia, 2007). The strains of *Yersinia enterocolitica*, pathogenic ones, cause Yersiniosis. Infection due to *Yersinia enterocolitica* is indicated by abdominal pain, diarrhea along with fever and occasionally vomiting. These bacteria are linked with pork, contaminated milk, and water (Adams and Moss, 2008). The serogroup O157:H7 belonging to Enterohemorrhagic *E. coli* is responsible for hemorrhagic uremic syndrome and hemorrhagic colitis (severe bloody diarrhea). Foods associated with its infection include ground beef, salads, raw milk, sprouts, apple cider, uncooked sausages, and some fruits (Ray and Bhunia, 2007). Listeriosis is the disease resulting due to *L. monocytogenes*. The foods linked with various outbreaks of this include soft cheese, deli, and other ready-to-eat products. Shigellosis occurs due to *Shigella* species which find resemblance with *E. coli* regarding their DNA. Contaminated foods such as shellfish, milk, chicken, salads, and some fresh produces have been implicated in the disease (Bintis, 2017).

4.7.2 VIRUSES

These obligatory intracellular parasites need vulnerable host cells for infection and proliferation. The extracellular infectious particle or virion simply consists of a nucleic acid (single or double-stranded DNA or RNA) enclosed in a protein coat. The envelope of virus consists of viral proteins and a lipid bilayer, derivative of cell membrane of host cell and may be absent or present in viruses and this feature makes basis for the classification of viruses as non-enveloped and enveloped. Viruses are grouped into families and orders on the basis of their shape, size, structure of the genome, nucleotide composition, and mode of replication (King et al., 2012).

A great number of diverse viruses can be there in the human GI tract causing a broad range of illnesses. Highly documented viral foodborne illnesses include hepatitis and gastroenteritis resulting from HAV and human

NoV respectively. However, there are numerous other viruses including astroviruses, enteroviruses, adenoviruses, sapoviruses, hepatitis E virus (HEV) and rotaviruses which have been associated with water and foodborne disease transmission. Patients experiencing gastroenteritis or hepatitis may shed an extremely large number of viruses in their stools, excreting up to 10^{13} and 10^{10} viral particles/g of stool, respectively (Bosch et al., 2018).

Gastroenteritis resulting from viral infection includes symptoms such as abdominal pain, nausea, and vomiting, infrequently headache and fever (Arness et al., 2000). Hepatitis initializing from nausea, headache, malaise, and fever, may lead to vomiting, abdominal pain, diarrhea, and jaundice resulting in devastating state. Worldwide, HAV is responsible for approximately half of the overall hepatitis cases. Its infection can lead to death or transplantation of liver in an emergency (O'Grady, 1992).

HEV infects with less frequency in industrialized countries as compared to HAV, however, its mortality rate is higher especially in pregnant females where infection due to 1 and 2 genotype may result in 25% deaths (Kumar et al., 2004). Water is the major source of transmission for HEV infection (commonly linked with 1 and 2 genotype) in Africa, the Middle East, and Asia (Wong et al., 1980). However, in developed countries, infection spreads zoonotically, principally from pigs where 3 and 4 genotypes are almost 100% seropositive in animals more than six months old (Ruggeri et al., 2013).

Tick borne febrile encephalitis is a rare foodborne zoonotic disease caused by the virus. The disease is transmitted from flavivirus infected cattle by means of non-pasteurized milk and cheese (Kríz et al., 2009). Febrile encephalitis is another disease of zoonotic transmission caused by Nipah virus. Human outbreak of this disease was reported in Malaysia (1998) with severity and high mortality rate. The source of transmission of this virus was recognized as contaminated meat of pig (EFSA, 2011).

Enteroviruses are transmitted with contaminated food or water. After entering the host body, these replicate in digestive tract. Often, there are moderate or slight symptoms of this viral infection but nearly all infections caused by this virus are without any symptom. However, these viruses can spread to other organs, resulting in serious or fatal diseases like aseptic meningitis, and infrequently paralysis (Vasickova et al., 2005).

4.7.3 FUNGI

These are eukaryotic organisms, including molds (multicellular) and yeasts (unicellular). Molds consist of branched, filamentous fungi and produce spores which serve as a means of reproduction. Yeasts are a little bigger in size than bacteria and used in bread making and alcoholic fermentations. Some yeast species (e.g., *Candida albicans*) cause diseases, so known as pathogenic ones. Most fungi grow at room temperature under high oxygen and prefer the acidic surroundings (Alcamo, 1996).

Several molds produce toxic metabolites under favorable moisture and temperature, which are toxic to human beings and are known as mycotoxins. AFs, a mycotoxin, M1, M2 and M4 present in the feed are transferred in the milk of cattle, consuming the contaminated feed. These toxins may be detected within 48 hours of consumption of contaminated feed. Mycotoxins are recognized as hepatotoxic, mutagenic, and carcinogenic (Zahoor et al., 2018).

Fungi, being heterotrophic, get their food through absorbing soluble nutrients. These can metabolize complex materials however, required to be broken down through secreting extracellular enzymes. Many fungus species are parasites of plants and animals, and a few of these relations have turned out to be obligate and complex. However, mycotoxins produced by some molds are also associated with a broad range of human diseases (Ray and Bhunia, 2007). Over 100,000 species of molds are reported, of which almost 24 species may cause health issues in human beings. Since ancient times, illnesses resulting from ergot alkaloids from the *Claviceps purpurea* mold have been documented. This mold grows on grains such as barley, rye, and others. Consumption of such flour with mold growth results in ingestion of ergot alkaloids and leads to the appearance of disease symptoms (Campbell, 2016).

Aspergillus flavus produces AFs B_1 and B_2, which have food associations with peanut, maize, and other food products while aflatoxin B_1, B_2, G_1, G_2 are produced by *Aspergillus parasiticus* and may be associated with peanut and maize. Similarly, Fumonisin B_1 toxin, associated with maize, is produced by *Fusarium moniliforme* and various other rarely found species. *Penicillium verrucosum* and *Aspergillus ochraceus* produce Ochratoxin A, which finds its association with various foods such as wheat, barley, etc. Zearalenone (ZEN) associated with wheat and maize is produced by *F. graminearum*, *F. culmorum*, and *F. crookwellense* species (Adeyeye, 2016).

Paecilomyces colonize food products, and result in human illness. *Paecilomyces variotii* is the chief food spoilage agent of this genus which is frequently connected with hyalohyphomycosis in human beings, with clinical signs of peritonitis. *Phoma* spp. can also be pathogenic for human beings. These have been detected in food, crops, and water sources. *Wallemia*, including the highly xerophilic, xerotolerant, halophilic species, have been found in dry highly sugary foods, dry salted foods, salt crystals and dry feed. These fungi have also been implicated in human health issues (Paterson, 2019). Numerous other genera of fungi related with foods include *Curvularia*, *Alternaria*, *Trichoderma*, *Acremonium*, *Lichtheimia*, *Penicillium*, *Amnesia*, *Achaetomium/Chaetomium/Botryotrichum/Dichotomopilus*, and *Scopulariopsis/Microascus* (Paterson and Lima, 2017).

4.8 FOOD SAFETY AND QUALITY (FSQ) MANAGEMENT SYSTEM

Development of food business and growing consumer awareness and expectations and their concern for quality and safety has motivated food industries to imply such management system that will assure quality and safety of their products. Food safety is the vital component of food quality that is regulated by food laws to assure consumers that the food they are purchasing is safe. The concept of food safety is stating that food will not cause any harm to consumer at the level of consumption, when it is being prepared and consumed according to its envisioned use (ISO 22000, 2005). A safe food is defined as: "one that does not go beyond an adequate level of risk and risk may be evaluated by using approaches either a science-based or value-based." Risk is "a function of the probability of an adverse health effect to be occur" while hazard is "severity of that effect" (Manning and Baines, 2004).

FSMS are designed mainly to control the particular food safety hazards that are primarily associated with the product and to endorse compliance with food safety legislation. Quality management systems (QMS) are principally designed to ensure compliance in between third party and retailer standards. Quality assurance (QA) schemes have developed out as a need to demonstrate due assiduousness in association with hygienic requirements, to target the access of quality food in the market and its traceability as well. QA systems are needed in the inspection and grading of quality products and also for development of purchase protocols. Effective QA systems include safety and quality of products as well as of processes (involves production methods). Quality can also be addressed in terms of intrinsic quality that covers the quality of the product and extrinsic quality that involves production and processing systems (Manning and Baines, 2004). The assurance and management can be distinguished as management "relates to company" while assurance "relates to product" (Sikora and Strada, 2005).

A food business entails a higher focus on quality, safety, and traceability system rather than traditional economic systems that is affecting the entire food chain from producer to retailer. General quality systems and standards of retailers often cover more parties in the chain like Good Agricultural Practices (GAPs), Hazard Analysis of Critical Control Points (HACCPs) and ISO, e.g., Quality management systems (QMS): ISO 9001 and FSMS: ISO 22000 (Trienekens, 2004) these systems are adopted to build quality into the product in order to improve food safety and increase consumer confidence toward the brand (Korada et al., 2008). Private safety control systems, standards, and certification programs have developed to respond advanced consumer expectation, because no longer quality is related alone to the product, but also to the physiognomies of the production and distribution processes. Examples of private quality and safety standards in food chains include EUREPGAP whose key element is Good Agricultural Practices (GAPs) and involved at primary production level, other is British retail consortium (BRC) and its key element is HACCP and involved at level of processing firm in supply chain and safe quality food (SQF) whose key element is HACCP and ISO 9000 and the part of chain involved is primary production, processing firms and retail (Trienekens and Zuurbier, 2008).

On the basis of requirements of the public sector, private safety and quality standards are emerging and implemented in the food industry where QA has become a reality. In the modification process, obligatory standards such as HACCP are a prerequisite for companies' behavior; however, the other disseminated standards are also implemented as well throughout the industry. The food safety point of view, the pressure to standardize quality has become significant within the food industry and for numerous producers of raw materials, disintegration, and low levels of concentration.

4.9 TOTAL QUALITY MANAGEMENT (TQM)

The term quality signifies the potential of all elements of an establishment to fulfill the stated and inferred needs; a quality product will satisfy the requirements of the system and services. Quality denotes "conformance to the purpose," "customer satisfaction," "fitness for both use and purpose," or has an expedient elucidation as non-inferior or superior to something. In both product and service cycles to attain satisfying quality levels, all three stages of cycle are significant, which include:

- Distinctly defined needs
- The product design and conformance, and
- Lifetime product support

Total quality management (TQM) comprises of combined efforts at the organizational level and along with system consolidation of principles, techniques, and best strategies to create a stable organizational environment, enabling the system to continuously improve in terms of delivering high-quality customer services and products. According to ISO, TQM can be defined on the basis of its three paradigms: Total, Quality, and Management. Total refers to "organization-wide," Quality signifies "usual definitions with all complexities," and Management defines "governing systems including leading, planning, organizing, and controlling." There are two prime quality aspects: (1) quality of design and (2) quality of conformance.

Quality of design refers to how effective process design is in establishing operational requirements of the product and incorporating them in the design requisites. This process requires grading in product/service quality based on construction material, reliability, and manufacturing tolerance. Quality of conformance refers to how effective the design and production systems are in terms of compliance to standards specified for product and design while meeting product tolerances and production goals with in-process control limits.

4.9.1 FIVE QUALITY APPROACHES

According to David Garvin there are five principal approaches that can define quality:

- Transcendent (quality cannot be defined in words but it is ethereal notion that you will know it when you see it)
- Product-based (quality is a quantifiable attribute/characteristic)
- User-based (quality is an individual matter and products that best satisfy their preferences are those with the highest quality)
- Manufacturing-based (quality is mainly concerned with engineering designs and production practices, i.e., conformance to requirements)
- Value-based (consumer perceive quality as desired attributes at best suited acceptable price)

4.9.2 PDCA CYCLE

Edwards Deming devised a four-step management strategy Plan-Do-Check-Act cycle. PDCA cycle provides business control and unceasing process and product improvement:

1. **Plan**: Identify and analyze problem, establish objectives, and process lines in accordance with quality requirements and output expectations.
2. **Do**: Implementing and executing the process plan to make the product.
3. **Check**: Measuring solution effectiveness and analyzing scope of improvement.
4. **Act**: Implementing improved solutions by process modification, taking corrective action in case of significant discrepancies in planned and actual results, conducting a root cause analysis.

The detail of PDCA cycle is described in Figure 4.1.

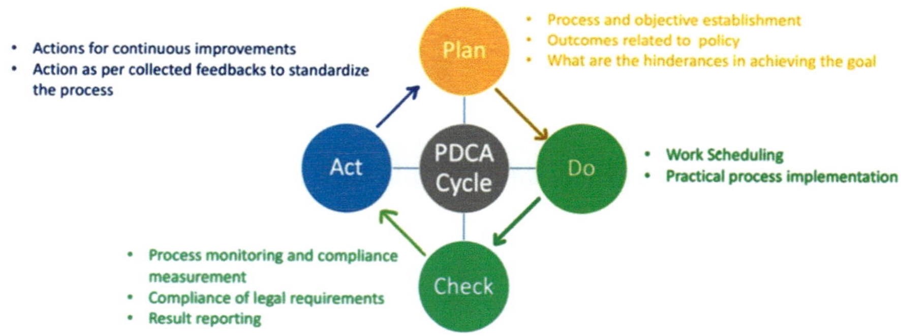

FIGURE 4.1 Elaboration of Deming management strategy.

4.9.3 DEMING'S GUIDELINES FOR QUALITY IMPROVEMENT

To signify the process of total quality control, Deming propounded the Deming philosophy, with 14 principles:

- Develop uniformity of purpose for product and service improvement
- Embrace novel philosophy
- Wind up mass inspection dependence to attain quality
- Do not award the business on the grounds of price tags
- Ensure continual improvement of service and production systems
- Develop proactive on job training and educating techniques
- Establish leadership
- Eradicate fear
- Disintegrate departmental barriers
- Annihilate zero-defect slogans and targets for the workforce
- Extirpate work standards endorsing numerical quotas, instead provide aids and supported supervision (for both workforce and managers)

- Eliminate barriers that hinder workmanship pride of managers and daily wagers
- Impalement vigorous learning and re-training approach
- Empower top management to attain the above goals

4.9.4 JURAN'S CONCEPT OF QUALITY TRILOGY

Joseph Juran gave the quality function concept of "cross functional management."

4.9.4.1 QUALITY PLANNING

Quality planning includes proceeding to develop process and product that complies consumer demands through following approach; establish clear quality goals, identify your target market, establish consumer requirement, design your product accordingly, set-up a process and prove process potential.

4.9.4.2 QUALITY CONTROL

Quality control activities establish system and process parameters by selecting a control subject and measurement unit, after establishing measurement and controls, transfer them to operations, set performance standards, compare actual results with standards, interpret non-conformance and take corrective actions.

4.9.4.3 QUALITY IMPROVEMENT

Quality improvement focuses on continuous improvement in performance by defining the improvement need, developing infrastructure to achieve annual improvement goals, identify specific projects for improvement, organize project guidelines and diagnostic system, prove effectiveness of operational solutions, and providing control to hold gains. Concept and implantation of TQM was first initiated in manufacturing industry, due to is benefits service and other industry also adopted this approach. The current challenge is to ensure that managers maintain a firm grip and vision regarding the basic principles of quality management and performance excellence.

4.10 ISO 9001

Implementation of quality management system is a strategic entity for meeting ultimate quality standards by an organization. The implementation of these quality standards has a key role in improving overall performance regarding sustainability of basic goals of an industry. The ISO is a powerful business improvement tool for attaining maximum level of consumer satisfaction. The basic focus of ISO implementation is to meet customer standards and requirements for exceeding customer expectations. The implementation of ISO standards provokes concept of continual improvement for seeking product of high priorities. The exercising implementation of ISO standards sets an appropriate criterion for executing QMS which is audited by third auditing bureau. The quality management principles are based on evidence-oriented decision making, process approach, engagement of team, leadership, customer focus, continual improvement, and relationship management.

4.10.1 CLAUSES OF ISO 9001:2015

4.10.1.1 SCOPE

The scope of quality standards based on two tenets of quality management which are tends to be achieved by implementing industry. Because scope of that quality standard covers consistency of product and service output by ensuring customer satisfaction and conformity.

4.10.1.2 NORMATIVE REFERENCES

Normative references ensure proper provision of better comprehension of terms or terminologies used in the implementation of ISO 9001:2015 by relatable vocabulary sources.

4.10.1.3 TERMS AND DEFINITIONS

That clause provides references for better understanding of terminologies and definitions used in ISO 9001:2015.

4.10.1.4 CONTEXT OF THE ORGANIZATION

It provides understanding about organizational context. Moreover, that clause outlines the general requirements of quality management system on wholly basis by developing business reviews and strategy plans.

4.10.1.5 LEADERSHIP

This clause deals with the role of "top management" which is a core group of people, who are responsible for directing the control of organization and decision making. The top management is involved in communication and execution of quality control policy; throughout the premises of an organization.

4.10.1.6 PLANNING

This clause is a remarkable addition to ISO 9001:2015 for introducing the concept of risk in business. It helps in provoking the understanding of risks and opportunities for maximizing consumer satisfaction. Besides this, continual improvement and progressive change in an organizational setup is a part of that clause.

4.10.1.7 SUPPORT

The productive implementation of quality management system cannot be achieved without the availability of suitable resources. That clause is based on awareness, competence, communication, resources, and documented information. It believes on appropriate maintenance of infrastructure particularly building, equipment, and transport, etc. The main intention behind this clause is to priorities working people for attaining maximum level of competency.

4.10.1.8 OPERATION

This clause especially deals with operational and production aspects of quality management standards; optimally, operational planning and control, requirement determination of products, design, and development of product, release of product or services and safety of product.

4.10.1.9 PERFORMANCE EVALUATION

Analysis and evaluation of data is an important phenomenon for assessing effectiveness and suitability of management systems. Business goals and objectives can easily be determined and assessed by continual performance evaluation programs. Internal auditing and management reviews fall under the umbrella of this clause.

4.10.1.10 IMPROVEMENT

Mostly, continual improvement is a rather complicated process to achieve on a continual basis. Reduction in non-conformance is a key priority in sustaining continual improvement process and corrective actions also results in proper implementation of ISO 9001:2015.

4.10.2 BENEFITS

Proper executions of quality standards have numerous benefits based on international standards described in briefly as below:

- Able to make products more proficient in consistency and sustainability of products with respect to consumer needs
- Make facilitation for meeting customer satisfaction
- Addressing risks and hazards interlinked with biological, physical, and chemical hazards
- Make able to demonstrate conformance with other quality management systems
- Versatility in the structure and application of different integrated quality executing systems
- ISO is suitable for both small and large industries
- Results in productivity, efficiency, and profitability
- Develops compatibility with other integrated quality-related systems
- Ameliorated consumer fidelity and retention with product
- Results in uniform and conforming outcomes, monitored, and précised
- Minimize mistakes to gross root level by improving reporting process and communications
- Improve marketing and sale-oriented company-based efforts
- Increase control over the business by strengthening organizational departments

4.11 ISO: 22000

From a number of decades, the governmental as well as non-governmental organizations (NGOs) from developed countries of the globe are using many FSMS to uplift the transparency and traceability of food in order to improve the national and international trading of food commodities (Hanak et al., 2000). At the same moment, owing to the exigencies of worldwide trading, manufacturing requirements, their self-imposed demands, global climate change and consumer awareness towards food safety has also forced the suppliers to follow all the food safety measurements throughout the food chain. Meanwhile, the insurance of food safety through these measurements along with maintaining the effectiveness of FSMS has always evoked as daunting challenge (Dumond, 1994; Redshaw, 2000; Vlachos, 2009). Nevertheless, various incidence related to food safety were caused by failure and shortcomings of FSMSs. To cope with these issues, ISO 22000 was promulgated in the form of international standards to establish same standards of FSMSs that has the ability to ensure the safety of food products from farm to the table throughout the chain without producing harmful effects and providing the consumer safe and healthy, wholesome food (Theodoros and Ioannis, 2008). A number of scientists depicted that the scientific as well as the technical research work has been started to investigate all the hazards associated with the cause of shortcoming and failures of FSMSs that further suppressed the effectiveness of system (Henroid and Sneed, 2004; Taylor and Taylor, 2004; Azanza and Zamora-Luns, 2005). In this context amid various companies, a few revealed insufficient knowledge of relevant food safety concepts, constructing an over-reliance of CCPs (critical control point) to note the hazards in a pipeline where a prerequisite program (PRP) can be more appropriate. The point was investigated by Al-Busaidi et al. (2017) who worked on the determination of major barriers in FSMSs for processed foods with reference to effect on cost. Although, with all the possible cost-related problematic issues, the supervisory author of officials of ISO, 2000 also lacked in different capabilities such as employee commitment, competence, and knowledge of every aspect. For that reason, the implementation of methodology of ISO 22000:2018 along with HACPP (hazard analysis critical control point) principles approved by Codex Alimentarius Commission (CAC) was introduced in various food categories (Purwanto et al., 2019).

Previously the food factories needed a number of standards to be implemented at the same time to adhere various FSMSs according to the standards of all the globes based on the requirement of private, regional-based market-specific standards either national or international level. Although, it is very difficult for industries to meet all the audits (internal and external) and managing all the requirements of FSMS simultaneously. Generally, ISO declared to review the standards after every five years to find out if any revision is required. Hence, ISO officials can ensure that the present standard is up to the mark and up to the date in terms of legality and applicability. Furthermore, ISO 22000 was based on the integration of ISO 9001, ISO 15161 over and above to HACCP. In addition to the implementation of the previously present ISO along with principles for food hygiene in the food chain, it was essential to combine all the critical control points (CCP) with hazard analyzes and further identification of any risk that can lead to the hazard at any point in food chain. The establishment and design of PRPs for ISO 22000 were required owing to the reason that GFSI (global food safety initiative) did not mark ISO 22000 as references standard in the start. Therefore, latest PRPs were formulated in 2018 to overcome the shortness of earlier ISO 22000 (Stefan et al., 2018).

ISO 22000 was supposed to construct as FSMS for a given food industry and constituted on the basis of the PDCA approach, i.e., Plan, Do, Check, and Act. ISO 22000:2018 confirmed the principle 7 principles and 12 application steps of HACCP. The ISO 22000:2018 specifically emphasis on the integration of risk-based thinking with risk assessment and enhanced it to the highest level of management commitment and participation towards implementation of food safety policies (Purwanto et al., 2019).

4.11.1 ISO 22000 BENEFITS

ISO 22000 as FSMS endow with the guidelines for any of the organization to develop a plan, then to implement it, maintain it and update it according to the FSMS followed to improve the effectiveness. However, with the implementation of ISO 22000, an organization can have more orders, become more trustworthy in customers regarding the product safety aspects and successfully get their satisfaction level. The companies following ISO 22000, directly or indirectly bound to follow all the laws and regulations that are applicable to that particular industry. ISO 22000 also provides outlines to the organization to assess and cover the mutually agreed demands of customers towards food safety (Stefan et al., 2018).

Previously, the requirements of ISO 22000:2005 standard were used for the selection and corrective measurements of the control measures over and above to conduct the assessment of hazard analysis. Proper planning should be required to determine which hazard needs to be controlled firstly. With the passage of time, ISO 22000 shared the common elements with other existing FSMS standards *viz*., policy, implementation, planning, operation, improvement, performance assessment and management review (Surak, 2007).

Recently, the ISO 22000:2018, evoked as a FSMS containing all the requirements and demands of the food chain as well as organization. This FSMS translated the food safety system as a process that has the capability of improvement on a continuous basis. ISO 22000:2018 focuses on the preventive action of food safety that includes identification of hazard, prevention from that hazard and reduction in the food as well as feed chain. Furthermore, ISO 22000:2018 applies pragmatic approach as an essential step to define distinctions between strategic level risk and operational level risk of the management system (ISO 2018). The ISO defines ISO 22000:2018 as a standard that offers complete power on hazards dealing with the safety of food. All the standard of ISO 22000:2018 combined together to form key elements such as PRPs, HACCP, interactive communications, and management system. These combined standards are applicable to the organizations that deal with feed production, harvesting the crop, food processors and production, packaging, distribution, storage, and food catering services (ISO, 2018). Hence, ISO 22000 has a number of benefits in safety concern as well as in verification of any organization such as:

- All the resources are optimally allocated within the food chain by the help of ISO 22000
- All the legal requirements are effectively communicated between relevant institution, supervisory authorities, customers, and suppliers
- There is more focus on PRPs, health measurements and conditions along with the preventive measurements to mitigate the possible failures
- ISO 22000 provides better documentation
- ISO 22000 creates and build trust hence, improving the credibility of management system that is further based on the reliable results of clear management processing and prerequisite of visual checks (Theodoros and Ioannis, 2008)

4.12 CONCLUSION

The implementation of FSQ standards have become a major concern for food manufacturing sector in order to assure wholesomeness of the processed product. The utmost care during handling, processing, and storage of food items has become inevitable in order to produce high quality and hazard free products. Hence, implementation of quality assuring standards can lead to the production of hazard free products. The outcomes of these quality-assuring measures ultimately would lead to a shrinkage of the global burden of diseases. The building of barriers either in the form of physical or chemical entity against the biological hazard is a key factor in limiting the proliferation of food quality degrading elements.

KEYWORDS

- **food quality**
- **food safety**
- **food terrorism**
- **food hazard**
- **foodborne microorganisms**
- **food quality assurance**
- **food quality standards**

REFERENCES

Adams, M. R., & Moss, M. O., (2008). *Food Microbiology*. London: Royal Society of Chemistry.
Addis, M., & Sisay, D., (2015). A review on major foodborne bacterial illnesses. *Journal of Tropical Diseases and Public Health*, *3*(4), 176–183.
Adeyeye, S. A., (2016). Fungal mycotoxins in foods: A review. *Cogent Food and Agriculture*, *2*(1), 3121–3127.
Al-Busaidi, M. A., Jukes, D. J., & Bose, S., (2017). Hazard analysis and critical control point (HACCP) in seafood processing: An analysis of its application and use in regulation in the Sultanate of Oman. *Food Control*, *73*, 900–915.
Alcamo, E., (1996). *Cliffs Quick Review Microbiology*. Wiley Publishing Inc.
Al-Dobaib, S. N., & Mousa, H. M., (2009). Benefits and risks of growth promoters in animal production. *Journal of Food, Agriculture and Environment*, *7*(2), 202–208.
Antunes, P., & Gil, O., (2004). PCB and DDT contamination in cultivated and wild sea bass from Ria de Aveiro, Portugal. *Chemosphere*, *54*(10), 1503–1507.
Arness, M. K., Feighner, B. H., Canham, M. L., et al., (2000). Norwalk-like viral gastroenteritis outbreak in US army trainees. *Emerging Infectious Diseases*, *6*(2), 204.
Azanza, P., & Zamora-Luna, M., (2005). Barriers of HACCP team members to guideline adherence. *Food Control*, *16*, 15–22.

Barton, M. D., (2000). Antibiotic use in animal feed and its impact on human health. *Nutrition Research Reviews, 13*(2), 279–299.

Bintsis, T., (2017). Foodborne pathogens. *AIMS Microbiology, 3*(3), 529.

Bosch, A., Gkogka, E., Le Guyader, F. S., et al., (2018). Foodborne viruses: Detection, risk assessment, and control options in food processing. *International Journal of Food Microbiology, 285*, 110–128.

Bozoglu, F., (2012). Food allergies, intolerances and foodborne intoxications. In: Alpas, H., Smith, M., & Kulmyrzaev, A., (eds.), *Strategies for Achieving Food Security in Central Asia* (pp. 93–108). The Netherlands: Springer International Publishing.

Campbell, A. W., (2016). Molds and mycotoxins: A brief review. *Alternative Therapies in Health and Medicine, 22*(4), 8.

Chen, H., Liu, S., Chen, Y., Chen, C., Yang, H., & Chen, Y., (2020). Food safety management systems based on ISO 22000: 2018 methodology of hazard analysis compared to ISO 22000: 2005. *Accreditation and Quality Assurance, 25*(1), 23–37.

Crews, C., (2014). *Natural Toxicants: Alkaloids, Encyclopedia of Food Safety Hazards and Diseases*, 251–260.

Danaher, M., & Jordan, K., (2013). Identification of existing and emerging chemical residue contamination concerns in milk. *Irish Journal of Agricultural and Food Research, 17*(4), 173–183.

Davidson, P. W., Myers, G. J., & Weiss, B., (2004). Mercury exposure and child development outcomes. *Pediatrics, 113*(3), 1023–1029.

Dumond, E. J., (1994). Making best use of performance measures and information. *International Journal of Operations and Production Management, 14*, 16–31.

Edgar, J. A., Colegate, S. M., Boppré, M., & Molyneux, R. J., (2011). Pyrrolizidine alkaloids in food: A spectrum of potential health consequences. *Food Additives and Contaminants: Part A, 28*(3), 308–324.

EFSA Panel on Biological Hazards (BIOHAZ), (2011). Scientific Opinion on an update on the present knowledge on the occurrence and control of foodborne viruses. *EFSA Journal, 9*(7), 2190.

Esterbauer, H., Schaur, R. J., & Zollner, H., (1991). Chemistry and biochemistry of 4-hydroxynonenal, malonaldehyde and related aldehydes. *Free Radical Biology and Medicine, 11*(1), 81–128.

Francesconi, K. A., (1994). Determination of arsenic and arsenic species in marine environmental samples. *Cycling and Characterization, 18*(3), 263–281.

Gupta, A., (2006). *Pesticide Residues in Food Commodities*. Agrobios (India).

Hanak, E., Boutrif, E., Fabre, P., & Pineiro, M., (2000). *Food Safety Management in Developing Countries*. CIRAD-FAO, Montpellier.

Henning, R. J., Johnson, G. T., Coyle, J. P., & Harbison, R. D., (2017). Acrolein can cause cardiovascular disease: A review. *Cardiovascular Toxicology, 17*(3), 227–236.

Henroid, D. J., & Sneed, J., (2004). Readiness to implement hazard analysis critical control point (HACCP) system in Iowa schools. *Journal of the American Dietetic Association, 104*, 180–185.

Islam, G. M. R., & Hoque, M. M., (2013). Food safety regulation in Bangladesh, chemical hazard and some perception to overcome the dilemma. *International Food Research Journal, 20*(1), 47.

Jun-Shi, C. H. E. N., (2009). What can we learn from the 2008 melamine crisis in China? *Biomedical and Environmental Sciences, 22*(2), 109–111.

Kannan, K., Tanabe, S., Giesy, J. P., & Tatsukawa, R., (1997). Organochlorine pesticides and polychlorinated biphenyls in foodstuffs from Asian and oceanic countries. *Reviews of Environmental Contamination and Toxicology*, *2*(5), 1–55.

King, A. M., Lefkowitz, E., Adams, M. J., & Carstens, E. B., (2011). *Virus Taxonomy: Ninth Report of the International Committee on Taxonomy of Viruses*. Elsevier.

Kiran, D. R., (2016). *Total Quality Management: Key Concepts and Case Studies*. Butterworth-Heinemann.

Korada, S. K., Yarla, N. S., Putta, S., Hanumakonda, A. S., Lakkappa, D. B., Bishayee, A., & Peluso, I., (2018). A critical appraisal of different food safety and quality management tools to accomplish food safety. In: Grumezescu, A. M., & Holban, A. M., (eds.), *Food Safety and Preservation* (pp. 1–12). Academic Press.

Kríz, B., Benes, C., & Daniel, M., (2009). Alimentary transmission of tick-borne encephalitis in the Czech Republic (1997–2008). Epidemiologie, mikrobiologie, imunologie: Casopis spolecnosti pro epidemiologii a mikrobiologii Ceske lekarske spolecnosti. *JE Purkyne*, *58*(2), 98–103.

Kumar, A., Beniwal, M., Kar, P., Sharma, J. B., & Murthy, N. S., (2004). Hepatitis E in pregnancy. *International Journal of Gynecology and Obstetrics*, *85*(3), 240–244.

Lungarini, S., Aureli, F., & Coni, E., (2008). Coumarin and cinnamaldehyde in cinnamon marketed in Italy: A natural chemical hazard? *Food Additives and Contaminants*, *25*(11), 1297–1305.

Luning, P. A., & Devlieghere, (2006). *Safety in the Agri-Food Chain*. Wageningen Academic Publishers, The Netherlands.

Manning, L., & Baines, R. N., (2004). Effective management of food safety and quality. *British Food Journal*, *106*(8), 598–606.

Nguyen, H. T., Fels-Klerx, H. J. V. D., & Van, B. M. A. J. S., (2014). N ϵ-(carboxymethyl) lysine: A review on analytical methods, formation, and occurrence in processed food, and health impact. *Food Reviews International*, *30*(1), 36–52.

O'Grady, J., (1992). Management of acute and fulminant hepatitis A. *Vaccine*, *10*, 21–23.

Paterson, R. R. M., & Lima, N., (2017). Filamentous fungal human pathogens from food emphasizing *Aspergillus*, *Fusarium* and *Mucor*. *Microorganisms*, *5*(3), 44.

Paterson, R. R. M., (2019). Human pathogenic filamentous fungi from food/water and mycotoxins from water. *Immunology*, *77*(19), 78–101.

Purwanto, A., Putri, R. S., Ahmad, A. H., et al., (2019). The effect of implementation integrated management system ISO 9001, ISO 14001, ISO 22000 and ISO 45001 on Indonesian food industries performance. *Engineering Management Journal*, *82*, 14054–14069.

Rasenack, R., Muller, C., Kleinschmidt, M., Rasenack, J., & Wiedenfeld, H., (2003). Veno-occlusive disease in a fetus caused by pyrrolizidine alkaloids of food origin. *Fetal Diagnosis and Therapy*, *18*(4), 223–225.

Ray, B., & Bhunia, A., (2007). *Fundamental Food Microbiology*. Florida: CRC Press.

Ruggeri, F. M., Di Bartolo, I., Ponterio, E., Angeloni, G., Trevisani, M., & Ostanello, F., (2013). Zoonotic transmission of hepatitis E virus in industrialized countries. *New Microbiology*, *36*(4), 331–344.

Sapkota, A., Sapkota, A. R., Kucharski, M., et al., (2008). Aquaculture practices and potential human health risks: Current knowledge and future priorities. *Environment International*, *34*(8), 1215–1226.

Scallan, E., Hoekstra, R. M., Angulo, F. J., et al., (2011). Foodborne illness acquired in the United States – major pathogens. *Emerging Infectious Diseases*, *17*(1), 7.

Schirone, M., Visciano, P., Tofalo, R., & Suzzi, G., (2017). Biological hazards in food. *Frontiers in Microbiology*, *7*, 2154.

Schmidt, R. H., & Rodrick, G. E., (2003). *Food Safety Handbook*. John Wiley & Sons.

Sikora, T., & Strada, A., (2005). Safety and quality assurance and management systems in food industry: An overview. In: *The Food Industry in Europe* (pp. 85–95). Agricultural University of Athens, Athens.

Simoes, M., Simoes, L. C., & Vieira, M. J., (2010). A review of current and emergent biofilm control strategies. *LWT-Food Science and Technology*, *43*(4), 573–583.

Staples, P., (2016). *Advantage and Disadvantage of Monosodium Glutamate*, *4*, 17.

Stefan, Z. P., Raluca, D., & Carmen, V., (2018). Comparative study of certification schemes for food safety management system in the European Union context. *Amfiteatru Economic*, *47*, 9–29.

Stevens, J. F., & Maier, C. S., (2008). Acrolein: Sources, metabolism, and biomolecular interactions relevant to human health and disease. *Molecular Nutrition and Food Research*, *52*(1), 7–25.

Surak, J. G., (2007). A recipe for safe food ISO 22000 and HACCP. *Quality Progress*, *40*, 21–27.

Taylor, E., & Taylor, J., (2004). Using qualitative psychology to investigate HACCP implementation barriers. *International Journal of Environmental Health Research*, *14*, 53–63.

Theodoros, H. V., & Ioannis, S. A., (2008). Application of ISO 22000 and comparison to HACCP for processing of ready to eat vegetables: Part I. *International Journal of Food Science and Technology*, *43*, 1729–1741.

Trienekens, J. H., (2004). Quality and safety in food supply chains. In: *Bridging Theory and Practice* (pp. 253–267). Reed Business.

Trienekens, J., & Zuurbier, P., (2008). Quality and safety standards in the food industry, developments and challenges. *International Journal of Production Economics*, *113*(1), 107–122.

Van, A. E. D., Fels-Klerx, H. J. V. D., Marvin, H. J. P., Bokhorst-Van De, V. H. V., & Groot, M. N., (2017). Overview of food safety hazards in the European dairy supply chain. *Comprehensive Reviews in Food Science and Food Safety*, *16*(1), 59–75.

Vasickova, P., Dvorska, L., Lorencova, A., & Pavlik, I., (2005). Viruses as a cause of foodborne diseases: A review of the literature. *Veterinární Medicína*, *50*(3), 89–104.

Walker, R., (1985). Sulphiting agents in foods: Some risk/benefit considerations. *Food Additives and Contaminants*, *2*(1), 5–24.

Wong, D., Purcell, R., Sreenivasan, M. A., Prasad, S. R., & Pavri, K., (1980). Epidemic and endemic hepatitis in India: Evidence for a non-A, non-B hepatitis virus aetiology. *The Lancet*, *316*(8200), 876–879.

World Health Organization (WHO), (1995). Coumarin: A strong association with hepatotoxicity. *WHO Drug Information*, *9*, 159.

Xiong, Z. T., & Wang, H., (2005). Copper toxicity and bioaccumulation in Chinese cabbage (*Brassica pekinensis* Rupr.). *Environmental Toxicology: An International Journal*, *20*(2), 188–194.

Zahoor, T., Liaqat, A., & Azhar, N., (2018). Safety and quality aspects of animal origin foods. In: Khan, M. I., &. Sameen, A., (eds.), *Animal Sourced Foods for Developing Economies: Preservation, Nutrition, and Safety* (pp. 207–224). Florida: CRC Press.

CHAPTER 5

Foodborne Illness: Causes and Preventive Measures

BINDU SADANANDAN, GANGARAJU DIVYASHRI, and PRIYA ASHRIT

ABSTRACT

Foodborne illness that results from the exposure of toxins and/or pathogens from the food are causing severe threat to public health. Bacterial and viral causes of foodborne illness are highlighted. Microbial toxins contributing to food intoxications are clearly stated. Thus, food safety has become a matter of concern across the globe with an aim to prepare, handle, and store food to reduce the risk from foodborne illness during the entire cycle of food production for its consumers. Even though there exists a critical requirement for strong preventive measures, the present chapter highlights the available background to these problems to illustrate the cause of various foodborne illnesses.

5.1 OVERVIEW OF FOOD SAFETY AND FOODBORNE ILLNESSES

Food safety is a matter of concern across the globe. It aims to prepare, handle, and store food to reduce the risk from foodborne illness by either chemical or biological hazards (BIOHAZ) during the entire cycle of food production for its consumers. According to DALYs (Disability Adjusted Life Year) measurements, there are an estimated 420,000 deaths, with 125,000 deaths of children less than 5 years of age recorded every year because of the consumption of contaminated food. The increase in world population

Food Microbial and Molecular Biology: From Fundamentals to Applications. Saher Islam, Devarajan Thangadurai, Jeyabalan Sangeetha, Zaira Zaman Chowdhury (Eds.)

presents an increasing challenge in ensuring food safety (Schirone et al., 2019) on food industry. This emphasizes the need for food safety guidelines to be imposed by regulatory authorities on food industries to prevent any undesired consequences and reduce the incidences of foodborne illness.

Foodborne illnesses are reported in many countries every year, and they form the main source of public health troubles globally (Heredia and Garcia, 2018; Schirone et al., 2019). Microbial pathogens *viz*. viruses, bacteria, and/or parasites represent the BIOHAZ. These pathogens become part of the food production cycle through their entry into the food chain, causing detrimental effects or by producing toxins that may act as food contaminants on consumption (Bintsis, 2017). The major foodborne pathogens are listed in Table 5.1. The majority of foodborne illnesses may result in discomfort of the digestive tract which sometimes may be fatal.

TABLE 5.1 Major Microorganisms Identified As Foodborne Pathogens

Type	Microorganism
Bacteria	*Arcobacter butzleri*
	Campylobacter jejuni
	Campylobacter fetus
	Cronobacter sakazakii
	Escherichia coli O157:H7
	E. coli non-O157 STEC
	Listeria monocytogenes
	Vibrio cholerae O139
	V. vulnificus
	V. parahaemolyticus
	Yersinia pseudotuberculosis
	Y. enterocolitica
Virus	Astrovirus
	Caliciviridae
	Hepatitis E
	Nipah virus
	Rotavirus
Parasite	*Cryptosporidium*
	Cyclospora cayetanensis
	Sarcocystis
	Trypanosoma cruzi

Food associated microorganisms are generally classified into three categories based on their activity:

- Beneficial microorganisms which aid in the process of making new foods; one such example is the conversion of milk sugar to an acid in the process of cheese making
- Spoilage microorganisms which are the reasons behind food spoilage and are not harmful to humans; one such example is souring of milk, and
- Pathogenic microorganisms are the actual disease-causing live micro-organisms and/or their toxins which when consumed cause various foodborne illness

Foodborne illness can be broadly classified as:

1. **Foodborne Infection:** Ingestion of live pathogen present in the food that gets established within the human intestinal tract
2. **Foodborne Intoxication:** Consumption of toxins produced by the pathogens present in the food

Typically, foodborne infections require a much longer time to produce symptoms than food toxins.

5.2 FOODBORNE ILLNESSES DUE TO PATHOGENS

In recent years, foodborne illnesses caused by microbial contamination of food have gained significant public concerns and awareness. Under the right conditions, some of these microbes thrive on food leading to food spoilage and human illness.

5.2.1 BACTERIAL CAUSES OF FOODBORNE ILLNESS

Bacterial foodborne illness results when food contaminated with bacteria is consumed resulting in bacterial growth within the intestinal cavity and setting up an infection.

5.2.1.1 SALMONELLOSIS

Salmonellosis results when food containing *Salmonella* bacteria are consumed. *Salmonella enteritidis*, *S. typhosa*, *S. cholerasuis* and *S. typhimurium* are the most common species that accounts for half of the infections. Most *Salmonella* spp. are known to cause (a) Enteric fever caused by *S. typhosa* (ingestion of *S. typhosa* with food, find its way into the bloodstream and then gets excreted out); (b) Septicemia by *S. cholerasuis*, causes blood poisoning; and (c) Gastroenteritis caused by either ingestion of *S. typhimurium* and *S. enteritidis* are known to persist for 2–3 days. The overall symptoms of the above diseases are mild headache, vomiting, nausea, restlessness, chills, abdominal pain, diarrhea, muscular weakness, drowsiness, and moderate fever.

Salmonellosis outbreaks due to consumption of contaminated meat, eggs, and egg products have been well documented (Kumar et al., 2019). The most prevalent serovars among *Salmonella* sp. are *S. enteritidis*, *S. typhimurium* and *S. gallinarum*. They are the primary cause of morbidity and mortality in commercial poultry resulting in significant economic losses, with the highest proportions observed during 2011–2016 contributing to nearly 43.7% of the total *Salmonella* isolates of poultry origin.

5.2.1.2 CAMPYLOBACTERIOSIS

Consumption of *Campylobacter jejuni* contaminated water or food causes Campylobacteriosis. Healthy animals ideally inhabit *C. jejuni* in the intestinal tracts and they are found in untreated water surfaces. However, non-chlorinated water and inadequately cooked foods (and even raw food) like undercooked chicken, raw milk, raw shellfish, and hamburger of animal origin are the most potent source of infections in human beings. The survival rate of *C. jejuni* is maximum under hypoxic environment and is heat-labile (120°F). However, the growth of this bacterium is hampered by the presence of acid and salt, and they fail to propagate at temperatures below 85°F. Symptoms of Campylobacteriosis embrace abdominal pain, headache, fever, nausea, and muscle pain. Onset of the symptoms usually takes 2–10 days to occur after the consumption of contaminated food. Furthermore, Campylobacteriosis also cause complications of reactive arthritis and urinary tract infections. Acute cholecystitis, recurrent colitis, meningitis, Guillain-Barre syndrome are also reported to occur, but at the lower complication rate. However, there have been rare cases of deaths reported. Campylobacteriosis can be prevented by milk pasteurization, avoiding raw poultry, meat and fish, and most importantly by averting the cross-contamination of raw food and cooked/ready-to-eat foods.

5.2.1.3 LISTERIOSIS

Listeria monocytogenes, the causative agent of Listeriosis, was discovered prior to the 1980s. Initially, it was associated with encephalitis in cattles with extensive environmental distribution. *Listeria* spp. has the ability to survive under adverse conditions and are reported to survive and grow at temperatures between 2–4°C. Humans and animals also carry *L. monocytogenes*. The organism survives and grows under optimal pH between 4.4 to 9.6. These organisms thrive under environments with high salt content and that are relatively dry. Pregnant women, new born infants, immune-compromised, and elderly individuals are most commonly affected. Listeriosis symptoms are common and mild, like fever, headache, and nausea in a healthy non-pregnant woman. Listeriosis in pregnant women causes infection of the uterus and cervical tract, which may result in still birth or abortions. However, infants born with infection reportedly develop meningitis. Onset of symptoms ideally take anywhere from 7 days to several weeks. Maintenance of proper sanitation, milk pasteurization and avoiding contamination after pasteurization are few ways to prevent listeriosis.

5.2.1.4 SHIGELLOSIS

The organisms of the genus *Shigella* causes Shigellosis, which is commonly known as bacillary dysentery. *Shigella* can be killed by heat during food processing or cooking and they fail to survive in acidic food conditions (pH below 4.6). However, they have demonstrated a high survival rate for extended periods only in certain food types. Symptoms of diarrhea, vomiting, fever, and abdomen pain are common in shigellosis. Incubation time for shigellosis ranges from 7–36 h, nonetheless the symptoms persist for nearly 13–14 days. Most outbreaks have resulted from contamination of raw or previously cooked foods in home or at foodservice stations.

5.2.1.5 YERSINIOSIS

Yersinia enterocolitica causes Yersiniosis, and is found in variety of food and water sources. Meat, poultry, raw milk, shellfish, vegetables, and tofu are the well-known food sources of *Y. enterocolitica*. Only certain species belonging to genus *Yersinia* are reported to produce disease, and it is found that most of them are demonstrated to be non-pathogenic. Survival rate of *Y.*

enterocolitica is best at room temperature and even can grow at refrigerator temperatures. However, the organism is sensitive to heat and can be easily destroyed with adequate cooking procedures and by milk pasteurization process. Thus, in line with this, preventive measures such as strict adherence to procedures for keeping perishable products (tofu and milk) in processing plants need to be employed.

5.2.2 VIRAL CAUSES OF FOODBORNE ILLNESS

Viruses belonging to several families may transmit foodborne illness (Shaheen, 2017). Foodborne viruses get transmitted through one-to-one contact and by consumption of contaminated food. For over a decade, most of the foodborne illnesses have been attributed to bacterial pathogens only. But, with the progression in diagnostic technology, few enteric viruses have also been proven to be transmitted by food occasionally and hence have become the major contributors to foodborne illness (Todd and Grieg, 2015). Viruses belonging to 24 families are known to infect humans. Of these, there have been at least 10 families known to cause foodborne illness. The severity of these infections may range from mild diarrhea to a severe case of encephalitis (Koopmans, 2012). Some of the viruses are occasionally transmitted by food. Scallan et al. (2011) has reported foodborne illness caused by Norovirus (NoV-26%) and Hepatitis A virus (HAV-7%) are greater compared to Human astrovirus, Human rotavirus (HRV) and Sapovirus which are responsible for less than 1% of illness.

5.2.2.1 NOROVIRUS (NoV)

NoV is reported to be the most frequent and prominent foodborne pathogen in many countries (Belliot, 2014). The Center for Disease Control and Prevention (CDC) in 2014 has estimated about 50% of foodborne outbreaks due to Notavirus. Human NoV belongs to the Caliciviridae family and is divided into two genera *viz.*, NoV and sapovirus (Koopmans, 2012). NoV causes upper respiratory tract infections, oral lesions, systemic disease with hemorrhagic syndromes in animals while it is known to cause gastroenteritis in humans (Martella et al., 2011). Foodborne transmission of NoV occurs due to widespread exposures, thereby resulting in much larger outbreaks. Shellfish, leafy greens, and fresh fruits are the foods that are frequently involved in NoV outbreaks. Furthermore, touching of food by infected food workers with their inadequately washed bare hands can transmit NoV (CDC, 2014).

5.2.2.2 HEPATITIS A VIRUS (HAV)

The HAV belonging to Hepatovirus genus of the Picornaviridae family affect both humans and primates. HAV of only genotypes I-III are known to infect humans (Koopmans, 2012). HAV causing foodborne infections are mainly transmitted through vegetables, fruits, juices, cheese, bread, milk, ice cream, cookies, and other raw or undercooked foods. Food establishments with poor sanitation, poor personal hygiene along with their improper waste disposal systems contributes more effectively to food contamination with HAV (Sattar et al., 2000). It is noteworthy to report that the world's first largest viral foodborne disease outbreak dates back to 1988 in Shanghai, China. There were 300,000 cases of infections that were related to the consumption of clams harvested from HAV sewage-polluted waters (Halliday et al., 1991). In France (2010), 59 cases of illness with HAV belonging to genotype I having no travel history were reportedly caused by the consumption of semi-dried tomatoes that were bought from one of the sandwich shops out of the three chains. Later it was successfully tracked that the original supplier in France imported the frozen semi-dried tomatoes from Turkey (Gallot et al., 2011). There were three other outbreaks of HAV infection which were highly similar to strains of genotype I that were reported in Australia in 2009 and in the Netherlands in 2010 that were also caused due to the consumption of semi-dried tomatoes. The investigation showed a pattern of spread where it pointed out the contamination originating from the food production chain. As described in the green onion outbreak, the use of contaminated water for tomato cultivation resulted in the outbreak (Wheeler et al., 2005). There was a possibility of all the four major outbreaks to have been originated from a farm operation in Turkey. In 2013, about 165 individuals in 10 US states were infected with HAV genotype I subtype B that are most common in Turkey. The infection was caused due to the consumption of pomegranate arils that were imported from Turkey (Collier et al., 2014). Infected individuals (69) were hospitalized, one required a liver transplant and two had fulminant hepatitis.

5.2.2.3 HEPATITIS E VIRUS (HEV)

HEV has a wide range of host and geographical distribution and is grouped into four genotypes (1–4). Genotype 1 and 2 are endemic to Asia-Africa and Mexico-Western Africa, respectively causing infections exclusively in humans with genotypes 3 and 4 as exceptions as they infect animal species (Tesse et al., 2012). HEV is waterborne infection accompanying low quality drinking water. The exact mode of transmission needs to be established. Person-to-person transmission also appears to be rare in case of HEV (Kumar et al., 2011). Epidemiological studies have now shown the evidence for foodborne HEV outbreaks resulting from the consumption of HEV-infected contaminated products (raw or undercooked meat), but only very few systematic studies have been accomplished so far (Colson et al., 2010; Lewis et al., 2010). HEV infected pork liver sold in retail shops have been reported in Netherlands and USA (Berto et al., 2012).

5.2.2.4 HUMAN ROTAVIRUS (HRV)

Group A HRV is the chief cause of dehydration and diarrhea in children, whereas diarrhea due to Group B HRV is an epidemic that affects only adults. Group C HRV causes rare and sporadic cases of diarrhea in children, in most countries (FDA, 2015). Humans are the main reservoir of HRV. However, cases of infection with group A rotaviruses have been observed in animals as well. Rotaviruses persist for several weeks in the environment due to the presence of three protein layers that act as protective coats. Person-to-person spread is a known means of HRV transmission, but under improper hygienic conditions, foodborne, and waterborne transmissions are also expected to play a role (WHO, 2008). HRV infection is associated with a short incubation time of 3 days. The onset of the disease results in watery diarrhea, abdominal pain, vomiting, and fever and finally leading to dehydration with hospitalization in case of infants (Todd and Grieg, 2015).

5.2.2.5 HUMAN ASTROVIRUS (HAstV)

Astrovirus was first discovered in 1975, following an outbreak of diarrhea in infants. HAstV belongs to the family of Astroviridae which consists of the genus Mammoastrovirus. AstVs are known to cause gastroenteritis (Lorenzo et al., 2018). AstV are single-stranded positive RNA viruses that are non-enveloped and consist of two genera: Mammoastrovirus and Avastrovirus (Vu et al., 2017). The modes of transmission of HAstV may be different. Reports have shown that HAstV contaminated foods at the preharvest stage are found to be stable in fresh surface water and drinking water (Todd and Grieg, 2015).

5.2.2.6 HUMAN ADENOVIRUS (HAdV)

Numerous disease outbreaks have been associated with nearly 52 different serotypes of HAdV. Most HAdV types attack the upper respiratory tract; however, few types cause gastroenteritis (Maunula et al., 2017). HAdV are now considered as a marker for fecal contamination of water as they are present at increasingly higher concentrations in sewage (Wyn-Jones et al., 2011).

5.2.3 PARASITE CAUSES OF FOODBORNE ILLNESS

The most prevalent form of foodborne illness in the recent years are caused due to parasites, but are however underestimated (Dorny et al., 2009). The organisms being parasitic, therefore inhabit and reproduce within the infected host and mostly excreted as feces. Some of the organisms that are most commonly recognized as foodborne pathogens include *Cyclospora cayetanensis*, *Toxoplasma gondii* and *Trichinella spiralis* (Bintsis, 2017). The difficulty in the assessment of their route of transmission and the burden on public health is mainly attributed to their complex life cycles, mode of transmission and their incubation time to develop and show symptoms after developing the infection (Panel et al., 2018).

5.2.3.1 CYCLOSPORA CAYETANENSIS

C. cayetanensis is a protozoan parasite which causes a watery diarrhea, nausea, and vomiting in humans (Bintsis, 2017). The first case of *C. cayetanensis* infection was reported in the late 1970s among travelers in regions where infections were endemic. *Cyclospora* has thus been known to be the cause of traveler's diarrhea (Ortega and Sanchez, 2010). They inhabit in small intestine reportedly in their intermediary life cycle stages within the cytoplasm of enterocytes and eventually produce oocysts that contain two sporocysts which encapsulate the four sporozoites (Dorny et al., 2009). Oocysts ideally take 7 to 15 days for sporulation (23°C to 27°C) and causes infections in susceptible hosts. Ingestion of oocysts contaminated water or food contaminated by a susceptible host results in the release of sporozoites and oocysts excyst into the epithelial cells of the jejunum and duodenum. Furthermore, asexual reproduction produces meronts of type I and II;

the latter ones later differentiate into gametocytes which are fertilized by microgametocyte to produce a zygote. Oocysts are then excreted into the environment as unsporulated oocysts (Figure 5.1) (Ortega and Sanchez, 2010). Many species of chickens, ducks, and dogs are known to contain oocysts of *C. cayetanensis* (Carollo et al., 2001; Chu et al., 2004). Thus far, *C. cayetanensis* infections have been identified and reported only in humans with 2 genome sequences being successfully completed with a total genomic length of 44.2991 Mb size as per the data published by NCBI in 2017.

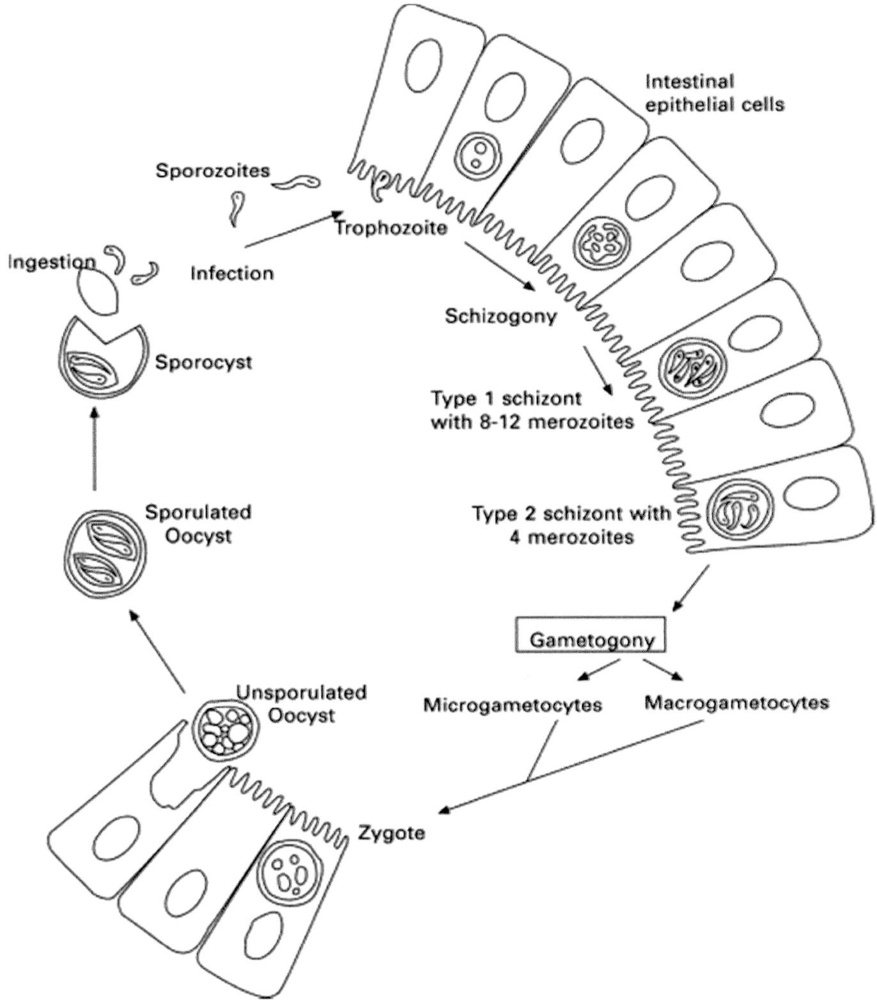

FIGURE 5.1 Different stages of *Cyclospora* life cycle.

Source: Reproduced with permission from: Smith and Evans (2009). © Elsevier.

5.2.3.2 TOXOPLASMA GONDII

T. gondii is an intracellular protozoan parasite and belongs to the phylum Apicomplexa (Bintsis, 2017). Cats are its primary hosts, but it is known to reside and complete its life cycle in almost all warm-blooded vertebrates (Bintsis, 2017), and cause toxoplasmosis in humans (Bacon and Sofos, 2003). *T. gondii* infections are caused due to the ingestion of cysts that are primarily present in beef and also to some extent in raw or undercooked meat, with fresh pork (Bacon and Sofos, 2003). Sometimes dairy and plant products along with water can also serve as vehicles for *T. gondii* infection as reported by Robert-Gangneux and Darde (2012) and Vieira et al. (2015), respectively.

5.2.3.3 TRICHINELLA SPIRALIS

T. spiralis of the phylum Nematoda is responsible for most trichinosis infections in humans. The diagnosis of trichinellosis in most patients is difficult as they exhibit nonspecific symptoms (Holzbauer et al., 2014). The unique property of *T. spiralis* is that it completes its life cycle from larval to the adult stage within the same animal but in different organs. The protozoan regulates host cell activities for its own intracellular survival (van Die and Cummings, 2010). Trichinosis is caused by the dissemination of new born larvae into lymph and blood and final invasion of various body tissues (Salvana et al., 2014). Two genomes of *T. spiralis* have been successfully sequenced and are available for retrieval from NCBI (NCBI, 2017). There were about 400 cases of trichinellosis reported every year by CDC in 1940, but the number of cases has dropped significantly as reported by Rawla and Sharma (2019) in the recent past. Trichinosis is characterized by headache, diarrhea, fever, myalgia, and facial edema. Sometimes aberrant larval migration into the central nervous system (CNS) might lead to various neurological diseases (Salvana et al., 2014). The most vulnerable individuals to trichinosis are the hunters and others who consume meat from wildlife. China stands highest in the world with maximum number of cases due to high consumption of pork. Polar bears, seals, and walrus have all been identified as possible vectors for *Trichinella* in the Arctic region. In recent years, the consumption of antibiotic-free meat by many consumers has led to the increase in trichinella cases especially in Europe (Rawla and Sharma, 2019).

The demonstration of Trichinellosis involves either the direct or indirect diagnosis of the infection. Compression and histological techniques are generally employed to detect larvae reaching musculature within the first 3 weeks of infection. Circulating antibodies can be detected in mild as well severely infected patients even after 2–4 weeks of infection. The excretory/secretory antigen is used to test the infection by enzyme-linked immunosorbent assay (ELISA). However, positive results of ELISA should be confirmed by performing an immunoblot (Bruschi and Gomez-Morales, 2017).

5.3 MICROBIAL TOXINS CONTRIBUTING TO FOODBORNE ILLNESS

Microbial toxins also cause foodborne illness. These toxins also contaminate food and are of concern today in addition to the diseases caused by foodborne pathogens (Hezam et al., 2019). Molds that show the ability to grow on nuts, dried fruits and cereals produce mycotoxins. Mycotoxins negatively impact human health as aflatoxins (AFs) are carcinogenic to humans and animals. Toxins produced by wild mushrooms, when consumed, cause severe illness leading to death. There have been 316 reported cases of mushroom poisoning by Health Protection Agency in 2010.

5.3.1 BACTERIAL TOXINS

Bacterial endotoxins and exotoxins contribute majorly for food intoxification (Hernandez-Cortez et al., 2017). Lipopolysaccharides (LPS) are the class of endotoxins that are present on the outer membrane of the Gram-negative bacteria. Endotoxic shock and tissue damage leads to the release of these toxins into the medium (Ramachandran, 2014). Exotoxins are proteins which are produced by the microorganism and later released into the medium. Exotoxins are classified as Type I, II, and III based on their mechanism of action (Hezam et al., 2019):

- Type I exotoxins cause host cell modification without internalization. Super-antigens of *Staphylococcus aureus* and *Streptococcus pyogenes* belong to this category.
- Type II exotoxins invade host cell by pore formation and membrane disruption. Type II exotoxins include glycerophospholipid cholesterol acyltransferase protein from *Aeromonas* spp. and aerolysin.

- Type III exotoxins consist of two fractions A/B (binary structure). Fraction A possesses enzymatic activity while fraction B helps in binding to cell surface receptors. Anthrax toxin from *Bacillus anthracis*, the cholera toxin (CT) from *Vibrio cholerae* and the Shiga toxin from *Escherichia coli* O157:H7 strain are the few type III exotoxins (Ramachandran, 2014).

Gram negative bacteria belonging to Enterobacteriaceae produce another class of toxin called enteric toxin. These include:

- Enterotoxins are mainly responsible for causing diarrheal disease leading to fluid hypersecretion without the actual destruction and demise of intestinal mucosal cells (Hezam et al., 2019)
- Cytotoxins which cause cell and/or tissue damage eventually leading to cell death (Badarau et al., 2017)
- Neurotoxins that alter the activity of smooth muscles in the intestine by releasing neurotransmitters from the enteric nervous system (Dong et al., 2019)
- Cytoskeleton-altering toxins which act on actin cytoskeleton, thereby causing cell dysfunctions, like changes in cell permeability and also the disruption of intercellular junctions (Richard et al., 1999)

5.3.2 FOODBORNE DISEASES CAUSED BY BACTERIAL TOXINS

The consumption of contaminated food containing bacterial toxins causes several important diseases. Notable ones are transmitted by *V. cholerae*, *S. aureus*, *Bacillus cereus*, *L. monocytogenes*, and *Clostridium botulinum*.

5.3.2.1 VIBRIO CHOLERAE

V. cholerae is ubiquitous in aquatic environments and more frequent in temperate waters. *V. cholerae* O1 and O139 are notable species which produce CT. The toxin causes profused diarrhea resulting in an electrolyte imbalance and eventual death. Rehydration therapies through oral solutions or IV fluids are effective treatment strategies. Antibiotics are suggested only in case of severe dehydration. Adequate sanitation, supply of pure drinking water and consumption of safe food are crucial to avoid cholera emergence.

5.3.2.2 *STAPHYLOCOCCUS AUREUS*

S. aureus produce toxins during their growth at permissive temperatures. Staphylococcal enterotoxins (SEs) are pyrogenic toxins (superantigens) consisting of nine thermostable enterotoxin serotypes. Pyrogenic toxins are stable at high temperatures and resistant to proteolytic enzymes even at low pH and hence fully functional in the digestive tract. Pyrotoxins cause immunosuppression and nonspecific T cell proliferation. Onset of disease is subjected to the quantity of the toxin ingested through the food (Zaghloul, 2015).

5.3.2.3 *BACILLUS CEREUS*

B. cereus is ubiquitous and, through their endospore-forming ability, they contaminate food production and processing system (Hezam et al., 2019). This bacterium survives pasteurization and cooking processes efficiently (Sanchez et al., 2016). *B. cereus* produces three enterotoxins [namely, cytotoxin K, hemolysin BL and nonhemolytic] along with along with emetic or cereulide toxin, which causes nausea and diarrhea. Enterotoxins and cereulide are produced during exponential and stationary phases of vegetative growth, respectively.

5.3.2.4 *CLOSTRIDIUM BOTULINUM*

Spores of *C. botulinum* have shown the ability to remain viable for longer periods of time (Rhoades et al., 2013). They produce botulinum neurotoxins of seven antigenic variants. Neurotoxins belonging to groups I and II are reported to cause botulism in humans. Group I neurotoxins producing *C. botulinum* strains are called *C. botulinum*-proteolytic bacteria, while group II neurotoxins producing *C. botulinum* strains are called *C. botulinum* non-proteolytic (psychrophilic) microbes.

Botulinum toxins possess endopeptidase activity and they require zinc for their function (metalloproteins). Structurally, they consist of two chains (150 kDa), and the double chain is further divided into a heavy and light structure. Light structure is mainly responsible for the catalytic function of the toxin. Heavy structure consists of a carboxyl-terminal and a nitrogen terminal domain. The carboxyl-terminal domain assists in binding of toxin to a presynaptic receptor for internalization and nitrogen terminal for translocation.

The fatality rate associated with botulism is very high. Flaccid paralysis in various muscles is a typical symptom associated with botulism. In addition, blurred vision followed by difficulty in speaking and swallowing with dry mouth is reported.

5.3.2.5 *LISTERIA MONOCYTOGENES*

L. monocytogenes, a facultative microorganism thrives in the cytosol of eukaryotic cells. The organism is classified into 13 serotypes out of which 4 serotypes (1/2a, 1/2b, and 4b) are responsible for more than 98% of the infectious cases. Listeriosis is caused due to the intake of food contaminated with listeriolysin O (LLO) toxin. Symptoms include fever and diarrhea which are treated using antibiotics (Morobe et al., 2012).

5.3.3 FUNGAL TOXINS

Many types of mycotoxins that are produced by certain molds have been identified. Most of these mycotoxins are a major cause of concern to the human health and livestock. Few of these toxins are AFs, ochratoxin A, patulin, fumonisins, zearalenone (ZEN), and nivalenol/deoxynivalenol. Foodborne fungal toxins cause severe illness which shows up immediately after consuming contaminated food products (Benedict et al., 2016).

5.3.3.1 AFLATOXINS (AFs)

Aflatoxins (AFs) are amongst the most poisonous mycotoxins produced by *Aspergillus flavus* and *Aspergillus parasiticus*. *Aspergillus* spp. frequently affects cereals, oilseeds, spices, and tree nut crops. Aflatoxin M1 is found in the milk of animals that are fed contaminated feed. Large doses of AFs cause acute poisoning (aflatoxicosis) leading to liver damage that can be life-threatening (Kumar et al., 2017).

5.3.3.2 OCHRATOXIN A

Ochratoxin A is a common food-contaminating mycotoxin produced by *Aspergillus* and *Penicillium* species. Ochratoxin A is produced by the organism during the storage of crops. It causes toxic effects in animals, the most potent of which is the kidney damage. In addition, it also affects fetal development and immune responses (Bui-Klimke and Wu, 2015).

5.3.3.3 PATULIN

Patulin is predominantly produced by *Aspergillus*, *Penicillium* and *Byssochlamys*. Patulin is genotoxic; however, a carcinogenic potential is yet to be demonstrated. Human dietary apples and apple juice made from affected fruit are the major dietary sources of the mycotoxin. Patulin causes damage to the liver, spleen, and kidney and presents toxic effects to the immune system (Zhong et al., 2018).

5.3.3.4 FUSARIUM TOXINS

Fusarium are most common inhabitants of the soil. They produce a variety of toxins which affect a variety of cereal crops. Deoxynivalenol and ZEN toxins affect wheat while T-2 and HT-2 toxins effect oats and fumonisins affect maize crops. Trichothecenes causes rapid irritation of skin and intestinal mucosa leading to diarrhea. High intake of ZEN can lead to hormonal changes with estrogenic effects and fertility problems, particularly in pigs. Fumonisins may cause esophageal cancer in humans. It causes toxic effects to liver and kidney in animals (Alshannaq and Yu, 2017).

5.4 STRATEGIES TO REDUCE/PREVENT FOODBORNE ILLNESS

Food contamination has become a major cause of public health concern. Foodborne microbes due to their persistent growth and toxin production are also responsible for contamination of the food supply chain. Scientific data hints at three major lines of defense that may be helpful in reducing the incidences of foodborne illness:

- The first line of defense focuses on the improvement of not only the quality of raw foodstuffs but also its hygiene
- The second line of defense employs hazard analysis and critical control point (HACCP) concepts and focuses on food processing technologies such as pasteurization and irradiations

- The third line of defense focuses on educating food handlers which also includes home-based food handlers (Scoot, 2003); the legislature for food hygiene applies to all food businesses, including caterers, farmers, manufacturers, and retailers.

5.4.1 FACTORS CONTRIBUTING TO THE FOODBORNE ILLNESS OUTBREAK

The factors contributing to the outbreak of foodborne illness include (a) improper holding temperatures during food production; (b) inadequate cooking; (c) use of contaminated equipment during food production; (d) poor personal hygiene; (e) cross-contamination between raw and cooked food; and (f) accidentally contaminated ingredients in cooked food.

5.4.2 PREVENTIVE MEASURES

Most foodborne illness can be prevented by avoiding cross-contamination. The prevalence of foodborne illness is due to the possible cross-contamination that can occur during food preparation. It has since then been the most important factor associated with most of the foodborne illness. Cross-contamination can be avoided by separate storage of raw and cooked food, by systematic cooking and maintenance of proper hygiene by washing hands. Thus, food handlers play a major role in ensuring food safety throughout the food chain by maintaining proper food hygiene practices. Such practices must be followed during production, storage, transportation, and preparation of food to minimize the emergence of foodborne pathogens. One of the efficient ways to reduce these risks is by providing proper training to food handlers. Food utensils and equipment are also prone to be contaminated with pathogens from the infected human, contaminated food, accidental escape from the faulty pipes and drains (FDA, 1978). These problems could be controlled by hiring trained microbiologists and food technologists who have immense knowledge on personal hygiene and food handling practices. Routine cleaning of food utensils and equipment should be in practice as per the guidelines and should be evaluated regularly. The U.S. Food and Drug Administration (FDA) recommends food to be cooked mostly without manual contact using suitable utensils and on surfaces that have been cleaned and sanitized before usage in order to avoid/prevent contamination

(FDA, 1978). Thus, continuous education of the process, practices, and perspectives of hygiene may contribute effectively to guarantee safety and quality of food to the consumer.

5.5 CONCLUSIONS

Safe and nutritious food is a key to sustaining good health. Consumption of unsafe food containing infectious agents and chemicals lead to more than 200 foodborne diseases. Foodborne diseases hamper the socioeconomic development of the country by disrupting health care systems. Thus, there is an intricate link between food safety, nutrition, and food security. Most foodborne illnesses, despite being complex, can be prevented by practicing effective prevention skills with sound knowledge across various disciplines. Quality education to improve food safety practices seems to be a key factor. Public health agencies, regulatory agencies, the food industry along with the consumers need to make constant effort to prevent food contamination in the farm, during processing and also at restaurants and homes. However, these days with proper farm practices and food processing regulations, some sort of national success is achieved in terms of reduction of microbial load in the foods. Foodborne illness can be minimized with suitable food safety education programs, which will lead to the reduction and occurrence of foodborne illness. This will in turn also set a platform for the population which could perhaps be better educated and equipped with knowledge as well as skills to fulfill the requirements of the food industry and foodservice sectors.

KEYWORDS

- **food consumers**
- **food intoxicants**
- **foodborne illness**
- **pathogens**
- **preventive measures**
- **toxins**

REFERENCES

Alshannaq, A., & Yu, J. H., (2017). Occurrence, toxicity, and analysis of major mycotoxins in food. *International Journal of Environmental Research and Public Health, 14*(6), 632.

Bacon, R. T., & Sofos, J. N., (2003). Characteristics of biological hazards in foods. In: *Food Safety Handbook* (pp. 157–195).

Badarau, A., Trstenjak, N., & Nagy, E., (2017). Structure and function of the two-component cytotoxins of *Staphylococcus aureus*–learnings for designing novel therapeutics. In: *Protein Reviews* (pp. 15–35). Springer, Singapore.

Belliot, G., Lopman, B. A., Ambert-Balay, K., & Pothier, P., (2014). The burden of norovirus gastroenteritis: An important foodborne and healthcare-related infection. *Clinical Microbiology and Infection, 20*(8), 724–730.

Benedict, K., Chiller, T. M., & Mody, R. K., (2016). Invasive fungal infections acquired from contaminated food or nutritional supplements: A review of the literature. *Foodborne Pathogens and Disease, 13*(7), 343–349.

Berto, A., Martelli, F., Grierson, S., & Banks, M., (2012). Hepatitis E virus in pork food chain, United Kingdom, 2009–2010. *Emerging Infectious Diseases, 18*(8), 1358.

Bintsis, T., (2017). Foodborne pathogens. *AIMS Microbiology, 3*(3), 529.

Bruschi, F., & Gomez-Morales, M. A., (2017). Parasites. In: *Foodborne Diseases* (pp. 305–324). Academic Press.

Bui-Klimke, T. R., & Wu, F., (2015). Ochratoxin A and human health risk: A review of the evidence. *Critical Reviews in Food Science and Nutrition, 55*(13), 1860–1869.

Carollo, M. C. C., Amato, N. V., Braz, L. M. A., & Kim, D. W., (2001). Detection of *Cyclospora* sp. oocysts in the feces of stray dogs in Greater Sao Paulo (Sao Paulo State, Brazil). *Magazine of the Brazilian Society of Tropical Medicine, 34*(6), 597–598.

Centers for Disease Control and Prevention, (2014). *Norovirus: Trends and Outbreaks*. Available from: http://www.cdc.gov/norovirus/trends-outbreaks.html (accessed on 11 February 2022).

Chu, D. M. T., Sherchand, J. B., Cross, J. H., & Orlandi, P. A., (2004). Detection of *Cyclospora cayetanensis* in animal fecal isolates from Nepal using an FTA filter-base polymerase chain reaction method. *The American Journal of Tropical Medicine and Hygiene, 71*(4), 373–379.

Collier, M. G., Khudyakov, Y. E., Selvage, D., et al., (2014). Outbreak of hepatitis A in the USA associated with frozen pomegranate arils imported from Turkey: An epidemiological case study. *The Lancet Infectious Diseases, 14*(10), 976–981.

Colson, P., Borentain, P., Queyriaux, B., et al., (2010). Pig liver sausage as a source of Hepatitis E virus transmission to humans. *The Journal of Infectious Diseases, 202*(6), 825–834.

Dong, M., Masuyer, G., & Stenmark, P., (2019). Botulinum and tetanus neurotoxins. *Annual Review of Biochemistry, 88*, 811–837.

Dorny, P., Praet, N., Deckers, N., & Gabriel, S., (2009). Emerging foodborne parasites. *Veterinary Parasitology, 163*(3), 196–206.

Food and Drug Administration, (1978). *Food Service Manual*. U.S. Department of Health, Education and Welfare, Public Health Service. DHEW Publ. N° FDA 78-2081, Washington, DC.

Gallot, C., Grout, L., Roque-Afonso, A. M., et al., (2011). Hepatitis A associated with semidried tomatoes, France, 2010. *Emerging Infectious Diseases, 17*(3), 566.

Gould, L. H., Walsh, K. A., Vieira, A. R., et al., (2013). Surveillance for foodborne disease outbreaks - United States, 1998–2008. *Morbidity and Mortality Weekly Report: Surveillance Summaries*, *62*(2), 1–34.

Guo, M., Lambertini, E., Buchanan, R. L., et al., (2017). Quantifying the risk of human *Toxoplasma gondii* infection due to consumption of fresh pork in the United States. *Food Control*, *73*, 1210–1222.

Halliday, M. L., Kang, L. Y., Zhou, T. K., et al., (1991). An epidemic of hepatitis A attributable to the ingestion of raw clams in Shanghai, China. *Journal of Infectious Diseases*, *164*(5), 852–859.

Heredia, N., & Garcia, S., (2018). Animals as sources of foodborne pathogens: A review. *Animal Nutrition*, *4*(3), 250–255.

Hernandez-Cortez, C., Palma-Martínez, I., Gonzalez-Avila, L. U., Guerrero-Mandujano, A., Solís, R. C., & Castro-Escarpulli, G., (2017). Food poisoning caused by bacteria (food toxins). *Poisoning: From Specific Toxic Agents to Novel Rapid and Simplified Techniques for Analysis*, *33*.

Hezam, A. M., Al-Jasimme, A. S., & Emran, F. K., (2019). A review on bacterial foodborne disease. *International Journal of Research in Pharmaceutical Sciences*, *10*(4), 3223–3228.

Holzbauer, S. M., Agger, W. A., Hall, R. L., et al., (2014). Outbreak of *Trichinella spiralis* infections associated with a wild boar hunted at a game farm in Iowa. *Clinical Infectious Disease*, *59*(12), 1750–1756.

Hussain, M. A., Stitt, V., Szabo, E. A., & Nelan, B., (2017). *Toxoplasma gondii* in the food supply. *Pathogens*, *6*(2), 21.

Kamar, N., Garrouste, C., Haagsma, E. B., Garrigue, V., et al., (2011). Factors associated with chronic hepatitis in patients with hepatitis E virus infection who have received solid organ transplants. *Gastroenterology*, *140*(5), 1481–1489.

Koopmans, M., (2012). Foodborne viruses from a global perspective. In: *Improving Food Safety Through a One Health Approach: Workshop Summary* (p. 225). National Academies Press.

Kumar, P., Mahato, D. K., Kamle, M., Mohanta, T. K., & Kang, S. G., (2017). Aflatoxins: A global concern for food safety, human health and their management. *Frontiers in Microbiology*, *7*, 2170.

Kumar, Y., Singh, V., Kumar, G., Gupta, N. K., & Tahlan, A. K., (2019). Serovar diversity of *Salmonella* among poultry. *Indian Journal of Medical Research*, *150*(1), 92.

Lewis, H. C., Wichmann, O., & Duizer, E., (2010). Transmission routes and risk factors for autochthonous hepatitis E virus infection in Europe: A systematic review. *Epidemiology and Infection*, *138*(2), 145–166.

Lorenzo, J. M., Munekata, P. E., Dominguez, R., Pateiro, M., Saraiva, J. A., & Franco, D., (2018). Main groups of microorganisms of relevance for food safety and stability: General aspects and overall description. In: *Innovative Technologies for Food Preservation* (pp. 53–107). Academic Press.

Martella, V., Pinto, P., & Buonavoglia, C., (2011). Canine noroviruses. *Veterinary Clinics: Small Animal Practice*, *41*(6), 1171–1181.

Maunula, L., Rönnqvist, M., Åberg, R., Lunden, J., & Nevas, M., (2017). The presence of norovirus and adenovirus on environmental surfaces in relation to the hygienic level in foodservice operations associated with a suspected gastroenteritis outbreak. *Food and Environmental Virology*, *9*(3), 334–341.

Morobe, I. C., Obi, C. L., Nyila, M. A., Matsheka, M. I., & Gashe, B. A., (2012). Molecular characterization and serotyping of *Listeria monocytogenes* with a focus on food safety and disease prevention. *Biochemical Testing*, *8*, 197–216.

NCBI. National Centre for Biotechnology Information. (2017). Available at: https://www.ncbi.nlm.nih.gov/genome (accessed on 11 February 2022).

Ortega, Y. R., & Sanchez, R., (2010). Update on *Cyclospora cayetanensis*, a foodborne and waterborne parasite. *Clinical Microbiology Reviews*, *23*(1), 218–234.

Panel, E. B., Koutsoumanis, K., Allende, A., et al., (2018). Public health risks associated with foodborne parasites. *EFSA Journal*, *16*(12), e05495.

Ramachandran, G., (2014). Gram-positive and gram-negative bacterial toxins in sepsis: A brief review. *Virulence*, *5*(1), 213–218.

Rawla, P., & Sharma, S., (2019). *Trichinella spiralis (Trichnellosis)*. In StatPearls [Internet]. StatPearls Publishing.

Rhoades, J., Nastou, A., Smirniotis, P., & Likotrafiti, E., (2013). Effect of relative humidity and storage temperature on the behavior of *Listeria monocytogenes* on fresh vegetables. *Journal of Food Safety*, *33*(4), 545–551.

Richard, J. F., Petit, L., Gibert, M., Marvaud, J. C., Bouchaud, C., & Popoff, M. R., (1999). Bacterial toxins modifying the actin cytoskeleton. *International Microbiology*, *2*(3), 185–194.

Robert-Gangneux, F., & Darde, M. L., (2012). Epidemiology of and diagnostic strategies for toxoplasmosis. *Clinical Microbiology Reviews*, *25*(2), 264–296.

Salvana, E. M. T., Salata, R. A., & King, C. H., (2014). Parasitic infections of the central nervous system. In: *Aminoff's Neurology and General Medicine* (pp. 947–968). Academic Press.

Sanchez, J. A., Correa, M., & Castaneda-Sandoval, L. M., (2016). *Bacillus cereus* and important pathogens in the microbiological control of food. *Journal of the National School of Public Health*, *34*(2), 230–242.

Sattar, S. A., Tetro, J., Bidawid, S., & Farber, J., (2000). Foodborne spread of hepatitis A: Recent studies on virus survival, transfer and inactivation. *Canadian Journal of Infectious Diseases and Medical Microbiology*, *11*(3), 159–163.

Scallan, E., Hoekstra, R. M., Angulo, F. J., et al., (2011). Foodborne illness acquired in the United States major pathogens. *Emerging Infectious Diseases*, *17*(1), 7.

Schirone, M., Visciano, P., Tofalo, R., & Suzzi, G., (2019). Foodborne pathogens: Hygiene and safety. *Frontiers in Microbiology*, *10*, 1974.

Scott, E., (2003). Food safety and foodborne disease in the 21st century. *Canadian Journal of Infectious Diseases and Medical Microbiology*, *14*(5), 277–280.

Shaheen, M., (2017). Viral infections as a cause of foodborne illness. *Virology and Immunology Journal*, *1*(6), 000132.

Smith, H., & Evans, R., (2009). Parasites: *Cryptosporidium*, *Giardia*, *Cyclospora*, *Entamoeba histolytica*, *Toxoplasma gondii* and pathogenic free-living amoebae (*Acanthamoeba* spp. and *Naegleria fowleri*) as foodborne pathogens. In: *Foodborne Pathogens* (pp. 930–1008). Woodhead Publishing.

Tesse, S., Lioure, B., Fornecker, L., et al., (2012). Circulation of genotype 4 hepatitis E virus in Europe: First autochthonous hepatitis E infection in France. *Journal of Clinical Virology*, *54*(2), 197–200.

Todd, E. C., & Grieg, J. D., (2015). Viruses of foodborne origin: A review. *Virus Adaptation and Treatment*, *7*, 25.

US Food and Drug Administration. (2012). Bad bug book. *Foodborne Pathogenic Microorganisms and Natural Toxins Handbook*. Available from: http://www.fda.gov/Food/FoodborneIllnessContaminants/CausesOfIllnessBadBugBook/ (accessed on 11 February 2022).

Van, D. I., & Cummings, R. D., (2010). Glycomics in unraveling glycan-driven immune responses by parasitic helminths. In: *Handbook of Glycomics* (pp. 367–396). Academic Press.

Vieira, F. P., Alves, M. D. G., Martins, L. M., & Rangel, A. L. P., Dubey, J. P., & Hill, D., (2015). Waterborne toxoplasmosis investigated and analyzed under hydrogeological assessment: New data and perspectives for further research. *Memorials do Instituto Oswaldo Cruz*, *110*(7), 929–935.

Vu, D. L., Bosch, A., Pinto, R. M., & Guix, S., (2017). Epidemiology of classic and novel human astrovirus: Gastroenteritis and beyond. *Viruses*, *9*(2), 33.

Wheeler, C., Vogt, T. M., Armstrong, G. L., et al., (2005). An outbreak of Hepatitis A associated with green onions. *New England Journal of Medicine*, *353*(9), 890–897.

WHO & FAO, (2008). Viruses in food: scientific advice to support risk management activities: meeting report. World Health Organization, Geneva, Switzerland, p. 58. https://apps.who.int/iris/handle/10665/44030.

Wyn-Jones, A. P., Carducci, A., Cook, N., et al., (2011). Surveillance of adenoviruses and noroviruses in European recreational waters. *Water Research*, *45*(3), 1025–1038.

Zaghloul, M. Z., (2015). *Staphylococcus aureus* toxic shock syndrome. *Tropical Medicine and Surgery*, *3*(2), 1–2.

Zhong, L., Carere, J., Lu, Z., Lu, F., & Zhou, T., (2018). Patulin in apples and apple-based food products: The burdens and the mitigation strategies. *Toxins*, *10*(11), 475.

PART II

Foodborne Pathogens: Emerging and Reemerging

CHAPTER 6

Foodborne Pathogens: An Introduction

FAIZA ANUM, ARIFA TAHIR, and SAIRA KHAN

ABSTRACT

An excessive number of diseases with noteworthy effects on the health of humans as well as economy are caused by foodborne microorganisms. Around one by three people are affected by such food-related bacteria in developed countries every year, whereas WHO also stated that food-related microorganisms are the main risk to food safety. Toxins are the cause of illness spread due to food; these may be either microbial toxin or chemical toxins. Toxins in food cause different diseases which are often mild and self-controlled. But in some cases, the end results are serious disease that leads to mortality. A large number of foodborne diseases are because of microbes that enter into the chain of food at any step from farm to fork. Foodborne pathogens such as bacteria (*Bacillus cereus*, *Campylobacter jejuni*, *Clostridium botulinum*, *Clostridium perfringens*, *Cronobacter sakazakii*, *Escherichia coli*, *Listeria monocytogenes*, *Yersinia enterocolitica*, *Salmonella* spp., *Staphylococcus aureus*, *Shigella* spp., *Campylobacter jejuni*, *Vibrio* spp., as well as *Yersinia enterocolitica*); viruses (Noroviruses (NoVs) as well as Hepatitis A); and foodborne parasites such as *Cyclospora cayetanensis*, *Toxoplasma gondii*, and *Trichinella spiralis* are discussed in this chapter.

6.1 INTRODUCTION

Food-related microorganisms, which include parasites, bacteria along with viruses are major biologically active agents that are responsible for diseases

Food Microbial and Molecular Biology: From Fundamentals to Applications. Saher Islam, Devarajan Thangadurai, Jeyabalan Sangeetha, Zaira Zaman Chowdhury (Eds.)

related to food occasionally. Such diseases are caused in result of ingestion of their specific microorganisms with some food product and then takes a place inside human host where they can multiply easily and express its characteristic symptoms or if any toxigenic microorganism get enter into food product at any stage in food chain and produce its toxin that will contaminate the food, and is eaten by human host will cause disease. Food-related illness epidemic is expressed as the incidence of two or more than two cases of alike illness because of ingesting a common food (Vieira et al., 1999). Usually, foodborne illness classification is expressed as intoxication and infection.

6.1.1 FOODBORNE INTOXICATION

Foodborne intoxication is the consumption of toxins produced by a pathogen in food. So far, the household is the most commonplace of contact. The most convincing evidence related to foodborne epidemics, *Salmonella* spp. was prevalently common pathogen identified in private households, however, "toxins produced from bacteria other than *C. botulinum* toxins," viruses as well as other causative agents were more commonly stated in public situations like cafeterias, catering service stations, eateries as well as bars (ECDC, 2014).

6.1.2 FOODBORNE INFECTIONS

Incubation period is typically included, the time from ingestion until side effects happen is any longer than that of food-related toxin. In excess of 200 distinctive foodborne sicknesses have been recognized (Mead et al., 1999). The most extreme cases will in general happen in the exceptionally old, in the youthful, in the individuals who have undermined resistant framework work, and in sound individuals presented to an exceptionally high portion of a living being (ECDC, 2014).

6.2 FOODBORNE BACTERIA

They are the vast group of organisms that are playing a major role in food-related diseases and have diverse shapes, kinds, and characteristics:

- Some of these are pathogenically spore-forming microbes that are resistant to high temperatures (heat), examples include *Clostridium botulinum*, *C. perfringens*, *Bacillus subtilis* as well as *Bacillus cereus*

- Some have ability to synthesis high temperature-resistant toxins (e.g., *Staphylococcus aureus* as well as *Clostridium botulinum*)
- Variety of them are mesophiles that have ideal growing temperature ranges of from 20 to 45°C
- Another characteristic class is of Psychrotrophs that include *L. monocytogenes*, as well as *Yersinia enterocolitica* that have the capability to survive in extremely cold conditions, i.e., below 10°C (Bintsis, 2017)

6.2.1 *BACILLUS CEREUS*

This species belongs to the family *Bacillaceae*; their cell wall is gram-positive and are rod shapes usually motile and are spore formers (Bintsis, 2017). Various *Bacillus* species are seen throughout the surrounding environment, including muds, freshwater, and marine environment. *B. cereus* produces spores that bear pili or other outgrowths that are hydrophobic in nature as compared to spores of other species. Such properties support their adherence to surfaces and avoid their elimination and decay during cleanliness/washing (Bacon and Sofos, 2003). Vegetative growth of this species is favorable at 4 to 5°C and 35 to 55°C temperature, but growth is more supported at temperature ranges between 30 to 40°C which varies according to strain (Rajkowski and Smith, 2001). pH required for their growth ranges between 4.9 to 9.3 and inhibition is growth observed at pH 4.35 as observed in meat (Andersson et al., 1995; Bintsis, 2017). The water activity (aw) values were observed as 0.93 however it is optimized at 0.921 that is required for growth as minimum. Fried rice possesses aw value 0.912–0.961 and *B. cereus* grows fast on it (Stenfors et al., 2008). *B. cereus* have the ability to synthesis 2 sorts of toxins, one is emetic (vomiting) while the other is diarrheal toxin, which in turn cause two types of diseases. Former one is due to emetic toxin synthesis by microbe while going through growth step in food. Later one is because of diarrheal toxins synthesized during bacterial growth in the small intestine. Former one (rapid) symptom include nausea as well as vomiting whereas symptoms of late diarrheal type is diarrhea as well as abdominal pain. Vomiting and diarrhea are the diseases that are caused as an after effect of *B. cereus* endospores enduring the cooking procedure, after that germination and resulting expansion of vegetative cells happens sooner or later during storage step. Foods that are normally concerned in *B. cereus* causing diarrheic food poisoning contain meat products, puddings, soups, sauces, vegetables, milk as well as milk products. Symptoms are characterized by diarrhea, abdominal pain, and nausea; mostly symptoms are shown after 8 to 16 hours. Diarrheal syndrome symptoms generally persist not more than 12 to 24 hours. Following an incubation period (1–5 h), emetic syndrome symptoms include primarily vomiting and nausea and persist for 6–24 h. Food items linked to intoxication by this species are pasta, pastry, fried, cooked rice, and noodles (Stenfors et al., 2008). Between two diseases, diarrhea is caused due to a complex that is thermolabile and vomiting is caused as a result of reaction by thermostable toxin. *B. cereus* has the ability to form adhesive endospore that is why it is usually exist in food-producing settings where it gets spread to all types of foods.

They produce series of circulating factors that can cause disagreeable illness in people when found in gastrointestinal (GI) tract or in food, it is the main foodborne pathogenic microbes, although mostly ailment is gentle and of brief-term (Scallan et al., 2011; Sayers et al., 2021). Pasta salad as well as spaghetti remains were the reason of two epidemics, from where clinical data as well as rapid onset of symptoms, altogether with the microbiological as well as molecular study, shows *B. cereus* as causative agent (Schaffter et al., 2004; Humphrey et al., 2007).

6.2.2 CAMPYLOBACTER JEJUNI

These species are members of family *Campylobacteriaceae* from which *C. jejuni* is greatest common reason of diarrheal disease and is widely found in the nature. This species is found both in the intestine of mammals as well as birds where they can colonize easily, and transmitted to human beings happened through contaminated food items. The systemic diseases lead to more severe infections. Around 932 genomes have been studied up till now conferring to the data recovered through NCBI. The genome length is measured as 1.686 Mb (median) (Altekruse et al., 1999). These organisms are small, spirally shaped, having Gram-negative cell walls, it has 18 species, of which six are the subspecies and two are biovars (Stern et al., 1994). This species varies from other microorganisms in the sense that it causes diseases that are basically microaerophilic that grows in temperature constitutes on Carbon dioxide (10%) and Oxygen (5%). It causes diseases in human and have a temperature range between 30°C to 46°C for their growth that is why they are classified as thermophilic campylobacters (Stern et al., 1994). In 2015, *Campylobacter* spp. are a piece of the typical intestinal vegetation of a wide assortment of sound local and wild creatures, including geese, pigs, dairy cattle, sheep, ducks, goats, rodents, wild flying creatures, chickens, hounds, felines, rodents, and marine well evolved creatures, and are regularly found related with waterways, for example, water troughs and streams. Most instances of campylobacteriosis are related with eating crude or half-cooked poultry meat, unpasteurized kind, debased water, or from cross-sullying of different nourishments by these things. All creatures utilized for food can be *C. jejuni* positive as can many friend species (household pets) (Carter and Peck, 2015). Freshly made contaminated with disease-causing microbes offers a hazard to consumers (Grass et al., 2013; Jaradat et al., 2014).

6.2.3 *CLOSTRIDIUM BOTULINUM*

These are spores producing bacteria and belongs to *Bacillaceae* family and constitutes of anaerobic bacteria (obligatory), rod-shaped and ability to produce spore in absence of air, do not produce spores in air presence specially at their initial growth stages and are mostly gram-positive in nature. Vegetative cells look as straight, sometimes curved rods in a few species. They are seen all over the atmosphere but predominantly seen in the soil as well as in the intestine of animals. There is a distinctive characteristic structure called endospores present that grow under situations not favorable for vegetative growth and enlarge single cells terminally or sub-terminally. These bacteria are motile by the use of a specific structure called peritrichous flagella as well as produce botulinum neurotoxins, the deadliest poison identified so far. It has seven types of neurotoxins, A through G, has the ability to recognize an antigen specifically, a toxin synthesized through every strain. Types A, B, E, as well as F causing botulism in human beings, types C as well as D causing botulism in birds as well as mammals, and type G (Stenfors et al., 2008; Bintsis, 2017). One of the methods used most commonly is "thermal processing" which increases shelf life, decrease acid, and increases moisture of food that disables *Clostridium botulinum* spores.

Evolutionary studies gave these bacteria a specific position the most primitive bacteria. Widely distributed strains as well as serovars are of *Clostridium botulinum* which synthesis toxin type A that is also used in numerous treatments, i.e., Botox. Genome sequencing of about 177 genomes according to NCBI and genome size is measured as 3.898 Mb (Bintsis, 2017). This bacterium is found in marine, freshwater, and soil also in GIT of animals. Also found in food, i.e., vegetables, meat, and honey (children less than one year should not be fed honey).

Diversity of food products, for example, canned beets, corn, mushrooms, peppers, green beans, soups, spinach, asparagus, ready olives, meats, ham, wiener, stuffed eggplant, lobster, as well as salted/smoked fish have been related with botulinum poison (Andersson et al., 1995). Customarily, food-borne botulism has been related with under processed as well as mishandled wieners or home-canned food products; such as botulism caused through the utilization of defiled food sources, for example, potato plate of mixed greens, sauteed onions, garlic sauce, cheddar, yogurt, bean glue, and olives. Indications appear after an incubation period of 12 to 36 hours because of ingesting botulinum neurotoxin in some food that would be contaminated. Symptoms are nausea and vomiting followed by advanced neurological symptoms which include visual damage as well as acute flaccid paralysis which initiates from facial muscles, head, and pharynx, plunging to involve thorax muscles as well as extremes then leading to probable death from respiratory arrest.

Botulinum toxin is a protein that is neurotoxin which produced during cell growth, it is released through cell lysis and activated by proteolytic cleavage. Botulinum has four types, i.e., produced from assimilation of preformed toxin in foods, produce as a result of microorganism growth in wound, production in infant's intestine by toxin produced, in adult's intestinal tract. Botulinum produced by any of these methods is moved through the bloodstream to neuromuscular junctions where it attached to receptors and disturb the normal functioning of nerve cells (Healy et al., 2010; Bintsis, 2017).

6.2.4 *CLOSTRIDIUM PERFRINGENS*

Its old name was *Clostridium welchii* that is member of *Bacillaceae* family and responsible for diseases spread by food intake. These are immotile, rod-shaped bacteria. They are encapsulated and produces spores that show resistance to various ecological stresses like radiation, desiccation, as well as heat. Vegetative cells grow at temperatures beginning from 6–50°C but favor an ideal temperature flanked by 43°C and 47°C. AW-value required for *Cl. perfringens* is 0.93. pH range for its growth ranges between 5.0 to 9.0 however optimized at 6.0–7.2 as ideal also required salt concentration ($NaCl_2$) below 5 to 8% (Stenfors et al., 2008; Bintsis, 2017). *Cl. perfringens* spores are widely dispersed through air and dust particles in it therefore easily contaminate food items, i.e., fish and meat products. *Cl. perfringens* are predicted to be the second most communal bacterial reasons of food-related diseases in the US, producing illness in 1 million people each year (Kandhai et al., 2004). Foodborne illness nearly at all times a result of temperature mishandling, and many times food item such as meat remains partially cooked or potentially cool too gradually or has experienced inadequate warming, permitting enduring spores to grow, prompting vegetative cell expansion. An incubation time of 7 to 30 hours start after ingestion, indicative signs mostly result in pain in the abdominal area, nausea, cramping, and may lead to vomiting, and this situation may last for 24 to 28 hours. Almost 5 types of *Cl. perfringens* have been identified that produces toxin (A-E), besides each one of them synthesize phospholipase which is an α-Toxin and causes a muscular disease "myonecrosis." B type strains synthesis beta- as well as epsilon-toxins, D type strains also synthesis ε-toxin, E type synthesize iota-toxin and the sporulation stage produces enterotoxin all through cell break down. After release, these toxins get attached to cells of epithelium, initiating damage to cytotoxic cell membrane as well as successive modification in permeability that will lead to diarrhea and intestinal cramping (Bintsis, 2017).

6.2.5 *CRONOBACTER SAKAZAKII*

The genus *Cronobacter* is comprised of seven different species, i.e., *Cr. turicensis*, *Cr. universalis*, *Cr. sakazakii* (previously known as *Enterobacter sakazakii*), *Cr. dublinensis*, *Cr. muytjensii*, *Cr. condiment*, *Cr. malonaticus* (Andersson et al., 1995). From these species *Cr. sakazakii* is linked with septicemia of infants, meningitis, necrotizing enterocolitis and brain abscesses in neonates. As per data received from NCBI almost 49 genomes have been sequenced till now and genome size is observed as 4.5475 Mb (Altekruse et al., 1999). It is also found responsible for different infections in adults (Croxen et al., 2013) and possible causes of its spread are house environment, i.e., cleanliness (Croxen et al., 2013) and processed food such as dried spices, milk, and meat products (Wells et al., 1983; Croxen et al., 2013). Many species of *Cronobacter* are found in a variety of food products and causing contamination but *Cr. sakazakii* is found mostly causing infections and epidemics in infants by intake of powdered infant formula caused due to inadequate good manufacturing practice (GMP) and contamination in cans during usage (Hochel et al., 2012).

6.2.6 *ESCHERICHIA COLI*

These are rod-shaped bacteria having gram-negative cell wall, non-spore-forming, can be motile or not and can be found in both flagellated and not flagellated form. As per the data received from NCBI 5351 genome sequencing has been completed till now and genome size is measured as

5.171 Mb (Altekruse et al., 1999). *E. coli* has many disease-causing variants known as pathotypes are the reason of abundant disease as well as death world all over; several of these pathotypes are a main public health apprehension as they have little infectious doses plus, they are spread via universal mediums, such as food as well as water. *E. coli* is spread by intake of contaminated food and water contaminated with feces of infected individuals and animals. Such contaminations happen through butchery and processing of animals.

The utilization of excrement from dairy cattle or different animals as manure crops can pollute produce as well as irrigation water. This microbe has an ability to grow for so long in surrounding and can multiply in foods products as well as vegetables. Pathogenic *E. coli* have been divided into six groupings according to the disease-causing mechanism:

- Enterotoxigenic *E. coli* (ETEC) or Shiga toxin-producing *E. coli* (STEC)
- Attaching and Effacing *E. coli* (A/EEC)
- Enteroaggregative *E. coli* (EAggEC)
- Enterohemorrhagic *E. coli* (EHEC)
- Enteropathogenic *E. coli* (EPEC)
- Enteroinvasive *E. coli* (EIEC) (Jemmi and Stephan, 2006 Buchanan et al., 2017)

ETEC strain of *E. coli* (O157:H7) can cause up to 63,000 diseases, 2,100 isolated hospitalizations, as well as 20 deaths each year. The major reservoir for this zoonotic disease-causing microbe is the intestine of farm animals, however other animal can also be the host. *E. coli* produce toxins that are analogous to Shiga toxin (caused by *Shigella dysenteriae*) as well as different from formerly defined *E. coli* heat-stable and heat-labile toxins.

6.2.7 *LISTERIA MONOCYTOGENES*

This species is the main reason of mortality via foodborne microorganisms, particularly in pregnant women, neonates, adult age groups, as well as immuno-compromised people. Infections in expecting women can be disturbing to the fetus that leads to miscarriages, stillbirths, as well as birth defects. Reservoirs of this organism in surroundings are manure, water, decaying vegetables. It can tolerate high and low temperature both ranges between 1–45°C, can survive in salt concentration thus known as most harmful foodborne pathogen. It is mostly grown in the food items that are consumed without heating properly and found in numerous food items that are uncooked, for example, uncooked meats and vegetables, just as in foods that get contamination subsequent to cooking or handling. It may transfer the disease in the digestive organs to the central nervous system (CNS) as well as to the placenta of humans. This species might become the cause of meningitis (aggravation of the layer encompassing spinal cord as well as cerebrum), gastroenteritis (irritation of mucous films of stomach and digestive tract), and septicemia (foundational spread of microscopic organisms and poisons in the blood) can result from disease (Andersson et al., 1995; Mossel et al., 1995; Vargas et al., 1999). As per the data received from NCBI 1243 genomes are sequenced and genome size is measured as 2.974 Mb (Altekruse et al., 1999). *L. monocytogenes* is found causing a serious infectious disease "Listeriosis" (Janda et al., 1988; Kaper and Morris, 1995).

6.2.8 *SALMONELLA* SPP.

This bacterium is famous for most commonly causing worldwide enteric infections (food poisoning) that belongs to *Enterobacteriaceae* group. The name *Salmonella* was given to these species after Dr. Daniel Salmon, who identified for the first time this organism from the pig intestine (*Salmonella choleraesuis*) (Bintsis, 2017).

This genus is mainly comprised of two species that causes diseases in human being: *S. bongori* and *S. enterica*. *Salmonella* is further sectioned into serotypes, grounded on the Kaufmann-White typing scheme first printed in 1934, that distinguishes *Salmonella* strains via their surface as well as flagellar antigenic characteristics. *Salmonella* spp. are generally mentioned by their serotype names (Andersson et al., 1995). Few *Salmonella enterica* serovars are accountable for extreme serious disorders like Typhoid fever. As per data received from NCBI total 5323 genomes of *S. enterica* are sequenced till now and genome size is observed as 4.783 Mb (Altekruse et al., 1999). *Salmonella* spp. reside in the intestine of animals (wild as well as livestock) and its transmission happen by eating contaminated food products due to the feces of these animals, who are an intestinal carriers of these organisms.

Shigella (genus) is affiliated with family and it owns four serogroups that have been conventionally expressed as species: serogroup A expressed as *Shigella dysenteriae*, serogroup B expressed as *Shigella flexneri*, serogroup C expressed as *Shigella boydii*, and serogroup D expressed as *Shigella sonnei*. However, serogroups A, B, as well as C comprise of 38 serotypes, serogroup D keeps only one. Member of this genus are unable to move, lack ability of spore-forming, anaerobes (facultative) Gram-negative cell wall and are rod-shaped bacteria. Temperatures range for growth is from 6 to 48°C, but prefers to grow at 37°C.

Whereas *S. sonnei* can grow on low temperature as compared to other serogroups. Ideal growth occurs between pH six and pH eight, even though growth is observed at pH range from 4.8 to 9.3. *Shigella* spp. are found associated with *E. coli* species genetically and also found similar to some antibodies biochemically though they are significantly different from other bacterial species, which can be identified by the symptoms they cause in humans after they cause infection. *Shigella* spp. are found most frequently in surroundings of compromised cleanliness as well as poor sanitation, moreover their mode of spread is mostly contact with the infected person and can cause the diseases such as shigellosis spread by intake of polluted water and food items. Food products related to outbreaks of shigellosis have comprised of milk and products related to milk, salad, poultry meat, marine shellfish, as well as many other foods consumed commonly at cafes, homes, hostel mess and schools (Stenfors et al., 2008; Bintsis, 2017). This species is mostly found associated with causing GIT infections in humans which causes symptoms after 12 to 50 hours and can result into anxiety, fever, abdominal cramps, and diarrhea and can also causes dysentery (Wu et al., 2014)

6.2.9 STAPHYLOCOCCUS AUREUS

Staphylococcus aureus (*S. aureus*) are non-motile and rounded shaped (Cocci) that sometimes express themselves in single, make pairs or in chain form and sometimes express themselves as tetrads or sometime as a specific grape-like structure, whose cell wall is gram-positive. In addition, they are facultative anaerobes also. They are prevalent throughout nature as well as seen on the skin and mammalian skin glands as well as birds, in addition they are also very much prevalent in areas of host body during infection such as in the mouth, upper respiratory tracts, mammary glands, intestinal GU (genitourinary) and bloodstream. They can stay alive outside of our body in dry condition, that is why they can be easily isolated from air, dust, sewage, as well as water, presenting them as one the most resistant as well as non-spore-forming pathogens. Along with the causes of infection in the environment, some reported *S. aureus* containing foods consist of minced beef, sausages of pork meat, minced turkey, steaks of salmon's meat, marine oysters, marine shrimps, creamy pies, milk, as well as delicatessen salads. *S. aureus* grows at temperature range 7–47.8°C which varies from one strain to other and produce enterotoxins flanked by 10 to 46°C but prefer an ideal temperature range from 40 to 45°C. The bacterium grows at pH 4.5–9.3, with an ideal pH between 7.0 as well as 7.5, and is very accepting to high levels of salt (greater than 10% to NaCl); production of enterotoxin requires a least 0.86 water activity, although growth is observed at an *aw* 0.83 (Andersson et al., 1995; Stenfors et al., 2008; Bintsis, 2017).

S. aureus major repositories are humans, and food contamination happens through direct contact or sometimes indirectly via skin fragments, and via respiratory tract droplets that cause food contamination while making. Other reason is insufficient refrigeration, insufficient cooking or else heating, and poorly managed personal hygiene. After enterotoxin ingestion, 6 to 10 hours incubation time, visible symptoms are nausea, unsettled stomach, perspiration, normal weakness, abdominal cramps, headache along with dizziness, chills, muscle cramping and prostration, as well as diarrhea might or might not be bloody. These enterotoxins are part of large toxins family "pyrogenic toxins" that can act as superantigens, in that way presenting a higher number of T-Cells. They cannot be deactivated by heat as the temperature required for their killing is high than required for killing the microorganism (Bintsis, 2017).

6.2.10 VIBRIO SPP.

The very important genus *Vibrio*, representing the *Vibrionaceae* family, having a number of 35 species above in which one by two of the species have been explained in the last two decades and those that are humanly pathogenic are one-third of the ratio. These specific morphological types of organism are curves rods basically that do not contain spores and facultative anaerobes, usually motile and Gram-negative rods (straight or curved). These organisms are medically important in diseases production, and species that are its pathogenic forms include *Vibrio cholera* specifically then *Vibrio parahaemolyticus* plus *Vibrio vulnificus*, which are pathogenic forms that need sodium for optimum growth.

These are motile bacteria that move with the help of locomotory structure called flagellum. They are monotrichous polar sheathed bacteria. If we talk about its primary ecological reservoirs, they are usually found as a normal resident of estuaries. *V. cholerae* has been recovered from zones not related with a marine or brackish water supply, freshwater lakes as well as rivers and from birds as well as herbivores (Bintsis, 2017).

According to epidemiological studies, it is clearly understood that there are two biogroups of *V. cholera*, that has been isolated during previous pandemics as well as from the current pandemic. The former is for *Vibrio cholerae* O1 and the later is for El Tor (Ma et al., 2014). *V. cholerae* ideal temperature for growth is between 30°C and 37°C, though growth might occur between 10°C and 43°C. For every bacterium, there is a specific temperature and pH range which those particular bacteria required for its growth. In such ranges, there is a specific point at which bacteria growth at its maximum, this is called the optimum level either it is for temperature or it is for pH. Similarly, for these particular bacteria, the pH range for growth is 5.0–9.6 but its optimum pH is 7.6. When we talk about its growth at a specific width, then it would be around 0.97 but preference is of 0.984. When we talk about its optimal growth occurs in a salty environment, then NaCl concentration of point 5%, though it might also be seen at concentrations of point 1% to 4% (Stenfors et al., 2008). There are some diseases in which transmission route is an animal source such diseases are called zoonotic diseases, and this microorganism is called zoonotic organism. Here is an example of zoonotic disease in which way of its transmission is likewise over ingesting contaminated food product, for example, mollusks (raw oysters) as well as crustaceans. Here there are many situations that will lead towards disease if we acquire these microbes either be eaten raw, underdone cooking, or later cooking contamination, as well as upon contact of wound that might not be dressed properly or open may lead to get contaminated with such raw water. Some clinical manifestations ensuing infection from *V. cholerae* O1 have a large number of symptoms from asymptomatic or some may lead to the most chronic form called "cholera gravis" of the classic biogroup that will lead towards asymptomatic infections (Bintsis, 2017). Moreover, the El Tor biogroup will express chronic disease in 2% of the already diseased people or you can say immunosuppressant people and insignificant or moderate disease in 23% whereas the classic biogroup produces chronic disease in 11%, of individuals and mild or moderate disease in 30% (Cary et al., 1999). Inoculum size as well as the amount of food taken are the two basic factors, in fact later in an incubation period of numerous hours to five days characteristic symptoms include excess dehydrative cramping of muscles in which amount of fluid loss is around 500–1,000 ml/h (Cuthbert et al., 2001). The disease produced by *V. cholerae* O139 Bengal is clinically similar to the symptoms shown by *V. cholerae* O1-diseased individuals. Other *V. cholerae* serogroups, in addition to *V. cholerae* O1 as well as *V. cholerae* O139 Bengal, are known as non O1, non-agglutinating vibrios or noncholera vibrios and are not known to cause epidemic disease. Still, many vibrios are considered to be as noncholera and are recognized as a culprit of self-limiting gastroenteritis and also may lead to infectious wounds, bacteria in bloodstream cause bacteremia, as well as their toxin in high amount cause septicemia when accompanying former hepatic disorder (Koff et al., 1967). Every bacterium carries an infectious dose (ID dose) which helps particular bacteria to cause disease, ID dose in host of *V. cholerae* is nearly 10^{11}, but the ID does is abridged to about 10^6 reliant on food's buffering capacity with the ingestion of food. There are a variety of food sources considered to be vector for *Vibrio parahaemolyticus* spread comprise of crabs, seaweed, oysters, prawns, scallops as well as clams. Growth temperatures lies between 5 to 44°C, by an ideal temperature similarly growth requires proper pH that lies in range of 30 and 37°C and at pH 7.6 as well as 8.6, the organism will grow in an environment at pH 4.8 to 11.0, in NaCl concentrations of 0.5 to 10.0%, and in environments with a least water activity of 0.94; however, it prefers a concentration of sodium chloride in the range of 2% to 4% as well as water activity of 0.981 (Stenfors et al., 2008; Bintsis, 2017). A very important *Vibrio* species is *V. parahaemolyticus* that is most commonly isolated from clinical samples observed in the US. Gastroenteritis is characteristically linked with ingesting of raw, imperfectly cooked, or cooked but contaminated seafood. *V. parahaemolyticus* lead to cause gastroenteritis, in which after incubation period of 4-to-96-hour symptoms appear like nausea, vomiting, headache that are usually very nonspecific symptoms along with very specific clinical manifestation like abdominal cramps, slight fever, chills plus watery diarrhea that is often bloody. More symptoms along with these may initiate later contact to contaminated water, may comprise of infected wounds, eyes, as well as ears. These visible symptoms are self-limiting and last for just two to three days, in adverse cases, conditions like dysentery, primary septicemia, or cholera disease occur with the possibility of death (Cary et al., 1999; Longenberger et al., 2014).

6.2.11 *YERSINIA ENTEROCOLITICA*

Genus *Yersinia* is a representative of the family *Enterobacteriaceae* which also contain 10 recognized species though only three are termed as pathogenic to humans as well as animals. Plague's chief contributing pathogen is *Yersinia pestis*, whereas *Yersinia pseudotuberculosis* is mainly a pathogen of animals even then it might affect humans by taking contaminated food/water, besides all these *Yersinia enterocolitica* is a foodborne pathogen that in humans causes gastroenteritis (Cuthbert, 2001; Bintsis, 2017). They are rod-shaped under the microscope, non-spore-forming, Gram-negative microorganisms that grow under both oxygen-rich as well as suppressed oxygen circumstances nevertheless regarded as facultative anaerobic bacteria. These are motile species because of the presence of a specialized peritrichous flagella. Their motility is temperature specific that is they are only motile at 22°C to 30°C but not at 37°C.

Because of its ability to act as psychrotroph, *Y. enterocolitica* might express a health risk in refrigerated foods products that are contaminated, even in refrigerating temperatures the microbe is generally expanded instead of other opposing psychrotrophs (Koff et al., 1967). This species grows between 0 and 45°C, but then again for growth optimal temperature range is flanked by 25°C and 30°C. *Y. enterocolitica* can endure alkaline conditions just as some other gram-negative bacterium however does not endure well in acidic situations, since growth is seen well at pH ranges between 4.0 to 10.0, but optimally grow at pH 7.6. Additionally, growth is also seen in concentration above 5% NaCl by *Y. enterocolitica* (Stenfors et al., 2008; Bintsis, 2017). Ingesting of contaminated food or having water that is contaminated or simply raw or not cooked properly seems to be underdone pork, is a source of foodborne infection in humans, that will show symptoms after few days to week of incubation period. History of intestinal yersiniosis in adults age group people show one to two weeks of persistence where as in children age group it is about 4 weeks with visible symptoms of diarrhea (watery or bloody) in combination with fever, pain abdominal cavity along with vomiting (Andersson et al., 1995; Bintsis, 2017). Immunocompromised people as well as kids younger than 15 are mostly affected, and yersiniosis's extraintestinal infections incorporate septicemia, meningitis, Reiter condition, myocarditis (inflammation of the heart muscle), glomerulonephritis (inflammation of the tiny filters in your kidneys (glomeruli), thyroiditis (inflammation, of the thyroid gland), and erythema nodosum (skin inflammation that is located in a part of the fatty layer of skin) (Cuthbert et al., 2001; Longenberger et al., 2014). *Y. enterocolitica* poison is heat stable, opposes enzymatic denaturation, stays stable during delayed capacity, and is of comparative pH steadiness as the thermostable enterotoxin delivered by ETEC (Koff et al., 1967).

6.3 FOODBORNE VIRUSES

The specific nature of growing only in living cells is seen in particulate matter, i.e., viruses. Thus, they do not survive for a longer period of time outside living cell. Enteric viruses have more than 100 types that are the leading cause of foodborne illness. From these types, Hepatitis A and norovirus (NoV) are major pathogenic forms related to food whose transmission is through food; species of phylum Mollusca, e.g., bivalve mollusks, in which most common clams, cockles, mussels, as well as oysters, are particularly susceptible to spread viruses. The waters wherein they develop are progressively dependent upon human fecal defilement, here and there from sewage releases and now and again from contamination harvesters of shellfish. The shellfish gather infections over the span of their channel taking care of movement. Human infections do not taint these species; however, they are harbored for a considerable length of time or weeks in the shellfish stomach related tract and are obviously harder to expel than microscopic organisms during forms expected to wash down the shellfish (for example depuration). In contrast to numerous different seafood, shellfish are typically eaten with their stomach-related tracts in place. They are frequently eaten crude or delicately cooked. Shellfish, in contrast to different foods, may likewise protect such viruses from warm inactivation during cooking (Wait and Sobsey, 1983).

6.3.1 HEPATITIS A

These are environmentally tough microorganisms that usually spread via water as well as food that are contaminated, contaminated daily usage surfaces in our environment (tabletops, cooking utensils) as well as via direct or indirect individual to individual contact. They are highly stable organisms with a huge range of environmental circumstances, however they cannot grow freely in the environment, for example, freezing, heating, chemicals, as well as desiccation (Andersson et al., 1995). In spite of the fact that Hepatitis A offer some significant attributes from picornavirus family various other genera, as it is adequately unique that it is named the main species in genus Hepatovirus (WHO, 2015). Very advanced technique RNA sequencing examination is being used to report six Hepatitis A genotypes (I-VI). Human diseases are linked with strains of Genotypes I, II, and III, with most of human strains gathered inside genotypes I and III. Structurally these are single-stranded having positive charge genome carries 7.5 kb (kilobases). They are usually hydrophobic, non-enveloped viruses having size of 22 to 30 nanometer (Andersson et al., 1995). Shellfish has also been observed as a major cause for viral sporadic diseases (Iturriza-Gomara and Brien, 2016). The waters that are specifically used to grow Shellfish are routinely examined for indicator organism which includes fecal coliforms group as well as *Escherichia coli* (Estes et al., 2006).

6.3.2 NOROVIRUSES (NoVs)

These viruses are basically involved in causing acute viral gastroenteritis. This virus is very much related to cause disease (acute gastroenteritis (AGE)) among children less than 5 years of age who seek medical care (Glass et al., 2009). Structurally these are non-enveloped with a full width of 30 to 35 nanometers and RNA single-stranded genome of around 7.5 kilobases. Classification of viruses is done on the basis of genogroups because of their diversified forms: in which only 3 of the groups are causing diseases in human beings out of 30 genogroups (Verhoef et al., 2011). Latest upgrades to diagnostic methods have permitted scientists to portray the huge commitment of this RNA virus as exceptionally infectious to the weight of foodborne ailment, especially as the reason for various outbreaks of foodborne sickness in network settings, for example, nursing homes, clinics, the military, and voyage ships. Fecal-oral spread is the essential method of transmission. The infection's capacities to withstand a wide scope of temperatures (from sticking to 60°C) and to persevere on natural surfaces and food things add to quick dispersal, especially by means of auxiliary spread (by means of food handlers or to relatives). Food can be debased at the source (by means of contaminated water) or while preparing them. Avoidance of disease is troublesome on the grounds that these infections can continue on natural surfaces and food things. Correlation of NoV arrangements gathered from around the globe over the previous decade have raised the likelihood that pandemic strains of NoV are blown out through nourishments sold globally, or through individual-to-individual contact when explorers convey the infection (Koopmans, 2008; McCarter, 2009; Müller et al., 2015). Latest reports propose likelihood of reservoirs of animal, but direct zoonotic transmission is seen as infrequent. Latest technologies have confirmed the extensive nature of infections through virus in the food chain, be that as it may, this does not really connect with the danger of infection (McCarter, 2009).

6.4 FOODBORNE PARASITES

Parasites are microorganisms that are singled and nucleated celled not having a tough cell wall and are larger in size to that of bacteria. They have the mutual quality as that of viruses that they do not increase their number in food products but multiply in their specific living host. Their communicable stages are termed as a cyst. Parasites are life forms that get sustenance and security from other living beings known as hosts. Their transmission is very simple that is from animals to humans, from humans to humans, or might be from humans to animals. Several parasites have emerged as significant causes of foodborne as well as waterborne illness. Parasites living and reproduction areas are inside the organ as well as tissues of humans that are infected as well as host (animal) that might shed it off through fecal material. Most widely recognized foodborne parasites are *Cyclospora cayetanensis*, *Toxoplasma gondii* and *Trichinella spiralis*.

6.4.1 *CYCLOSPORA CAYETANENSIS*

Majority of parasites are involved in foodborne diseases in which *Cyclospora cayetanensis* is also a protozoan parasite, have its place in a family *Eimeriidae*, that lives in the small intestine of host because parasite requires a host for its development, nutrition, and shelter, where they present intermediary life cycle stages in the cytoplasm of cells of digestive tract and afterward produce oocysts containing two sporocysts encapsulating four sporozoites. These oocysts present subsequent shedding afterwards, maximum of 7–15 days are needed for sporulation to perform. *C. cayetanensis* genomes nearly two in number have been reported until now according to the data retrieved from NCBI. Genome total median length is 44.2991 Mb. The groups of people that is immunocompromised and immunocompetent are likely to bear prolonged disease nearly around six weeks or longer, with nonspecific symptoms including nonbloody diarrhea, nausea, vomiting, anorexia (an eating disorder characterized by an abnormally low body weight, an intense fear of gaining weight and a distorted perception of weight), bloating, abdominal cramping, malaise, fever, as well as fatigue (Altekruse et al., 1999; Bintsis, 2017).

6.4.2 *TOXOPLASMA GONDII*

It is a parasitic protozoan of Apicomplexa (Phylum) that causes human toxoplasmosis. It is an obligate intracellular pathogen. The parasites have the ability to complete their lifecycle in the host. There are various types of hosts. Mainly two hosts are common in the majority of humanly parasites. These are definitive and intermediate host.

In *T. gondii* cats are definitive host and an intermediate host is any other warm-blooded animal (Bintsis, 2017). Its life cycle three stages are tachyzoites, bradyzoites, or sporozoites. The first two stages occur in body tissues, where the former stage increase in number and destroy infected host cells whereas on the other hand, bradyzoites reproduce inside tissue cysts. Shedding of sporozoites inside oocysts, in cat's feces where they produce spores after 15 days and can survive for months as they are resistant to cold, dryness, and disinfectants (Bintsis, 2017). As per the data received from NCBI 17 genomes are sequenced till now of *T. gondii* and the length measured is as 64.1936 Mb (Altekruse et al., 1999). In human beings, it is transmitted easily through multiple ways such as eating food or taking water contaminated with oocyst, transplacental transmission, blood transfusion, accidental tachyzoite inoculation, and organ transplantation. The infection is caused due to cysts uptake in meat (raw/undercooked), with fresh pork and beef appearing to be the primary sources. Toxoplasmosis is from the ingestion of as few as 100 tissue cysts or oocysts, at which time cyst walls rupture, releasing the sporozoites or bradyzoites to move through epithelium (intestinal) and circulate body all over. Sporozoites and bradyzoites transform into tachyzoites and begin to rapidly multiply intracellularly, and after host cell death, the tachyzoites enter head-to-head cells as well as repeat the reproduction procedure; taking aid of the host immune response, are forced for transforming back in bradyzoites as well as make cysts in the local tissue of host organism, where they can live their entire life. Fever, rash, headache, muscular pain, and edema of lymph nodes are its symptoms. The most widely recognized bases of toxoplasma are half-cooked meat, creature defecation, and transplacental transmission. Many of the people that are ill with toxoplasma carries no sign and symptoms, unborn kids (who contract it from their moms) and grown-ups with bargained safe framework's chance genuine reactions. The results of inherent toxoplasmosis extend from gentle to extreme to deadly and include: mental hindrance, seizures, visual impairment, and demise (Andersson et al., 1995; Altekruse et al., 1999; Gandhi and Chikindas, 2007; Bintsis, 2017).

6.4.3 *TRICHINELLA SPIRALIS*

This parasite is basically involved in most human trichinosis infections. It is a roundworm belongs to Nematoda (Phylum). Not only in humans, parasite is also involved in infections of mammals that are carnivores. *T. spiralis* have 2 genomes that have been completed till conferring to the data retrieved from *National Center for Biotechnology Information*. The total genomic median length is 56.7757 Mb. Worms (adult) are 1.4 to 1.8 millimeter in size and located in the small intestine epithelium of host, adult worms (male) lays larvae in bloodstream as well as they spread into muscle fiber the place of their encystation; where they remain operable for a longer period of time. Pathogeneses is mainly due to the migrating as well as encystment phase that led to pain, fever, edema, neurological disorders that are symptoms even expiry.

Nematodes usually undergo various stages in their life cycle in the host body at any of their anatomical site, their adult stage usually reside in the epithelial mucosa of duodenum and jejunum. Here these nematodes stay for up to eight weeks in host body earlier they are ejected out from the body. Adult female nematodes can produce about 1,500 larvae into the bloodstream to travel around the body during this transient period, and afterward enter muscular tissue, where they can subsist for many years. In skeletal muscle, larvae develop, mature, and undergo encapsulation in a calcified wall several months later. Both the larval and the adult stages are passed from the same host. The stages described above in which there is larval and adult stage both are found in the same host. Those larvae that have cyst around them persists their viability for up to 10 years, and they get rid of the stomach enzymes of the host after eating flesh carrying cyst stage. Period of 3 to 14 days later incubation, symptoms comprise of nonspecific symptoms of gastroenteritis, nausea, vomiting, headaches, fever, visual deficiencies, difficulty breathing, chills, night sweating, low level of eosinophils, myalgia, and circumorbital edema nematode can be thermally inactivated, and therefore the USDA recommends cooking pork products to an internal temperature of 76.7°C (Altekruse et al., 1999; Bintsis, 2017).

6.5 CONCLUSION

Foodborne diseases are a global issue, and a unified and joint approach by all countries and the relevant international organizations is a prerequisite for the identification of causative organism and its control of all emerging foodborne problems that hover human health and international trade. The toxins also become the reason for food-related diseases; these may be either microbial toxin or chemical toxins. Most of the diseases with respect to food contamination are mild and the majority of them do not require treatment as they get recovered by themselves such conditions are called self-limiting, most of them show serious complications that results into mortality, some are preventable even they have intricate biology, analysis as well as epidemiology. Various health-related agencies that are dealing with our general public, some of the regulatory agencies, the food-related industries, and their consumer need to make unremitting efforts to eliminate the factor of food contamination not only on farms but also in processing units, restaurant, and homes. If proper education be given to people regarding the safety of food, much of the cases could be minimized.

KEYWORDS

- **diarrhea**
- **gram-positive**
- **incubation period**
- **intestinal cramping**
- **nausea**
- **pathogen**
- **respiratory tracts**
- **stimulus**
- **toxins**
- **vegetative cells**

REFERENCES

Altekruse, S. F., Stern, N. J., Fields, P. I., & Swerdlow, D. L., (1999). *Campylobacter jejuni* – an emerging foodborne pathogen. *Emerging Infectious Diseases*, *5*(1), 28.

Andersson, A., Ronner, U., & Granum, P. E., (1995). What problems does the food industry have with the spore-forming pathogens *Bacillus cereus* and *Clostridium perfringens*?. *International Journal of Food Microbiology*, *28*(2), 145–155.

Bacon, R. T., & Sofos, J. N., (2003). Characteristics of biological hazards in foods. *Food Safety Handbook* (pp. 157–195). New Jersey: John Wiley and Sons.

Bintsis, T., (2017). Foodborne pathogens. *AIMS Microbiology*, *3*(3), 529.

Buchanan, R. L., Gorris, L. G., Hayman, M. M., Jackson, T. C., & Whiting, R. C., (2017). A review of *Listeria monocytogenes*: An update on outbreaks, virulence, dose-response, ecology, and risk assessments. *Food Control*, *75*, 1–13.

Carter, A. T., & Peck, M. W., (2015). Genomes, neurotoxins and biology of *Clostridium botulinum* group I and group II. *Research in Microbiology*, *166*(4), 303–317.

Cary, J. W., Linz, J. E., & Bhatnagar, D., (1999). *Microbial Foodborne Diseases: Mechanisms of Pathogenesis and Toxin Synthesis*. CRC Press, Florida.

Croxen, M. A., Law, R. J., Scholz, R., Keeney, K. M., Wlodarska, M., & Finlay, B. B., (2013). Recent advances in understanding enteric pathogenic *Escherichia coli*. *Clinical Microbiology Reviews*, *26*(4), 822–880.

Cuthbert, J. A., (2001). Erratum: Hepatitis A: Old and new. *Clinical Microbiology Reviews, 14*(3), 38–58.

ECDC, E., (2014). EFSA (European Food Safety Authority) and ECDC (European Centre for Disease Prevention and Control). *The European Union Summary Report on Trends and Sources of Zoonoses, Zoonotic Agents and Foodborne Outbreaks in 2015, 13*, 4036.

Estes, M. K., Prasad, B. V., & Atmar, R. L., (2006). Noroviruses everywhere: Has something changed?. *Current Opinion in Infectious Diseases, 19*(5), 467–474.

Gandhi, M., & Chikindas, M. L., (2007). *Listeria*: A foodborne pathogen that knows how to survive. *International Journal of Food Microbiology, 113*(1), 1–15.

García, A., Fox, J. G., & Besser, T. E., (2010). Zoonotic enterohemorrhagic *Escherichia coli*: A one health perspective. *ILAR. Journal, 51*(3), 221–232.

Glass, R. I., Parashar, U. D., & Estes, M. K., (2009). *Norovirus gastroenteritis. New England Journal of Medicine, 361*(18), 1776–1785.

Grass, J. E., Gould, L. H., & Mahon, B. E., (2013). Epidemiology of foodborne disease outbreaks caused by *Clostridium perfringens*, United States, 1998–2010. *Foodborne Pathogens and Disease, 10*(2), 131–136.

Hakwenye, H. L., (2019). *Knowledge, Attitudes and Practices on Food Safety Among Food Handlers in* Opuwo district, Kunene region, Namibia (Doctoral dissertation, University of Namibia).

Healy, B., Cooney, S., O'Brien, S., et al., (2010). *Cronobacter* (*Enterobacter sakazakii*): An opportunistic foodborne pathogen. *Foodborne Pathogens and Disease, 7*(4), 339–350.

Hochel, I., Růžičková, H., Krásný, L., & Demnerová, K., (2012). Occurrence of *Cronobacter* spp. in retail foods. *Journal of Applied Microbiology, 112*(6), 1257–1265.

Humphrey, T., O'Brien, S., & Madsen, M., (2007). Campylobacters as zoonotic pathogens: A food production perspective. *International Journal of Food Microbiology, 117*(3), 237–257.

Iturriza-Gomara, M., & O'Brien, S. J., (2016). Foodborne viral infections. *Current Opinion in Infectious Diseases, 29*(5), 495–501.

Janda, J. M., Powers, C., Bryant, R. G., & Abbott, S. L., (1988). Current perspectives on the epidemiology and pathogenesis of clinically significant *Vibrio* spp. *Clinical Microbiology Reviews, 1*(3), 245–267.

Jaradat, Z. W., Al Mousa, W., Elbetieha, A., Al Nabulsi, A., & Tall, B. D., (2014). *Cronobacter* spp. – opportunistic foodborne pathogens. A review of their virulence and environmental-adaptive traits. *Journal of Medical Microbiology, 63*(8), 1023–1037.

Jemmi, T., & Stephan, R., (2006). *Listeria monocytogenes*: Foodborne pathogen and hygiene indicator. *Rev. Sci. Tech., 25*(2), 571–580.

Kandhai, M. C., Reij, M. W., Van, P. K., Guillaume-Gentil, O., Beumer, R. R., & Van, S. M., (2004). A new protocol for the detection of *Enterobacter sakazakii* applied to environmental samples. *Journal of Food Protection, 67*(6), 1267–1270.

Kaper, J. B., Morris, J. G., & Levine, M. M., (1995). Cholera. *Clinical Microbiology Reviews, 8*(1), 48–86.

Koff, R. S., Grady, G. F., Chalmers, T. C., Mosley, J. W., Swartz, B. L., & Boston Inter-Hospital Liver Group, (1967). Viral hepatitis in a group of Boston hospitals: Importance of exposure to shellfish in a nonepidemic period. *New England Journal of Medicine*, 276(13), 703–710.

Koopmans, M., (2008). Progress in understanding norovirus epidemiology. *Current Opinion in Infectious Diseases, 21*(5), 544–552.

Longenberger, A. H., Gronostaj, M. P., Yee, G. Y., et al., (2014). *Yersinia enterocolitica* infections associated with improperly pasteurized milk products: Southwest Pennsylvania. *Epidemiology and Infection, 142*(8), 1640–1650.

Ma, C., Deng, X., Ke, C., et al., (2014). Epidemiology and etiology characteristics of foodborne outbreaks caused by *Vibrio parahaemolyticus* during 2008–2010 in Guangdong province, China. *Foodborne Pathogens and Disease, 11*(1), 21–29.

McCarter, Y. S., (2009). Infectious disease outbreaks on cruise ships. *Clinical Microbiology Newsletter, 31*(21), 161–168.

Mead, P. S., Slutsker, L., & Dietz, V., (1999). Food-related illness and death in the United States. *Emerging Infectious Diseases, 5*(5), 607.

Mitscherlich, E., & Marth, E. H., (2012). *Microbial Survival in the Environment: Bacteria and Rickettsiae Important in Human and Animal Health.* Springer Science and Business Media.

Mossel, D. A. A., Corry, J. E., Struijk, C. B., & Baird, R. M., (1995). *Essentials of the Microbiology of Foods: A Textbook for Advanced Studies*. John Wiley and Sons.

Müller, L., Schultz, A. C., Fonager, J., et al., (2015). Separate norovirus outbreaks linked to one source of imported frozen raspberries by molecular analysis, Denmark, 2010–2011. *Epidemiology and Infection, 143*(11), 2299–2307.

Raheem, D., (2016). Outbreaks of listeriosis associated with deli meats and cheese: An overview. *AIMS Microbiol., 2*, 230–250.

Rajkowski, K. T., & Smith, J. L., (2001). Update: Food poisoning and other diseases induced by *Bacillus cereus*. In: Hui, Y. H., Pierson, M. D., & Gorham, J. R., (eds.), *Foodborne Disease Handbook* (pp. 61–76).

Sayers, E. W., Beck, J., Bolton, E. E., et al., (2021). Database resources of the national center for biotechnology information. *Nucleic Acids Research, 49*(1), 10.

Scallan, E., Hoekstra, R. M., Angulo, F. J., et al., (2011). Foodborne illness acquired in the United States – major pathogens. *Emerging Infectious Diseases, 17*(1), 7.

Schaffter, N., Zumstein, J., & Parriaux, A., (2004). Factors influencing the bacteriological water quality in mountainous surface and groundwaters. *Acta Hydrochimica et Hydrobiologica, 32*(3), 225–234.

Stenfors, A. L. P., Fagerlund, A., & Granum, P. E., (2008). From soil to gut: *Bacillus cereus* and its food poisoning toxins. *FEMS Microbiology Reviews, 32*(4), 579–606.

Stern, N. J., Jones, D. M., Wesley, I. V., & Rollins, D. M., (1994). Colonization of chicks by non-culturable *Campylobacter* spp. *Letters in Applied Microbiology, 18*(6), 333–336.

Vargas, M., Gascon, J., De Anta, M. T. J., & Vila, J., (1999). Prevalence of *Shigella* enterotoxins 1 and 2 among *Shigella* strains isolated from patients with traveler's diarrhea. *Journal of Clinical Microbiology, 37*(11), 3608–3611.

Verhoef, L., Kouyos, R. D., Vennema, H., et al., (2011). An integrated approach to identifying international foodborne norovirus outbreaks. *Emerging Infectious Diseases, 17*(3), 412.

Vieira, K. C. D. O., Silva, H. R. A. D., Rocha, I. P. M., Barboza, E., & Eller, L. K. W., (2021). Foodborne pathogens in the omics era. *Critical Reviews in Food Science and Nutrition*, 1–16.

Wait, D. A., & Sobsey, M. D., (1983). Method for recovery of enteric viruses from estuarine sediments with chaotropic agents. *Applied and Environmental Microbiology, 46*(2), 379–385.

Wells, J. G., Davis, B. R., Wachsmuth, I. K., et al., (1983). Laboratory investigation of hemorrhagic colitis outbreaks associated with a rare *Escherichia coli* serotype. *Journal of Clinical Microbiology*, *18*(3), 512–520.

World Health Organization, (2015). *WHO Estimates of the Global Burden of Foodborne Diseases: Foodborne Disease Burden Epidemiology Reference Group 2007–2015*. World Health Organization.

Wu, Y., Wen, J., Ma, Y., Ma, X., & Chen, Y., (2014). Epidemiology of foodborne disease outbreaks caused by *Vibrio parahaemolyticus*, China, 2003–2008. *Food Control*, *46*, 197–202.

CHAPTER 7

Foodborne Pathogens and Food Safety Regulations

MAHESH PATTABHIRAMAIAH and SHANTHALA MALLIKARJUNAIAH

ABSTRACT

Food safety is a global issue, and consumers deserve the benefit of healthy and nutritious food. A substantial percentage of diseases that have dire implications on the environment and human health are caused by food-borne pathogens (FBP). Pathogenic bacteria, viruses, and parasites cause human disease. Global interventions by food producers, companies, and policymakers in recent decades have ensured that food supply achieves the high standard demanded by health authorities, protecting public health, and reducing monetary losses from foodborne diseases. To fulfill this aim, many projects and strategies have been established globally by both government departments and the private sector. Foodborne diseases caused by bacteria, environmental pollutants, and chemical contaminants pose a major problem for public health, as new risks are increasingly arising while others are being managed. In this context, to detect possible threats, perform a risk analysis, and an effective sanitary surveillance system is immediately required to monitor foodborne disease outbreaks globally.

7.1 INTRODUCTION

Food is any commodity and/or substance that people and/or animals eat or drink. It retains and boosts the standard of living or is used as a nutritional or medical supplement (FAO/WHO, 1996). Food is produced in the principal

Food Microbial and Molecular Biology: From Fundamentals to Applications. Saher Islam, Devarajan Thangadurai, Jeyabalan Sangeetha, Zaira Zaman Chowdhury (Eds.)

industry and will continue to be produced and consumed by the public through production and retailing (agricultural, aquaculture, livestock, and fisheries). Food plays a significant role in the population's health, nutritional condition, and productivity. Unhygienic food is the principal reason for foodborne diseases. Contaminated or adulterated foods cause major health problems worldwide and foodborne illness is the foremost problem of community health apprehension.

National or local authorities must enforce food control laws to protect the public and ensure that all foods during the entire food production, handling, processing, storage, and distribution process are nutritious and safe for human consumption and that they adhere to safety and quality standards set by authorities. The goal of food control strategies is to improve population health, economic growth by reduction of spoilage, and minimizing food losses. When it comes to food production and processing on a massive scale, food safety and quality (FSQ) are crucial. Globally, foodborne illness can lead to hampered economic growth, impair tourism, unemployment, and fatality, hence FSQ have a significant impact worldwide. To safeguard national food supplies, and to exercise suitable food control strategies. All nations must guarantee that food is highly nutritious, is available in sufficient quantities, and is inexpensive. Since food is the most important aspect of health, adequate strategies have to be planned to control food safety.

The link between food consumption and human sickness was first established by Hippocrates (460 BC), who stated that there is a significant link between food consumption and human illness (Hutt and Hutt, 1984). Foodborne illness can be caused by a virus, bacteria, and parasites. In the event that there are two or more cases of a foodborne epidemic of a comparable illness arise as a result of the consumption of a common food (CDC, 2012). When a toxigenic pathogen establishes itself (and generally multiplies) in a food product, it generates a toxin, which is subsequently consumed by the human host. There are about 200 distinct foodborne illnesses that have been recognized (Mead et al., 1999). Quite elderly, relatively young and those with compromised immune systems as well as healthy people exposed to a very high dose of an organism are the most at risk (CDC, 2012).

Foodborne illness-causing organisms can invade this food production chain at any point. Several factors that can impute to foodborne disease, including unsuitable storage conditions, are rarely restricted to one sector alone, thus the term paddock to plate. As a result, its quality and safety are crucial for underdeveloped and developing nations. Furthermore, consumer safety is one of the most critical issues to solve in global food supply management.

7.2 FOOD SAFETY

Food safety is the sustaining hygienic conditions required and safeguarding the good and secure processing of food in all processes from raw materials to food production, storage, stockpiling, delivery, and use. It means a guarantee that food will not cause any damage to the consumers and is directly related to the regulation of potentially dangerous substances. Food safety is an issue for the foodservice industry, public health authorities, and consumers. Consumers have an important responsibility for food safety. Many consumers think foodborne diseases are most common in the home, although most infections occur in the home environment (Lahti et al., 2017). Improper food handling, contaminated raw food supply, and purposeful ingestion of meals that are uncooked or undercooked of animal origin are all factors that contribute to these illnesses.

Consumers enjoy fresh fruit, vegetables, and cereals, and these commodities can travel a long distance from one another within the context of a healthy lifestyle. The year-round demand for fresh produce has led to a rise in consumption of products imported from countries where production methods are less well-controlled and microbial contamination is more likely (Sean, 1999; Rajkowski and Smith, 2001). Consumers are more likely to consume commercial meal services – whether at home or away from home- and are more likely to purchase ready-to-eat take-out meals from grocery shops or restaurants. As a result of mass manufacturing, inadequate storage, or reheating of these meals, pathogens can grow in them (Sean, 1999).

Raw meats, raw fish, and seafood, as well as swift cooking methods that do not adequately heat food, have grown increasingly popular. Buyers must be made aware of the hazards of infection associated with these approaches, as well as the importance of ensuring that the preparatory components are safe (Sean, 1999). The risk of contamination from organic produce is the same as it is from nonorganic produce, and in certain cases, the risk is much higher. Both sorts of products should be prepared with caution (Edwards et al., 2013).

Food irradiation using a regulated source of low-dose ionizing radiation improves food safety by eliminating the majority of microbial infections. Food that has been properly irradiated is both healthy and safe (Inns et al., 2010; CDC, 2013). In contrast, food that has been irradiated does not eliminate all microorganisms and hence is not a substitute for proper food production, processing, and preparation.

Many countries have made significant efforts to strengthen food control regimes. Food quality, as stated by the Food and Agriculture Organization of the United Nations (FAO), is "a multidimensional feature of food that controls its worth or appropriateness to consumers." Globally, 420,000 people died and 600 million were sickened in 2010 due to 31 significant food safety hazards (WHO, 2015). Consequently, food safety is a crucial factor in ensuring food security, which is debilitated by both the health and cost consequences of unhygienic food.

According to the FAO, quality is "a dynamic aspect of food determining its worth or acceptance by customers." The nutritional value of food contributes to its perceived consistency, as do its functional and organoleptic properties. Another quality factor is the protection of food. Healthy food is free of substances that may jeopardize a person's health.

The FSA (Food Standard Agency) states that "controlling pathogens is at the core of good food hygiene, which can cause serious disease." A common cause of food poisoning, cross-contamination happens when microorganisms from different sources are transmitted across food, surfaces, or kitchen equipment when raw food comes into contact with other foods, surfaces, or machinery. Cleaning the equipment, surfaces, and food properly may prevent dangerous bacteria from spreading. Cleanliness and disinfection of all surfaces and equipment are required after raw food handling. Food must be refrigerated to prevent the growth of dangerous microorganisms and the refrigerator is checked on a regular basis to maintain proper temperature. When food is thoroughly cooked, it kills bacteria and other microorganisms.

The word food quality relates to characteristics that influence the value of a commodity to consumers. Food safety covers all unwanted qualities, such as deterioration and waste as well as positive aspects like color, texture, and flavor. It is, therefore, a comprehensive term that incorporates factors such as nutritional characteristics, sensory properties (flavor and smell), social factors, health, and safety. Safety is a conceptual feature and a reference to consistency. For ensuring healthful and high-quality food, various governments, and international bodies around the world have laid down the food requirements that the supplier has developed. Therefore, every retail and food provider should act in accordance with suitable management practices, and guarantee food health. The essential factors to be considered are given below:

- Quality of raw materials and water
- Maintaining a clean environment in all parts of the business including the facility itself, personnel, supplies, food preparation, storage, and service
- Preservation and storage of food at acceptable temperatures
- Food hygiene
- Good service practices

Food safety and food quality are phrases that might be confusing. This includes both chronic and acute hazards that can also harm the consumer's health which is not acceptable. Quality includes all other features that influence the consumer's value of the product. This covers negative aspects such as spoilage, filth intrusion, discoloration, odorlessness, and positive aspects such as the food's origin, and processing technique. Factors responsible for spoilage include:

- Many people consume food outside their homes because of changing lifestyles and eating habits. There are ample risks in hotels and restaurants of food being contaminated, as food is cooked in large amounts and is touched by many people. Besides, food products can be spoiled if not properly stored, since they will be cooked several hours in advance. The food safety of processed and packaged food is especially significant.
- There will be a significant gap between foods processing as well as food consumption. These commodity foods are difficult to transport in large quantities, and there is a substantial danger of food deterioration during manufacturing and distribution.
- Foodborne infectious illnesses are becoming increasingly common due to adaptations/mutations of microbes, antibiotic resistance, it is therefore mandatory to detect, characterize, and diagnose emerging pathogens to provide vigorous management of these microbes at both international and national level.
- Employing physical and chemical contaminants such as preservatives, colorants, and flavoring agents is the root cause of foodborne illness. Other factors include pesticides, soil pollutants, and contaminated water.

Changes in diet and lifestyle have a trivial effect on the rising incidence of foodborne disease. Earlier, food was prepared and consumed locally. Traditional ancient food was simple and very well-cooked. Nowadays, the trend is that fewer meals are prepared at home and more depend on packaged, ready-to-eat, and take-away food. The urge to consume fresh, unprocessed food and packaged food lacking preservatives promotes the growth of foodborne germs (Altekruse and Swerdlow, 1996). Minimally processed and prolonged shelf-life food often bears inherent risks of increased contamination. Social cohesion around the world has led to a much wider variety of food options, including a wider range of ingredients. Eggs and egg products, as well as pig meat, are essential livestock foods (both accounting for 10% of all strong-evidence outbreaks), crustaceans (3%), bovine meat (4%), milk, and dairy products (5%), fish, and fish products (7%), cheese (8%), and, broiler meat (9%) are predominantly included in our diet. For example, raw fish, common in Japanese, Korean, and many other cultures, seems to have become increasingly popular with consumers who are ignorant of the various packaging and storage techniques used in the conventional preparation of such raw foods or the hazards of foodborne illness triggered by *Vibrio parahaemolyticus*, the leading cause of foodborne outbreaks in Japan (Osawa et al., 1996).

Food contamination is described as foods that have gone bad or have been contaminated by microorganisms (bacteria, viruses, or parasites) or toxic substances, rendering them unsuitable for human consumption. A food contaminant, which can be biological, chemical, or physical in origin, can infiltrate the food supply chain (farm to fork) and render a food product unsuitable for eating. Chemicals, microorganisms, and diluents are examples of contaminants, which can occur before, during, or after food preparation or storage.

Contagious bacterial outbreaks involving agents including *Salmonella*, *Escherichia coli*, and contaminants containing chemical illustrate food safety issues and increase public concern that modern farming, food processing, and marketing systems do not provide protection for public health. Potential risks in food are caused by a variety of factors, including inappropriate farming methods; improper sanitation at all levels of the food chain; absence of preventative measures in food processing and preparation processes; chemicals misuse; polluted raw materials, additives, water, and hazardous storage, etc. Consumers' health can be jeopardized by risks such as physical, chemical, and biological agents that have harmful/adverse consequences. Pertinent questions about food hazards have generally been based on the following.

7.2.1 PHYSICAL HAZARD

It is an external physical substance that may be present in food that causes ill health or damage and encompasses parts of the animals, wood, stones, rocks, nail fur, etc.

7.2.2 CHEMICAL HAZARDS

They are the contaminants or deleterious substances that may be deliberately or inadvertently introduced to food. This type of hazard contains pesticides, food colors, chemical products, polychlorinated biphenyls (PCB), radioactive metals, preservatives, and other additives. One of the greatest health concerns is toxins that occur naturally and pollutants to the ecosystem.

7.2.2.1 PERSISTENT ORGANIC POLLUTANTS (POPs)

They are substances that build up in human beings and the ecosystem. Examples that are undesirable by-products of industrial applications and waste annihilation include documented dioxins and PCBs. They are released into the environment globally, and bio-accumulate in the food web. Dioxins are extremely toxic which can have an ill effect on reproduction, development, harm the immune system, and meddle with hormones and cancer.

7.2.2.2 PESTICIDES

The word pesticides relate to all chemicals used to eliminate or control pests. In farming, these include herbicides (poisons for weed elimination), insecticides (insects), fungicides (fungi), nematodes, and rodenticides (vertebrates). The use of pesticides as a traditional form of pest management has been widespread since the 1950s, especially in areas where monoculture is practiced in large quantities. Regretfully, the advantages of chemistry have indeed jeopardized the long-term viability of global ecosystems leading to the destruction of ecological relationships and biodiversity loss, and there are significant human health implications (FAO, 2003).

Pesticides appear to have a significant impact on health across the world, although the precise toll is hard to assess given the wide range of chemicals and forms of exposure. Pesticides have considerably different levels of risk and can have implications that can be classified into two categories: acute effects that occur shortly after exposure and chronic impacts that may arise many years later and are sometimes difficult to detect. The impacts of exposure of pesticide on human health are mostly linked by various illnesses (cancer, birth abnormalities), neurotoxicity, endocrine abnormalities, and damage to kidneys or liver (Prüss-Ustün et al., 2011).

Pesticides are ingested mostly through touch with the skin, through the handling of pesticide-tainted products; by inhalation, through the pesticide dust inhaling or spraying; by consuming contaminated water or food. The worldwide disease incidence on agricultural and industrial pesticides and at least 1.2 million fatalities a year are caused by acute toxicity. This leads to 2% of all mortality and 1.7% of the overall disease burden worldwide (Prüss-Ustün et al., 2011).

7.2.2.3 HEAVY METALS

Even at low quantities, heavy metals, which have a relatively high density, can be hazardous or dangerous. Environmental pollution is a major cause of heavy metal contamination in the food supply, such as the presence of heavy metals such as Platinum and Mercury in food. As a result of bioaccumulation in the food chain, heavy metal pollution is a serious environmental health problem for animals and people. Because of this, humans have suffered from a variety of health problems including neurological and renal impairment due to the consumption of contaminated food.

7.3 MICROBIAL/BIOLOGICAL HAZARDS (BIOHAZ)

They are living organisms that contain bacteria, viruses, and parasites. Microorganisms that are food-related and cause disease are referred to as foodborne pathogens (FBPs). Among the many hazards, biological hazards (BIOHAZ) are a prominent source of foodborne diseases. Despite all food safety measures, microbial foodborne pathogens remain a major issue and emerging pathogens continue to develop. For example, there are two categories of foodborne illnesses caused by microbial pathogens (bacteria, viruses, and parasites) infections and poisoning.

7.3.1 BACTERIA

Bacteria are the most prevalent cause of foodborne illness and are available in a multitude of shapes, sizes, and characteristics. Some pathogenic bacteria may produce spores and are hence very heat resistant, e.g., *B. cereus*, *Bacillus subtilis*, *Clostridium botulinum*, *C. perfringens* (Bacon and Sofos, 2003). Very few bacteria can produce toxins which are heat-resistant (e.g., *C. botulinum*, *Staphylococcus aureus*). The majority of pathogens are mesophilic, with optimum growth temperatures ranging from 20°C to 45°C. Few psychrotrophic foodborne pathogens (FBP), such as *Listeria monocytogenes* and *Yersinia enterocolitica* may thrive under refrigerated environments or at temperatures less than 10°C.

Foodborne diseases are caused by *B. cereus*, a gram-positive bacterium that may be found in nature and various types of foods (Marrollo, 2016). *B. cereus* was isolated from several foods, including cooked rice, mixed salad, and ready to eat foods (RTE) (Tewari et al., 2015; Gao et al., 2018; Yu et al., 2019). Some reports show that RTE foods may be polluted by food storage surfaces with various bacterial pathogens (de Oliveira et al., 2011; Shiningeni et al., 2019). In the environment most of *Bacillus* spp. is found in soils, fresh, and marine waters. *B. cereus* may produce viscous spores that are very resistant to heat and drought. Spores of *B. cereus* include additional hydrophobic appendages and/or pili, allowing the spores to cling to numerous other diverse surfaces and defy elimination during sanitation and maintenance (Rajkowski and Smith, 2001). At pH 4.9–9.3, the organism grows, and the inhibitory impact of pH in foods is decreased, as demonstrated by restricted development at pH 4.35 on meat (FDA, 2012). *B. cereus* contains multiple virulence factors which, through ingesting, can invade the gastrointestinal (GI) tract causing diarrhea and vomiting (Song et al., 2019).

According to Tewari et al. (2015), among 860 RTE samples, 302 (35%) were infected with *B. cereus*, implying that retailed RTE meals pose a serious threat to consumers, accounting for 34% (220/656) of cooked meat samples. According to research, open-air booths increase the danger of environmental contamination (Ng et al., 2013).

Infection caused by *B. cereus* is characteristically linked with starch-containing foods producing emetic toxins causing nausea (Delbrassinne et al., 2015; Griffiths and Schraft, 2017). Around 50% (59/119) of the noodle/rice samples were contaminated with *B. cereus* (Chang et al., 2011; Delbrassinne et al., 2012; Merzougui et al., 2014). Cold vegetables in the sauce are the favored RTE food in China, also known as Chinese salads. *B. cereus* was found to be more prevalent in 22% of the (19/85) cold vegetable samples, however, in Korea, it is fewer than that documented in vegetable salads (Chon et al., 2015). *Campylobacter* spp. is an anaerobic, obligate gram-positive, spore-forming bacterium that is found in the natural gut flora of a diverse range of healthy wild and domestic animals, including sheep, pigs, goats, cows, cats, dogs, rodents, chickens, wild birds, geese, ducks, and marine animals, and is widely distributed with rivers and lakes. *Campylobacter jejuni* is a significant source of microbial foodborne diarrheal illness in people worldwide. It may inhibit the guts of both birds and mammals, and propagation to humans occurs through contaminated food products (Chlebicz and Śliżewska, 2018; Wieczorek et al., 2018). This organism can first invade the epithelial layer and then penetrate through it, attaching to epithelial cells. Diarrhea results from disruption to the cells of the epithelia. Systemic diseases that cause more serious illnesses may also occur. The majority of infections caused by campylobacteriosis are linked to ingesting unpasteurized milk, contaminated water, raw or undercooked chicken meat, or cross-contamination of other foods with these products and groundwater may also carry these bacteria (Schaffner et al., 2004; Hagos et al., 2019). Consumers are in danger if they ingest ready-to-eat food infected with enteric bacteria. The rising frequency of reported foodborne outbreaks is linked to unpasteurized fruit juices, raw fruits, vegetables, and poses a Campylobacteriosis threat to public health.

The key reservoir in developing countries is for *Campylobacter* infections in humans and food processing facilities (Rossler et al., 2019). Many species of animals can serve as reservoir hosts or *Campylobacter* carriers, making it difficult to track the source of infection (Skarp et al., 2016). *Campylobacter* colonizes rapidly and is present in animals (cattle, dairy cows, dogs, and swine), the GI tract of poultry and wild birds (Horrocks et al., 2009). *Campylobacter* is usually regarded as a commensal community in poultry, although certain measures may have a detrimental influence on chicken health (Humphrey et al., 2014). *Campylobacter* is rarely reported to cause pathological indications in poultry, pigs, and sheep, but can induce young stocks of diarrhea and often intermittent abortions (Hansson et al., 2018). Chickens are typical reservoir hosts, likely due to higher temperatures of the body (42°C) and the ideal circumstances for the colon and cecum *Campylobacter* growth (Hermans et al., 2011). *Campylobacter* can spread feces, dust, and soil on fleece, hair, and animal skin pre-processing. During the processing of poultry and cattle in a slaughterhouse, the meat might get contaminated, due to the spillage of the intestines and the feces might come in contact with machinery contaminating it (Facciola et al., 2017). Infections of *Campylobacter* in humans are frequently intermittent in nature. Outbreaks of milk unpasteurized, uncooked poultry pâté and raw drinking water have been classified as uncommon (Kaakoush et al., 2015).

Escherichia coli are gram-negative, non-spore-forming rod bacteria that can be motile or stationary; some rods are flagellated, while others are not (Mitscherlich and Marth, 1984). The bacterium is a facultative anaerobe that produces acetic, lactic, and formic acids by fermenting simple carbohydrates like glucose. The optimum pH range for growth is 6.0 to 8.0, although growth takes place at pH values as high as 9 to 10 and as low as 4.3.

The majority of *E. coli* strains are innocuous; but, certain strains have evolved characteristics that make them harmful to humans, such as the ability to produce toxins (Garcia et al., 2010). The major pathogenic bacteria in foods are *E. coli*. It is spread by a fecal or oral path and should not be found in any food in any circumstances. Shiga toxin-producing *E. coli* (STEC) is a pathotype that produces one or more Shiga-like toxins. Although the group is a sub-set of one genus, the serological and virulence variables described are complex. The most relevant serotype is *E. coli* O157:H7, which first came to eminence in the early 1980s and was known as the 'hamburger worm' in the USA. STEC has the potential, though rarely, to cause significant health results in humans, including renal failure and death. Infection results in diarrhea or bloody diarrhea more often [therefore, a subset of STEC is referred to as enterohemorrhagic (EHEC)]. Their reservoir is the intestinal content of ruminant animals and human infections are caused by direct or indirect absorption from this source through several routes. One such route is contaminated food, such as raw/undercooked meats, fresh food, and bean sprouts. The diarrheagenic effect is the most important symptom caused by this microorganism. Furthermore, skepticism, meningitis, and many enteric diseases are known to induce. One of the major health challenges is the inability to ensure food safety.

Food or water contaminated with feces of ill humans or animals can spread *E. coli* bacteria. Contamination of animal products is frequent in animal slaughter and processing (Mustafa and Inanc, 2018; Zelalem et al., 2019). Irrigated water can be contaminated by the use of bovine or other animal manure as a fertilizer for crops.

The genus *Listeria* organisms are gram-positive, motile, non-spore-forming, facultative anaerobic and rod-shaped bacteria (Lee et al., 2019). Especially in the elderly, pregnant women, infants, and immunocompromised patients, *Listeria monocytogenes* is one of the leading foodborne infections (Buchanan et al., 2017). Infections in pregnant women can be fatal to the fetus, leading to miscarriages, stillbirths, and birth abnormalities. Infections are present in decaying vegetable waste, soil, and sewage water, and it can tolerate extremes of temperature (1–45°C) and salt concentration, making it a highly hazardous foodborne pathogen, and it is carried asymptomatically by various animals. These microbes have been discovered in a range of raw foods, including uncooked vegetables and meats. Intestinal infection can travel to the brain and fetal placenta. As a result, infection can lead to meningitis (inflammation of the membrane that surrounds the spinal cord and brain), gastroenteritis (inflammation of the mucous membranes of the stomach and intestine), and septicemia (bloodborne transmission of bacteria and toxins) (Buchanan et al., 2017).

Salmonella spp. is a pathogenic genus of Enterobacteriaceae that is one of the most common causes of intestinal disease (food poisoning) across the world. They were called after Dr. Daniel Salmon, a scientist who discovered *Salmonella choleraesuis* from a pig's gut (Bacon and Sofos, 2003*)*. The *Salmonella* species consists of two distinct species viz., *S. enterica* and *S. bongori* that can cause disease in people (FDA, 2012). For instance, *S. enterica* subsp. is further divided into various serotypes, viz., *S. enteritidis* and *S. typhimurium* that can cause disease in both people and animals (Tegegne, 2019*)*. *Salmonella* infection develops as a result of kitchen staff consuming raw meats and eggs, vegetables, and fruits, and unpasteurized milk infected with *Salmonella*. It causes GI problems such as diarrhea, stomach pain, nausea, and vomiting (Salmonellosis). Few serovars of *S. enterica* are accountable for more serious diseases like typhoid fever.

Shigella is a gram-negative rod genus that is non-motile, non-spore-forming, and facultatively anaerobic. *Shigella* spp. are more common in regions with poor sanitation and hygiene, and although person-to-person contact is the major mode of transmission. Shigellosis can occur as a result of consuming fecally contaminated water or food. Fresh fruits and vegetables, salads, dairy, poultry, and various seafood, have been linked to shigellosis outbreaks in a range of environments, including homes, schools, restaurants, cruise ships, commercial airlines, and military mess dorms, and have been implicated in a substantial epidemic in southern and other parts of Asia, Africa, and Central America (ICMSF, 1996). These outbreaks have led to significant rates of illness and mortality, particularly among malnourished children, the elderly and immuno-compromised individuals and 188 million cases are recorded each year, with roughly 1 million fatalities (Kotloff et al., 2018).

S. aureus is non-motile, gram-positive cocci that occur individually or in short chains, pairs, tetrads, or distinctive "grape-like" clusters, and develop more quickly in aerobic environments (Bacon and Sofos, 2003). *Staphylococcus* spp. are prevalent in mammals, birds, and are present in the mouth, blood, GI, genitourinary (GU) and upper respiratory tracts, mammary glands, skin, and skin glands of infected hosts. One of the most resistant non-spore-forming bacteria, *S. aureus* can survive in a dry environment outside the body for prolonged periods of time and has been isolated from air, dust, sewage, and water (FDA, 2012). Infection by *S. aureus* foods infected by environmental sources of illness include contaminated meat sausage, turkey meatballs, salmon steaks, oysters, shrimp, dairy, and delicatessen salads (Bacon and Sofos, 2003; Wang et al., 2017). Despite their preference for

40°C to 45°C and a pH range of 4.5 to 9.3, with an optimum of 7.0 to 7.5, the bacteria can survive at temperatures ranging from 7°C to 47.8°C and generate enterotoxins at temperatures between 10°C and 46°C. It can withstand high salt concentrations (>10% sodium chloride) with ease (ICMSF, 1996). There are a number of diseases caused by *S. aureus* which includes infectious endocarditis, soft tissue infections (carbuncles and folliculitis), impetigo, and cellulitis, as well as gastroenteritis, meningitis, osteomyelitis, and toxic shock syndrome (TSS) (Tong et al., 2015).

The *Yersinia* genus is a member of the Enterobacteriaceae family and comprises 11 species; however, only three are recognized to be human or animal pathogens. *Yersinia* spp. is non-spore-forming, gram-negative, or gram-variable rods that thrive in both aerobic and anaerobic environments but are classified as facultative anaerobes. *Y. pestis* causes plague, *Y. pseudotuberculosis* is mostly an animal disease that can infect people following consumption of contaminated food or water, and *Y. enterocolitica* has emerged as a source of foodborne human gastroenteritis (Chlebicz and Śliżewska, 2018). Pork consumption, particularly uncooked pork products, is responsible for yersiniosis, and infection is spread primarily through the fecal-oral route. The illness manifests as aggressive diarrhea and mesenteric adenitis, terminal ileitis, and pseudo-appendicitis (Delibato et al., 2018; Raymond et al., 2019).

7.3.2 VIRUSES

Viruses only proliferate in live cells as they cannot survive outside the host for extended durations. It has been demonstrated that more than 100 kinds of enteric viruses cause foodborne diseases. Hepatitis A, norovirus (NoV), and rotavirus, including hepatitis E, are the most prevalent foodborne viral infections. NoVs and hepatitis A viruses (HAVs) are the major cause of viral gastroenteritis worldwide. Most of the time, these viruses are transferred through eating shellfish such as mussels as well as bivalve mollusks, and oysters. They proliferate in waters that are frequently contaminated with human feces, either from sewage discharges or by infected shellfish harvesters. Shellfish is occasionally consumed raw or minimally cooked and this cooking process may help protect the virus against heat inactivation (Digirolamo et al., 1970).

HAV is a single positive-stranded RNA (Heptovirus), hydrophobic, non-enveloped virus 22 to 30 nm in size, the genome size is approximately 7.5 kilobases (Cuthbert 2001; FDA, 2012). Disease can be transmitted by contaminated food, drinking water, and environmental surfaces (such as contaminated countertops and cooking utensils). It can also be transmitted through direct or indirect human contact (FDA, 2012). They are known to be highly stable under a wide variety of environmental conditions, including freezing, humidity, chemical, and desiccation. According to RNA sequence research, there are six Hepatitis A genotypes (I-VI), of which genotypes I, II, and III comprise strains linked with the majority of human illnesses. HAV may cause liver failure and usually progresses via raw or undercooked seafood. The introduction of the virus to the susceptible populations of adults can affect significant outbreaks (Lemon et al., 2018).

One of the first outbreaks of raw shellfish-associated viral disease was caused by storing clean oysters in an unsanitary port (Koff et al., 1967; Halliday et al., 1991). There were 292,301 instances of Hepatitis A linked to clams in Shanghai, which may be the largest documented epidemic of the foodborne sickness in history (Halliday et al., 1991).

In 2013 and 2014, there was Hepatitis A outbreaks in many European countries, with over 1,400 cases connected to strawberries which are fresh and frozen (Chiapponi et al., 2014; Montano-Remacha et al., 2014). The epidemic potential of a viral genotype, or even a subtype, is determined by its intrinsic genetic features as well as its ability to mutate, survive, and propagate via the food chain.

NoVs are single-stranded, nonenveloped RNA viruses with a diameter of 30–35 nm and a genomic size of about 7.5 kb. The viruses are divided into six genogroups, with only three of them causing human infection (Iturriza-Gomara and O'Brien, 2016). In children under the age of five, NoV is the most prevalent cause of acute viral gastroenteritis requiring medical care (Scallan et al., 2011; WHO, 2015). Nausea, severe vomiting, watery diarrhea, and stomach discomfort are all symptoms of NoV infections.

Investigators in diagnostic techniques recently reported the enormous influence of this highly contagious RNA virus on the severity of foodborne disease, notably because of several outbreaks of foodborne disease in community environments such as hospitals, armies, and sail ships (Estes et al., 2006; Glass et al., 2009).

The predominant route of transmission is fecal-oral spread. The virus's ability to withstand average temperatures from –40°C to 60°C, as well as its ability to survive on environmental surfaces and foodstuffs, in particular via secondary dissemination (by food handlers or family members) leads to a rapid spread in food (Glass et al., 2009). During preparation, food may be spoiled at the source (by polluted water). Infection deterrence is challenging as it may persist in the environment and foodstuffs. A comparison of NoV sequences gathered from throughout the world over the last decade has highlighted the likelihood that NoV pandemic strains are disseminated through foreign cuisines or person-to-person contact when travelers carry the virus (Koopmans, 2008; Glass et al., 2009). The research suggests that direct zoonotic transmission through animal reservoirs is uncommon.

NoV is the most prevalent enteric virus found in outbreaks and feces, indicating a high incidence of illness linked to raw or undercooked oyster consumption. Pathogen identification must be done quickly if control and preventive strategies are to be developed. Foodborne viruses cause significant morbidity and mortality. The role of asymptomatic food handlers in NoV outbreaks is significant, accounting for up to one-quarter of outbreaks; As far as eliminating viruses from fingers, soap, and water, hand washing is still the most effective approach (Tuladhar et al., 2015).

7.3.3 PARASITES

Parasites are organisms that dwell in or on a host to get nourishment. Parasites belong to three main classes which can cause disease in humans are protozoa, helminths, and ectoparasites. They are larger than bacteria, multiply only in hosts and transmit as a *cyst*. Other living species known as hosts provide these organisms with sustenance and protection. They survive and multiply in infected human and animal hosts' tissues and organs, and are frequently expelled in feces. They can be transferred from animals to people, humans to humans, or humans to animals. Several parasites represent significant reasons for food and waterborne disease. The most prevalent FBP documented include *Trichinella spiralis*, *Toxoplasma gondii*, and *Cyclospora cayetanensis*. FBP is a well-known yet neglected pathogen group (Robertson, 2018). A total of 156 trichinellosis and 41 congenital toxoplasmosis were reported in the European Union (EU) in 2015. The highest infection rate was observed in Lithuania, followed by Romania and Bulgaria.

In relatively low and middle-income countries, the load is particularly heavy, which has been noted since parasitic infection cycles are extremely specific to dietary sources all across the world (Murray et al., 2012). Mortalities of 33,900 and 2.94 million DALYs have been attributed to enteric protozoa, whereas 67.2 million cases, 5,560 deaths, and 492,000 DALYs are attributed to foodborne protozoa.

The parasitic protozoa *C. cayetanensis* belonging to the Family *Eimeriidae* lives in the human small intestine, where they complete their intermediate life cycle stages in enterocytes' cytoplasm, before developing into two-sporocyst-containing parasites with four sporozoites each (Bacon and Sofos, 2003). Because of the globalization of the human food supply, the prevalence of cyclosporiasis has been increasing in both food-producing and importing nations (Li et al., 2020). *C. cayetanensis* can induce severe sickness (6 weeks or more) in both immunocompromised and immunocompetent people, with symptoms such as fever, vomiting, nausea, anorexia, bloating, abdominal cramps, malaise, non-bloody diarrhea, and tiredness.

The protozoan parasite *Toxoplasma gondii* (phylum *Apicomplexa*) causes toxoplasmosis in humans. It is an obligate intracellular pathogen. In cats, *T. gondii* flourishes as its primary reservoir, and in any warm-blooded animal as a secondary host. *Tachyzoites*, *bradyzoites*, and *sporozoites* are three phases of the protozoan's life cycle. Tachyzoites and bradyzoites may be found in bodily tissues, where tachyzoites proliferate and kill infected host cells while bradyzoites reproduce inside tissue cysts. Oocysts of sporozoites are excreted by cats, where they sporulate after 1–5 days and can live for months by resisting disinfectants, freezing, and drying.

This bacterium can infect humans in a number of ways, including consumption of contaminated food or water containing the oocyst, accidental tachyzoite inoculation, or transplacental transmission, organ transplantation, or infected blood transfusion. *T. gondii* infections are often caused by the consumption of cysts found in raw meat of fresh pig and beef proving to be the most common causes. The intake of as little as 100 tissue cysts or oocytes may induce toxoplasmosis, during which time the cyst walls are ruptured, the sporozoites or the bradyzoites are released into circulation throughout the body, where they may remain in the host's organism throughout its life (Bacon and Sofos, 2003). Frequent fever, rashes, headache, and muscle pain and discomfort are signs of toxoplasmosis. Lymph node enlargement can continue for more than a month (FDA, 2012).

Among the most common parasites, *T. gondii* causes toxoplasmosis, an overlooked tropical disease associated with poverty (Rahman et al., 2018). This is primarily due to the close linkages between humans, cattle, and socio-cultural activities in the region of East Africa (John et al., 2020). While cats are the only known host where parasites may complete their life cycle, nearly all warm-blooded vertebrates, including humans, can be used as hosts. Approximately 87,000 diseases, 4,400 hospitalization, and 330 fatalities annually are reported due to *T. gondii* infections in the USA which make this the second-highest source of foodborne mortality, and the third major source of foodborne hospitalization (Scallan et al., 2011). Animal feces, undercooked meat, and mother-to-child transmission are the most prevalent sources of toxoplasma.

This is especially concerning for women of reproductive age who have not developed antibodies to this parasite, as it can be transferred to the embryo via the placenta (congenital toxoplasmosis). Congenital toxoplasmosis infections range from mild to severe, leading to mental impairments, convulsions, and blindness, including death.

Trichinella spiralis (phylum Nematoda), a parasitic roundworm is responsible for the majority of human trichinosis infections, which are induced by eating undercooked or raw meat (usually pork). Sized between 1 and 2 millimeters, adults are lodged in the epithelium of small intestines, where females and male's breed. Female adults transfer larvae into the bloodstream, where they encyst in muscle fibers; larvae encysted in muscle survive for a long period. The manifestations and pathogenicity are induced by the encystment and migratory processes, which result in pain, fever, edema, neurological problems, and even death. Adult nematodes can survive for 6–8 weeks in the mucosal epithelium of the duodenum and jejunum before being discharged into the body; during this transient period, nearly 500 larvae are released into the bloodstream by adult female nematodes around the body and then invade muscle tissue, where they can sustain for several years. The larval and adult stages are both transmitted by the same host. Encysted larvae can live for up to 10 years. Symptoms include nonspecific gastroenteritis, nausea, vomiting, vision impairments, nocturnal sweating, trouble breathing, chills, headaches, fever, eosinophilia, myalgia, and circumorbital edema following a 3–14-day incubation period. Considering that heat can inactivate the parasite, the USDA recommends cooking pig products to a temperature of 76.7°C (Bacon and Sofos, 2003). The work of Vutova et al. (2020) assessed 72 trichinellosis cases in Bulgaria between 2009 and 2011 for epidemiological and clinical features.

Fish-borne trematodes are parasites that are transmitted through food. *Taenia solium* and *Echinococcus* spp., for example, are tapeworms that can spread illness via food or close contact with cattle. Other parasites, such as *Entamoeba histolytica*, *Cryptosporidium*, *Ascaris*, or *Giardia*, enter the food web via contaminating food by water or soil.

7.3.4 MICROBIAL TOXINS (NATURALLY OCCURRING TOXINS)

Some microbial toxins such as aflatoxins (AFs), are carcinogenic in animals and probably humans. Some types of macroscopic fungal (for instance wild mushrooms) toxins can lead to illness or death when consumed. Mycotoxins are released by molds that grow on a variety of foods, including nuts and dried fruits, and have been linked to a number of acute and chronic illnesses in humans. Dietary staples including maize or cereals may typically contain mycotoxins, like aflatoxin and ochratoxin, developed on grain by molding which impacts immune modulation and normal development or even leads to cancer. Around 316 cases of mushroom poisoning were reported in 2010. Some strains of foodborne bacteria including *S. aureus*, *C. botulinum*, and *B. cereus* release toxins in food, causing a variety of symptoms, ranging from vomiting to a severe neuroparalytic condition.

7.3.5 PRIONS

Prions are infectious agents, exceptional in being correlated with different types of neurodegenerative disease. Pathological, infectious prion diseases are caused by the misfolding of a self-protein (PrPC) into a pathological, infectious conformation (PrPSc). During the 1980s bovine spongiform encephalopathy (BSE) outbreak, prion diseases of food animals acquired notoriety. The spread of a deadly, incurable illness to humans, in particular, shifted the focus of cattle prion infections from food production to food safety. Among the prion diseases in cattle is BSE, often known as "mad cow disease," also allied with the variant in humans is referred to as Creutzfeldt-Jakob disease (vCJD). Consumption of bovine products, such as brain tissue, is the primary method of prion agent transfer to humans. To prevent these infections/diseases, each government's priority task is to safeguard food supply security. Diarrheal diseases are transmitted by water and food, killing 2.2 million individuals per year, and are major causes of mortality in developing nations, particularly among adolescents (WHO, 2008).

7.4 THE EFFECTS OF FOODBORNE PATHOGENS (FBP)

Food, as an indispensable part of daily life, passes through several manufacturing phases before reaching the customer. The microbiota plays an imperative role in every food matrix, oscillating from fermentation, spoilage, and contamination. The pathogenic properties of foodborne microorganisms are strain-dependent. Besides, the epidemic potential of a foodborne strain within its population can diverge in the role of its genetic makeup and ecological items (EFSA, 2014). Foodborne infections are important causes of morbidity and death worldwide, and the ability to achieve epidemiological studies and intrude in foodborne diseases is an essential component of the existing public health infrastructure. Illness caused by foodborne microorganisms is a significant economic and public health burden for the entire world.

More than 200 diseases are triggered by ingesting bacteria, viruses, fungi, or chemical compounds like heavy metals from contaminated food. This rising issue of public health has major socio-economic impacts due to a loss of profitability in medical systems and damage to hospitality and commerce. The global incidence of sickness and death was greatly influenced by these diseases. Foodborne diseases are caused by food contamination and exist in the food production, supply, and consumption chain at any point. They are the result of numerous sources of contamination in the environment, including air, soil, or water pollution, inadequate food storage and processing.

Hundreds of outbreaks of foodborne pathogens (FBP) occur around the world. A total of 5,648 foodborne outbreaks involving 69,553 persons have been recorded in the EU in 2011 (EFSA, 2014). Most FBP are transmitted from the environment or animals through food products to humans. Knowledge of pathogens behavior and techniques of detection are necessary to minimize FBP incidence in the food production chain and therefore improve food safety. FBP pose a public health risk and present a challenge to the food industry to control. The occurrence and transcription of genes, which encode specialized functions, play a key role in the survival of pathogens, growth under harsh circumstances and their capability to vintage toxins, antimicrobial resistance, and pathogenicity.

For decades, there has been a documented risk of foodborne pathogenic microorganisms. Similarly, the application of heat to low acid foods in insulated containers handled botulism problems. In all countries, foodborne illness is the growing cause of morbidity, and the list of possible foodborne microbial pathogens continues to expand despite significant progresses in food science and technology. Besides, foodborne disease in most countries is a leading cause of preventable death and financial burden. Regrettably, there are limited data on and impact foodborne diseases in most countries.

Malnutrition and disease - account for the vast majority of child deaths in poor nations. Also, undernourished babies and children are far more vulnerable to foodborne threats and the danger of diarrheal disease. The high incidence of diarrheal disease in under-developed countries again emphasizes the significance of food safety issues (WHO, 2007a). In contrast, serious diseases frequently associated with non-hygienic services, often lead to infant mortality. Such outbreaks, in effect, worsen malnutrition, contributing to a negative feedback loop of illness and mortality (Prüss-Ustün et al., 2011). Reasonably high mortality and low quality of life for children in several parts of the globe are significant concern for the public health community, particularly nutritionists. The root source of infant mortality in underdeveloped countries can be attributed to pneumonia and diarrhea, both entirely avoidable and curable diseases. Diarrhea is, therefore, a significant public health issue. The leading causes of diarrhea are inaccessibility of drinking water, dearth of hygiene facilities, and rejection of admittance to sanitation, congested living conditions, consumption of tainted food or inadequate food storage, and consumption of raw fruit (Hénock and Dovie, 2007).

Even though not well established, developing countries are assumed to have the greatest influence on the foodborne disease issue due to the prevalence of a wide variety of foodborne diseases, as well as those triggered by parasites (WHO, 2008). As a consequence, the high prevalence of diarrheal diseases in many developed countries reveals significant fundamental food safety concerns (WHO, 2007b). Worldwide, potable water coverage rose from 77% in 1990 to 87% in 2008, with just 84% of the people in underdeveloped countries having better access to water supplies, contrary to almost 100% in developed nations. Nevertheless, only 60% of the population in sub-Saharan Africa (SSA) has admittance to safe, drinking water. Among infants under 5 years of age, 70% of fatalities are often attributed to biologically contaminated food. Such toxins are more extensively generated by foodborne parasites; although, mycotoxin contamination is more common.

A wide spectrum of disorders from diarrhea to cancer is caused by foodborne infections. These can cause neurological, gynecological, and immunological symptoms, most often as GI problems. Diarrheal diseases are an important issue for every country in the globe; however, they are disproportionately burdened by nations with low and moderate incomes and adolescents below the age of 5.

Almost one in 10 people worldwide are sick each year and cause over 420,000 deaths following consumption of contaminated food. Children suffer tremendously, and 125,000 fatalities in people below the age of five are recorded annually in which diarrheal disorders are the major instances. Kidney and liver damage, brain and nerve diseases, reactive arthritis, cancer, and deadness may have severe effects on foodborne diseases. Countries with low- and medium-income, foodborne illnesses are meticulously connected to poverty but represent an increasing global public health problem. Increased global trade and increasingly complicated food chains increase the movement of contaminated food across national boundaries and foodborne illness risk. Such problems are amplified and linked to rising towns, climatic change, migration, and increasing international travel.

7.4.1 FOODBORNE DISEASES (FBDs)

Foodborne diseases are generally infectious and toxic and are caused by food or water contaminants that enter the body. Many individuals may be at threat for foodborne diseases (WHO, 2007a). Foodborne diseases most commonly elicit GI symptoms, but these diseases can also cause neurological, gynecological, immunological, and other symptoms, which can lead to multi-organ failure, including cancer. In this way, foodborne diseases bear a major risk of injury, morbidity, and mortality (WHO, 2008).

FBP promote foodborne diseases (FBDs) directly (by pathogens) and indirectly (by bacterial mycotoxins and endotoxins) (Martinovic et al., 2016), which can have disastrous health and financial consequences in both developed and poor nations (Pires et al., 2012; EFSA, 2014; Henao et al., 2015). Diarrheal diseases are reason for a sizable proportion of FBDs, with a disproportionate effect on children (Pires et al., 2015). FBDs are typically caused by FBPs such as bacteria, viruses, parasites, and few fungi.

Important factors to the development of pathogens include human hosts, animal hosts, and their links with humans, the pathogen itself, and the environment, including food production, processing, and storage. For illustration, fluctuations in the proneness of the host due to malnutrition, age, and other conditions lead to new infections in vulnerable populations. Genetic exchange or mutations in species may lead to new strains with the ability to cause disease. Introduction to emerging pathogens through variations in eating habits, climate, mass production, food processing, and intensified globalization of food supply may trigger pathogens to evolve in new demographics or new geographical regions.

Foodborne diseases are a source of substantial sickness and mortality in human beings especially in immune-compromised individuals and also in infants (Stein et al., 2007; Tauxe et al., 2010). Viruses, bacteria, and parasites are liable for many foodborne diseases (Scallan et al., 2011). Therefore, in many nations, sporadic infections or eruptions are investigated continuously. Technological innovation has also led to an upsurge in the spread of FBP crossways international borders, adversely affecting heavily on trade and food safety (Frank et al., 2011; Bernard et al., 2014) that led to the establishment of intricate restrictions by individual governments and international organizations to strengthen food safety. As a result, various food production systems are burdened with the accountability of formulating food appropriate for human consumption. Foodborne pathogens are a major threat to food safety (Bintsis, 2017). Nearly each year, one-third of people in developing countries are infected by FBP reported by the WHO. This figure is considerably higher in developed countries. The price tag and obligation of foodborne illness are extraordinary and are of global concern. Global trade means that tainted food is likely to spread quickly.

Food should not contain any infectious micro-organisms (bacteria, viruses, amoebae, fungi, algae, yeast, or giardia), parasites (e.g., worm eggs, larvae, or cysts), toxins or poisons, or toxic chemicals such as pesticides. Bad hand hygiene habits and viruses (not just bacteria) are also considered significant risk factors for foodborne diseases (Rolando, 2011). Bacteria that flourish on food contamination before feasting can yield toxins that can yield illness.

7.4.2 AVOIDING FOODBORNE ILLNESS

Foodborne microbes are typically destroyed or managed by cooking or freezing. Many foodborne diseases can be avoided by preventing cross-contamination. This can be done by preserving cooked and raw food separately, preparing food precisely and cleaning hands earlier and subsequently processing raw food. These food hygiene activities must be implemented during the processing, transportation, and preparation of food, to lessen the growth and spread of pathogens. Food hygiene law shields all food companies, together with cooking, producers, distributors, and supermarkets. To control them, excellent personal, and food hygiene, sound agricultural practice, post-harvest measures, and appropriate human sewage treatment is still required (Tuladhar et al., 2015; Manuela et al., 2017). However, urea and citric acid augmented formulations for hand sanitizers are more efficient against viruses (Ionidis et al., 2016). Steady cleaning of hands with water and soap, and monitoring of pathogens in foods, will certainly be useful to reduce the incidence of foodborne illnesses (Pal, 2020).

7.5 GLOBAL FOOD SAFETY REGULATIONS

Foodborne diseases have been a significant concern in human history. International agencies such as FAO, The International Organization for Standardization (ISO), Codex Standards, Sanitary, and Phytosanitary (SPS) developed by the World Trade Organization (WTO), Food Safety Standards Authority of India (FSSAI) and the Hazard Analysis Critical Control Point (HACCP) are therefore critical in carrying out risk analysis and compliance to accomplish an acceptable level of protection for human health, the interests of consumers, including fair practices in all kinds of food trade. There are more nearly 20 frameworks for food safety regulation, and also several trade standards, wholesale trading systems, and certification support systems. Some widely applied systems are discussed below.

7.5.1 ISO 9000

The International Standardization Organization (ISO) is a global non-governmental organization (NGO) encompassing national standards organizations (ISO members). ISO 9000 is an international standard for quality standards. The quality assurance (QA) of ISO 9000 is a quality standard program with a key emphasis on identifying and preventing non-compliance during the manufacturing and supply cycle and preventing its recurrent presence. The goal of ISO is to facilitate the implementation of calibration and coupled events on the earth through a commitment to promoting the international exchange of products and services and improving coordination in the areas of cultural, scientific, technical, and economic activity. The ISO was initially adopted by the food industry. Subsequently, it was applied to other fields, such as the processing of fresh fruit and vegetables, packaging, and distribution. ISO includes a fair evaluation of organizational structure; administrative, operating procedures; personnel, equipment, material resources; working areas, facilities, and processes; compliance with requirements and specifications; documentation and archiving.

7.5.2 FOOD AND AGRICULTURE ORGANIZATION (FAO)

FAO emphasizes the food protection systems of all countries to guarantee that the world's citizens are food-safe in terms of consistency, quantity, safety, and competitiveness. This is vital for the advancement of public health. FAO is a renowned pioneer in designing and transforming global food safety policies into country-level practice. The FSQ Program promotes a holistic and interdisciplinary attitude towards food safety management and comprehensive and viable food chain approaches to particular food safety problems as set out in the FAO Strategy for Enhancing Food Safety Globally. When possible, the FSQ Units work in conjunction with other national and international organizations whose missions and guiding principles are congruent.

7.5.3 WORLD HEALTH ORGANIZATION (WHO)

The WHO's mission is to promote global prevention, identification, and response to public health risks connected with unhealthy food. Safeguarding consumer confidence in their jurisdictions and faith in the healthy supply of food is a consequence that perhaps the WHO seeks to accomplish. The WHO synchronizes with the World Organization for Animal Health (OIE), FAO, and other international organizations to guard food safety across the food chain, from production to consumption. To accomplish this, WHO encourages the Member States to develop the capacity to avoid, identify, and handle foodborne threats by:

- Microbiological and chemical risk assessments based on actual scientific research on the context of international food regulations, strategy, and recommendations, recognized as the Codex Alimentarius, to promote food safety wherever it emanates.
- Evaluating the protection of emerging technologies used throughout food processing, such as genetic engineering and nanotechnology.
- Attempting to build regulatory policies and national food operations and developing appropriate infrastructure to handle food safety risks.
- The FAO and the WHO created the International Food Safety Authorities Network (INFOSAN) to communicate information promptly during food safety emergencies; to encourage healthy food handling through comprehensive disease prevention and awareness-raising initiatives, through the Healthier Food policy and instructional materials.
- Integrate food safety in national policy in conjunction with International Health Regulations and promoting food safety as an integral element of health security.

The WHO Food Safety Program, in conjunction with several other WHO programs and agencies, involves improving food safety processes, encouraging good production techniques, and training distributors and customers on proper food handling. One of the most significant techniques in the prevention of foodborne illnesses is teaching consumers and food managers in safe food management.

7.5.4 CODEX ALIMENTARIUS COMMISSION (CAC)

The Codex Alimentarius Commission (CAC) drafted the Codex Alimentarius (Food Code), which is a compendium of food safety rules, instructions, and guidelines. The Commission, also recognized as the CAC, is a key module of the multiparty FAO/WHO Food Standards Program, to maintain public safety and rational trade practices. It had its first conference in 1963. CAC is an independent international agency established to achieve international standards to protect consumer health and facilitate trade in food and agriculture. Buyers can rely on the quality and safety of the food they purchase, and distributors can rely on that the food they order will comply with their requirements. In 2017, 187 participating countries and one-member organization (European Community) were members of Codex.

CAC has been the primary international point of reference for the implementation of food standards. 'Codex Alimentarius,' which means "Food Code," is the name of the handbook published by the CAC. The framework contains standards, codes of conduct, guidelines, and other suggestions intending to protect consumers and maintain fair trade practices in food. Codex specifications are used by various countries to establish national standards. The key goals are to ensure the safety of consumers and to create a fair business practice in food. The CAC is the most influential in implementing international convergence of food quality and safety requirements. Food standards were established and enforced by the FAO.

It has developed international guidelines for a broad variety of food items and unique specifications for food additives, food pollutants, sanitation, pesticide residues, veterinary drug residues, labeling, etc. Codex research has generated a global understanding of food health, quality, and consumer security issues and has established international consensus about how to handle them scientifically using a risk-based approach. The standards of food quality and health have been evaluated on an ongoing basis at the international level.

7.5.5 SANITARY AND PHYTOSANITARY MEASURES (SPS) AND TECHNICAL BARRIERS TO TRADE (TBT) AGREEMENTS

Sanitary and phytosanitary (SPS) programs are confinement and biosecurity intercessions introduced to safeguard plant, animal or human life or safety from threats emerging from the introduction, establishment, and dissemination of pests and diseases and hazards of pesticides, pollutants, and contaminants in food and feed. Such initiatives are regulated by the Agreement on the implementation of SPS measures of the WTO and its committee on SPS measures. The SPS Agreement lays the foundation for regulations to WTO members all through the conception, execution, and integration of sanitary (human or animal life or health) and Phytosanitary (plant life or health) processes that may affect trade.

The SPS strategy may be defined as any method used to safeguard plant, animal, or human life in contrast to a diversity of hazards. Engler et al. (2012) highlighted criteria (viz., hygiene, labeling, packaging, tolerance limits for pesticide residues and contaminants, etc.), items of SPS, and quality-related standards and regulations. According to WTO regulations, all the member nations are requisite to publish their regulations. The member countries must consider other country's SPS steps as equal. The agreement promotes the use of international standards, guidelines, or guidance wherever they arise and defines Codex requirements (regarding pesticide residues, food additives, veterinary drugs, pollutants, analytical, and sampling procedures and hygiene codes and guidelines) to be compliant with the provisions of the SPS.

7.5.6 TECHNICAL BARRIERS TO TRADE (TBT)

The motive of the Technical Barriers to Trade (TBT) contract guarantees that technical rules, requirements, and compliance testing processes are non-discriminatory and do not establish unwanted trade barriers. At the very same time, it respects the WTO members' right to take steps to reach concrete policy goals including the security of human health and safety or environmental protection. The TBT Agreement actively encourages participants to focus their actions on international standards as a way of promoting trade. It also aims, along with its regulations on transparency, to establish a reliable trading climate. The TBT Agreement specifies that the technical restrictions on conventional quality factors, unfair practices, packaging, labeling, etc., implemented by countries should not be more stringent on imported goods than they are on domestically manufactured products.

7.5.7 HAZARD ANALYSIS AND CRITICAL CONTROL POINT (HACCP)

HACCP is a management method that ensures food safety by identifying and monitoring biological, chemical, and physical risks throughout the manufacturing, processing, and management of raw materials, as well as the manufacture, distribution, and use of finished goods. This approach may be used throughout the entire food supply chain, from manufacturer to end-user, resulting in increased food safety, more productive use of resources and prompt responses to food safety issues. For the program to be successful, a team approach is required, as well as the engagement of management and the employees. The HACCP technique is compatible with all other quality management systems (QMS), including organizations, with the ISO 9000 series from the ISO (FAO, 1997). The following are the principles of HACCP:

- **Principle 1**: Execute risk evaluation
- **Principle 2**: Set Vital Control Points (CCPs)
- **Principle 3**: Develop critical guidelines
- **Principle 4**: Create a CCP control framework
- **Principle 5**: Establish disciplinary action to be undertaken when shows that a specific CCP is not to be controlled
- **Principle 6**: Develop testing procedures to ensure that the HACCP framework is functioning properly
- **Principle 7**: Create documents relating to and applying all procedures and records applicable to these principles

7.6 CONCLUSION

Food security, means an assurance that, by providing adequate hygienic conditions, food is not harmful to customers, ensuring high-quality food across the whole process, from raw ingredients to food manufacturing, stocking, and consumption. Food safety's economic influence on food enterprises has been emphasized. Foodborne illness outbreaks must be monitored for a long time to uncover trends in the foods, locations, and pathogens involved. Foodborne illness contributes to global health problems, putting everyone at risk. Diseases are more common among pregnant women, infants, and the elderly who have an underlying sickness. Foodborne illness feeds a rancorous cycle of diarrhea and malnutrition, endangering the nutritional status of the most liable.

Food safety a public health priority should be made by the governments they play a key role in rising policies and regulatory frameworks, and launching and implementing potent food safety systems. At any step of manufacturing and delivery, food will be contaminated and food producers will have the main responsibility. Evaluators should implement fundamental hygiene standards in order to preserve the health of their individuals during buying, selling, and preparing food.

Foodborne illnesses are a worldwide concern, and a unified and common strategy is a necessity for identifying and controlling any developing foodborne problems that affect the human health and international trade of all nations and of important international organizations. Despite their complexity in analysis, biology, and epidemiology, most foodborne diseases are preventable. A mix of knowledge and expertise is definitely needed across disciplines. Continued efforts to avoid food contamination on-farm, in processing, in restaurants, and in homes should be made by public health authorities, regulatory bodies as well as by the food sector and consumers. The number of incidents of foodborne diseases might be reduced if appropriate food safety education programs were implemented for all individuals concerned.

KEYWORDS

- **food safety**
- **foodborne diseases**
- **foodborne pathogens**
- **human illness**
- **polychlorinated biphenyls**
- **public health**

REFERENCES

Altekruse, S. F., & Swerdlow, D. L., (1996). The changing epidemiology of foodborne diseases. *American Journal of the Medical Sciences*, *311*(1), 23–29. doi:10.1097/00000441-199601000-00005.

Bacon, R. T., & Sofos, J. N., (2003). Characteristics of biological hazards in foods. In: Schmidt, R. H., & Rodrick, G. E., (eds.), *Food Safety Handbook* (pp. 157–195). John Wiley and Sons: New Jersey.

Bernard, H., Faber, M., Wilking, H., et al., (2014). Outbreak investigation team. *Euro Surveillance*, *19*(8), 20719.

Bintsis, T., (2017). Foodborne pathogens. *AIMS Microbiology*, *3*(3), 529–563. doi: 10.3934/microbiol.2017.3.529.

Buchanan, R. L., Goris, L. G. M., Hayman, M. M., et al., (2017). A review of *Listeria monocytogenes*: An update on outbreaks, virulence, dose-response, ecology, and risk assessments. *Food Control*, *75*, 1–13.

CDC (Centre for Disease Control and Prevention), (2012). *What is a Foodborne Disease Outbreak and Why Do They Occur*. Available from: http://www.cdc.gov/foodsafety/facts.html#whatisanoutbreak (accessed on 11 February 2022).

CDC (Centre for Disease Control and Prevention), (2013). Multistate outbreak of *Campylobacter jejuni* infections associated with undercooked chicken livers – northeastern United States. *Centers for Disease Control and Prevention Morbidity and Mortality Weekly Report*, *62*, 874–876.

Chang, H. J., Lee, J. H., Han, B. R., Kwak, T. K., & Kim, J., (2011). Prevalence of the levels of *Bacillus cereus* in fried rice dishes and its exposure assessment from Chinese-style restaurants. *Food Science and Biotechnology*, *20*, 1351–1359. doi: 10.1007/s10068-011-0186-3.

Chiapponi, C., Pavoni, E., Bertasi, B., et al., (2014). Isolation and genomic sequence of hepatitis A virus from mixed frozen berries in Italy. *Food and Environmental Virology*, *6*, 202–206.

Chlebicz, A., & Śliżewska, K., (2018). Campylobacteriosis, salmonellosis, yersiniosis, and listeriosis as zoonotic foodborne diseases: A review. *International Journal of Environmental Research and Public Health*, *15*(5), 1–28.

Chon, J. W., Yim, J. H., Kim, H. S., et al., (2015). Quantitative prevalence and toxin gene profile of *Bacillus cereus* from ready-to-eat vegetables in South Korea. *Foodborne Pathogens and Disease*, *12*, 795–799. doi: 10.1089/fpd.2015.1977.

Cuthbert, J. A., (2001). Hepatitis A: Old and new. *Clinical Microbiology Reviews*, 14, 38–58.

De Oliveira, M. A., De Souza, V. M., Bergamini, A. M. M., & De Martinis, E. C. P., (2011). Microbiological quality of ready-to-eat minimally processed vegetables consumed in Brazil. *Food Control*, *22*, 1400–1403. doi: 10.1016/j.foodcont.2011.02.020.

Delbrassinne, L., Andjelkovic, M., Dierick, K., Denayer, S., Mahillon, J., & Van, L. J., (2012). Prevalence and levels of *Bacillus cereus* emetic toxin in rice dishes randomly collected from restaurants and comparison with the levels measured in a recent foodborne outbreak. *Foodborne Pathogens and Disease*, *9*, 809–814. doi: 10.1089/fpd.2012.1168.

Delbrassinne, L., Botteldoorn, N., Andjelkovic, M., Dierick, K., & Denayer, S., (2015). An emetic *Bacillus cereus* outbreak in a kindergarten: Detection and quantification of critical levels of cereulide toxin. *Foodborne Pathogens and Disease*, *12*, 84–87. doi: 10.1089/fpd.2014.1788.

Delibato, E., Luzzi, I., Pucci, E., Proroga, Y. T. R., Capuano, F., & De Medici, D., (2018). Fresh produce and microbial contamination: Persistence during the shelf life and efficacy of domestic washing methods. *Annali dell'Istituto Superiore di Sanita*, *54*(4), 358–363.

Digirolamo, R., Liston, J., & Matches, J. R., (1970). Survival of virus in chilled, frozen, and processed oysters. *Applied and Environmental Microbiology*, *20*, 58–63.

Edwards, D. S., Milne, L. M., Morrow, K., et al., (2013). Campylobacteriosis outbreak associated with consumption of undercooked chicken liver pâte in the East of England, identification of a dose-response risk. *Epidemiology of Infections Journal*, *142*, 352–357.

EFSA (European Food Safety Authority), (2014). *Use of Whole Genome Sequencing (WGS) of Foodborne Pathogens for Public Health Protection*. Scientific Colloquium Summary Report, Parma: Italy. Available at: http://www.efsa.europa.eu/sites/default/files/corporate_publications/files/743e.pdf (accessed on 11 February 2022).

Engler, A., Nahuelhual, L., Cofre, G., & Barrena, J., (2012). How far from harmonization are sanitary, phytosanitary and quality-related standards? An exporter's perception approach. *Food Policy*, *37*(2), 162–170.

Estes, M. K., Prasad, B. V., & Atmar, R. L., (2006). Noroviruses everywhere: Has something changed?. *Current Opinion in Infectious Diseases*, *19*, 467–474.

Facciola, A., Riso, R., Avventuroso, E., Visalli, G., Delia, S. A., & Lagana, P., (2017). *Campylobacter*: From microbiology to prevention. *Journal of Preventive Medicine and Hygiene*, *58*, E79–E92.

FAO (Food and Agriculture Organization), (1997). *Hazard Analysis and Critical Control Point (HACCP) System and Guidelines for its Application*. Available at: http://www.fao.org/docrep/005/y1579e/y1579e03.htm (accessed on 11 February 2022).

FAO (Food and Agriculture Organization), (2003). *Assuring Food Safety and Quality: Guidelines for Strengthening National Food Control Systems*. Rome: Italy.

FAO/WHO, (1996). *Report of the FAO/WHO Workshop on Assessment of Fermentation as a Household Technology for Improving Food Safety*. World Health Organization, Geneva.

FDA (Food and Drug Administration), (2012). *Bad Bug Book, Foodborne Pathogenic Microorganisms and Natural Toxins*. Available from: https://www.fda.gov/Food/FoodborneIllnessContaminants/CausesOfIllnessBadBugBook/ (accessed on 11 February 2022).

Frank, C., Faber, M. S., Askar, M., et al., (2011). Large and ongoing outbreak of haemolytic uraemic syndrome, Germany. *Euro Surveillance*, *16*(21), 19878.

Gao, T., Ding, Y., Wu, Q., et al., (2018). Prevalence, virulence genes, antimicrobial susceptibility, and genetic diversity of *Bacillus cereus* isolated from pasteurized milk in China. *Frontiers in Microbiology*, *9*, 533. doi: 10.3389/fmicb.2018.00533.u.

Garcia, A., Fox, J. G., & Besser, T. E., (2010). Zoonotic enterohemorrhagic *Escherichia coli:* A one health perspective. *International League of Associations for Rheumatology*, *51*, 221–232.

Glass, R. I., Parashar, U. D., & Estes, M. K., (2009). Norovirus gastroenteritis. *New England Journal of Medicine*, *361*, 1776–1785.

Griffiths, M. W., & Schraft, H., (2017). *Bacillus cereus* food poisoning. In: Dodd, C. E. R., Aldsworth, T. R., Stein, A., & Cliver, D. O., (eds.), *Foodborne Diseases* (pp. 395–405). Academic Press: London. doi: 10.1016/b978-0-12-385007-2.00020-6.

Hagos, Y., Berhe, M., & Gugsa, G., (2019). Campylobacteriosis: Emphasis on its status as foodborne zoonosis in Ethiopia. *International Journal of Tropical Diseases*, *7*, 1–8.

Halliday, M. L., Lai, L. Y., Zhou, T. K., et al., (1991). An epidemic of hepatitis A attributable to the ingestion of raw clams in Shanghai, China. *Journal of Infectious Diseases*, *164*, 852–859.

Hansson, I., Sandberg, M., Habib, I., Lowman, R., & Engvall, E. O., (2018). Knowledge gaps in control of *Campylobacter* for prevention of Campylobacteriosis. *Transboundary and Emerging Diseases*, *65*(1), 30–48.

Henao, O. L., Jones, T. F., Vugia, D. J., Griffin, P. M., & Network, F. D. A. S., (2015). Foodborne diseases active surveillance network – 2 decades of achievements, 1996–2015. *Emerging Infectious Diseases*, *21*, 1529–1536.

Hénock, B. N. Y., & Dovie, D. B. K., (2007). Diarrheal diseases in the history of public health. *Archives of Medical Research*, *38*, 159–163.

Hermans, D., Van, D. K., Martel, A., et al., (2011). Colonization factors of *Campylobacter jejuni* in the chicken gut. *Veterinary Research*, *42*, 82.

Horrocks, S. M., Anderson, R. C., Nisbet, D. J., & Ricke, S. C., (2009). Incidence and ecology of *Campylobacter jejuni* and *E. coli* in animals. *Anaerobe*, *15*, 18–25.

Humphrey, S., Chaloner, G., Kemmett, K., et al., (2014). *Campylobacter jejuni* is not merely a commensal in commercial broiler chickens and affects bird welfare. *mBio*, *5*, e01364–1314.

Hutt, P. B., & Hutt, P. B., (1984). A history of government regulation of adulteration and misbranding of food. *Food Drug and Cosmetic Law Journal*, *39*, 2–73.

ICMSF (The International Commission on Microbiological Specifications for Foods), (1996). *Micro-organisms in Foods 5, Characteristics of Microbial Pathogens*. Kluwer Academic/ Plenum Publishers: New York.

Inns, T., Foster, K., & Gorton, R., (2010). Cohort study of a campylobacteriosis outbreak associated with chicken liver parfait, United Kingdom. *Euro Surveillance*, *15*, 19704.

Ionidis, G., Hubscher, J., Jack, T., et al., (2016). Development and virucidal activity of a novel alcohol-based hand disinfectant supplemented with urea and citric acid. *BMC Infectious Diseases*, *16*, 77.

Iturriza-Gomara, M., & O'Brien, S. J., (2016). Foodborne viral infections. *Current Opinion in Infectious Diseases*, *29*, 495–501.

John, M. M., John, M. K., David, M. K., et al., (2020). A review on the present advances on studies of toxoplasmosis in Eastern Africa. *BioMed Research International*, *2020*, 1–12. doi.org/10.1155/2020/7135268.

Kaakoush, N. O., Castano-Rodriguez, N., Mitchell, H. M., & Man, S. M., (2015). Global epidemiology of *Campylobacter* infection. *Clinical Microbiology Reviews*, *28*, 687–720.

Koff, R. S., Grady, G. F., Chalmers, T. C., et al., (1967). Viral hepatitis in a group of Boston hospitals – importance of exposure to shellfish in a nonepidemic period. *New England Journal of Medicine*, *276*, 703–710.

Koopmans, M., (2008). Progress in understanding norovirus epidemiology. *Current Opinion in Infectious Diseases*, *21*, 544–552.

Kotloff, K. L., Riddle, M. S., Platts-Mills, J. A., Pavlinac, P., & Zaidi, A. K. M., (2018). Shigellosis. *Lancet*, *391*(10122), 801–812.

Lahti, E., Löfdahl, M., Agren, J., et al., (2017). Confirmation of a Campylobacteriosis outbreak associated with chicken liver pâtè using PFGE and WGS. *Zoonoses and Public Health*, *64*, 14–20.

Lee, S., Cappato, P. L., Guimarães, T. J., et al., (2019). *Listeria monocytogenes* in milk: Occurrence and recent advances in methods for inactivation. *Beverages*, *5*(1), 1–14.

Lemon, S. M., Ott, J. J., Van, D. P., & Shouval, D., (2018). Type A viral hepatitis: A summary and update on the molecular virology, epidemiology, pathogenesis and prevention. *Journal of Hepatology*, *68*, 167–184. doi: 10.1016/j.jhep.2017.08.034.

Li, J., Cui, Z., Qi, M., & Zhang, L., (2020). Advances in cyclosporiasis diagnosis and therapeutic intervention. *Frontiers in Cellular and Infection Microbiology*, *10*, 43. doi: 10.3389/fcimb.2020.00043.

Manuela, C. F., Arisseto-Bragotto, A. P., & Jane, M. B., (2017). Review food quality, foodborne diseases, and food safety in the Brazilian food industry. *Food Quality and Safety*, *1*, 13–27. doi: 10.1093/fqs/fyx003.

Marrollo, R., (2016). *Bacillus cereus* foodborne disease. In: Savini, V., (ed.), *The diverse faces of Bacillus cereus* (pp. 61–72). Academic Press: London. doi: 10.1016/ b978-0-12–801474-5.00005-0.

Martinovic, T., Andjelkovic, U., Gajdosik, M. S., Resetar, D., & Josic, D., (2016). Foodborne pathogens and their toxins. *Journal of Proteomics*, *147*, 226–235. doi: 10.1016/j. jprot.2016.04.029.

Mead, P. S., Slutsker, L., Dietz, V., et al., (1999). Food-related illness and death in the United States. *Emerging Infectious Diseases*, *5*, 607–625.

Merzougui, S., Lkhider, M., Grosset, N., Gautier, M., & Cohen, N., (2014). Prevalence, PFGE typing, and antibiotic resistance of *Bacillus cereus* group isolated from food in Morocco. *Foodborne Pathogens and Disease*, *11*, 145–149. doi: 10.1089/fpd.2013.1615.

Mitscherlich, E., & Marth, E. H., (1984). *Microbial Survival in the Environment: Bacteria and Rickettsiae Important in Human and Animal Health*. Berlin: Springer-Verlag.

Montano-Remacha, C., Ricotta, L., Alfonsi, V., et al., (2014). Hepatitis A outbreak in Italy, 2013: A matched case-control study. *Euro Surveillance*, *19*, 20906.

Murray, C. J. L., Vos, T., Lozano, R., et al., (2012). Disability-adjusted life years (DALYs) for 291 diseases and injuries in 21 regions, 1990–2010: A systematic analysis for the global burden of disease study 2010. *Lancet*, *380*, 2197–2223.

Mustafa, S. A., & Inanc, L. A., (2018). Antibiotic resistance of *Escherichia coli* O157: H7 isolated from chicken meats. *Journal of Agriculture and Nature*, *21*(1), 7–12.

Ng, Y. F., Wong, S. L., Cheng, H. L., Yu, H. F., & Chan, S. W., (2013). The microbiological quality of ready-to-eat food in Siu Mei and Lo Mei shops in Hong Kong. *Food Control*, *34*, 547–553. doi: 10.1016/j.foodcont.2013.05.018.

Osawa, R., Okitsu, T., Morozumi, H., & Yamai, S., (1996). Occurrence of urease-positive *Vibrio parahaemolyticus* in Kanagawa, Japan, with specific reference to presence of thermostable direct hemolysin (TDH) and the TDH-related-hemolysin genes. *Applied and Environmental Microbiology*, *62*, 725–727.

Pal, M., (2020). Infectious hepatitis: A foodborne viral disease of public health importance. *Food and Drink Industry*, *1*, 49–50.

Pires, S. M., Fischer-Walker, C. L., Lanata, C. F., Devleesschauwer, B., Hall, A. J., Kirk, M. D., et al., (2015). Aetiology-specific estimates of the global and regional incidence and mortality of diarrhoeal diseases commonly transmitted through food. *PLoS One*, *10*, e0142927. doi: 10.1371/journal.pone.0142927.

Pires, S. M., Vieira, A. R., Perez, E., Wong, D. L. F., & Hald, T., (2012). Attributing human foodborne illness to food sources and water in Latin America and the Caribbean using data from outbreak investigations. *International Journal of Food Microbiology*, *152*, 129–138. doi: 10.1016/j.ijfoodmicro.2011.04.018.

Prüss-Ustün, A., Vickers, C., Pascal, H., & Roberto, B., (2011). Knowns and unknowns on burden of disease due to chemicals: A systematic review. *Environmental Health*, *10*, 9. doi: 10.1186/1476-069X-10-9.

Rahman, T., Rahman, A., & Chakraborty, S., (2018). Infection of *Toxoplasma gondii* in humans and livestock animals: An emerging silent threat for Bangladesh. *Open Journal of Medical Microbiology*, *8*(4), 109–117.

Rajkowski, K. T., & Smith, J. L., (2001). Update: Food poisoning and other diseases induced by *Bacillus cereus*. In: Hui, Y. H., Pierson, M. D., & Gorham, J. R., (eds.), *Foodborne Disease Handbook* (pp. 61–76). Markel Dekker: New York.

Raymond, P., Houard, E., Denis, M., & Esnault, E., (2019). Diversity of *Yersinia enterocolitica* isolated from pigs in a French slaughterhouse over 2 years. *Microbiology Open*, *8*(6), e00751.

Robertson, L. J., (2018). Parasites in food: From a neglected position to an emerging issue. *Advances in Food and Nutrition Research*, *86*, 71–113.

Rolando, R., (2011). *Hand Washing a Big Deal?* SA Food Review. http://www.foodreview.co.za/safety-andhygiene/32-hand-washing-a-big-deal (accessed on 11 February 2022).

Rossler, E., Signorini, M. L., Romero-Scharpen, A., et al., (2019). Meta-analysis of the prevalence of thermotolerant *Campylobacter* in food-producing animals worldwide. *Zoonoses and Public Health*, *66*, 359–369.

Scallan, E., Hoekstra, R. M., Angulo, F. J., et al., (2011). Foodborne illness acquired in the United States - major pathogens. *Emerging Infectious Diseases*, *17*(1), 7–15.

Schaffner, N., Zumstein, J., & Parriaux, A., (2004). Factors influencing the bacteriological water quality in mountainous surface and groundwaters. *Acta Hydrochimica et Hydrobiologica*, *32*, 225–234.

Sean, F. A., Stern, N. J., Fields, P. I., & Swerdlow, D. L., (1999). *Campylobacter jejuni*. An emerging foodborne pathogen. *Emerging Infectious Diseases*, *5*, 28–35.

Shiningeni, D., Chimwamurombe, P., Shilangale, R., & Misihairabgwi, J., (2019). Prevalence of pathogenic bacteria in street vended ready-to-eat meats in Windhoek, Namibia. *Meat Science*, *148*, 223–228. doi: 10.1016/j.meatsci.2018.05.014.

Skarp, C. P. A., Hanninen, M. L., & Rautelin, H. I. K., (2016). Campylobacteriosis: The role of poultry meat. *Clinical Microbiology and Infection*, *22*, 103–109.

Song, Z., Zhao, Q., Zhu, L., Zhang, Z., Jiang, L., & Huang, H., (2019). Draft genome sequence of multidrug-resistant β-lactamase-producing *Bacillus cereus* S66 isolated from China. *Journal of Global Antimicrobial Resistance*, *17*, 23, 24. doi: 10.1016/j.jgar.2019.02.019.

Stein, C., Kuchenmuller, T., Henrickx, S., et al., (2007). The global burden of disease assessments - WHO is responsible? *PLoS Neglected Tropical Diseases*, *1*, e161.

Tauxe, R. V., Doyle, M. P., Kuchenmüller, T., Schlundt, J., & Stein, C. E., (2010). Evolving public health approaches to the global challenge of foodborne infections. *International Journal of Food Microbiology*, *139*(1), S16–S28.

Tegegne, F. M., (2019). Epidemiology of *Salmonella* and its serotypes in human, food animals, foods of animal origin, animal feed and environment. *Journal of Food, Nutrition and Health*, *2*(1), 7–14.

Tewari, A., Singh, S. P., & Singh, R., (2015). Incidence and enterotoxigenic profile of *Bacillus cereus* in meat and meat products of Uttarakhand, India. *Journal of Food Science and Technology*, *52*, 1796–1801. doi: 10.1007/s13197-013-1162-0.

Tong, S. Y., Davis, J. S., Eichenberger, E., Holland, T. L., & Fowler, V. G., (2015). *Staphylococcus aureus* infections: Epidemiology, pathophysiology, clinical manifestations, and management. *Clinical and Microbiology Reviews*, *28*(3), 603–661.

Tuladhar, E., Hazeleger, W. C., Koopmans, M., Zwietering, M. H., Duizer, E., & Beumer, R. R., (2015). Reducing viral contamination from finger pads: Handwashing is more effective than alcohol-based hand disinfectants. *Journal of Hospital Infection*, *90*, 226–234.

Vutova, K., Velev, V., Chipeva, R., et al., (2020). Clinical and epidemiological descriptions from trichinellosis outbreaks in Bulgaria. *Experimental Parasitology*, *212*, 107874.

Wang, W., Baloch, Z., Jiang, T., et al., (2017). Enterotoxigenicity and antimicrobial resistance of *Staphylococcus aureus* isolated from retail food in China. *Frontiers in Microbiology*, *8*, 1–11.

WHO, (2007a). *Countries Urged to be More Vigilant About Food Safety*. https://www.who.int/news/item/19-07-2007-countries-urged-to-be-more-vigilant-about-food-safety (accessed on 18 February 2022).

WHO, (2007b). *Food Safety and Foodborne Illnesses*. Fact Sheet, 237. https://foodhygiene2010.files.wordpress.com/2010/06/who-food_safety_fact-sheet.pdf (accessed on 18 February 2022).

WHO, (2008). *WHO Initiative to Estimate the Global Burden of Foodborne Diseases*. A Summary Document. Geneva, WHO.

WHO, (2015). *WHO Estimates of the Global Burden of Foodborne Diseases, Technical Report*. Geneva. https://apps.who.int/iris/bitstream/handle/10665/199350/9789241565165_eng.pdf (accessed on 18 February2022).

Wieczorek, K., Wołkowicz, T., & Osek, J., (2018). Antimicrobial resistance and virulence-associated traits of *Campylobacter jejuni* isolated from poultry food chain and humans with diarrhea. *Frontiers in Microbiology*, *9*, 1–11.

Yu, P., Yu, S., Wang, J., et al., (2019). *Bacillus cereus* isolated from vegetables in China: Incidence, genetic diversity, virulence genes, and antimicrobial resistance. *Frontiers in Microbiology*, *10*, 948. doi: 10.3389/fmicb.2019.00948.

Zelalem, A., Sisay, M., Vipham, L. J., Abegaz, K., Kebede, A., & Terefe, Y., (2019). The prevalence and antimicrobial resistance profiles of bacterial isolates from meat and meat products in Ethiopia: A systematic review and meta-analysis. *International Journal of Food Contamination*, *6*(1), 1–14.

CHAPTER 8

Foodborne Organisms: General Characteristics, Virulence Factors, and Clinical Manifestations

HETTIPALA ARACHCHIGE DARSHANEE RUWANDEEPIKA, DENETHTHI NIMESH MADUSHANKA, and THOTAWATHTHAGE SANJEEWA PRASAD JAYAWEERA

ABSTRACT

Foodborne diseases play a significant threat in society, having tangible and intangible costs. Major impacts of foodborne diseases are on public health leading to gastrointestinal (GI) and other clinical symptoms in consumers. Ingestion of raw, inadequately processed or foods contaminated by microbial pathogens including bacteria, viruses, toxins, or other toxic agents. Due to the present lifestyle of people, where a variety of manufactured, ready-to-eat food products are daily consumed in large quantities, the knowledge on the occurrence of foodborne diseases (FBDs) by foodborne organisms has gained significant attention in public health protection. This chapter discusses different foodborne organisms focusing on general characteristics, virulence factors, clinical manifestation, etc. This highlights the importance of achieving better health education in terms of controlling or minimizing the occurrence of foodborne diseases.

8.1 INTRODUCTION

The growth of the global population, urbanization, and an increase in the purchasing power of the people, led to significantly high global demand for

Food Microbial and Molecular Biology: From Fundamentals to Applications. Saher Islam, Devarajan Thangadurai, Jeyabalan Sangeetha, Zaira Zaman Chowdhury (Eds.)

food in the last two decades. Due to the rising demand for food, the global food supply has become a booming business, and it has created a severe rush in food production chains. It started locally and then quickly expanded into a global-scale phenomenon. With an increasing demand for food, the global food supply has to undergo strict food safety standards. These safety standards measure preliminary based on common causes of foodborne diseases, varying from abusive food production to unhygienic storage of food products. Foodborne diseases are bound with public health, food security aspects. Not only that, it is a threat to the economy at global level and subsequently it affects badly the socio-economic growth. In 2010 there were 600 million cases of foodborne illnesses caused by 31 foodborne hazards (composed of hazards leading to diarrheal diseases; the most common one, non-diarrheal diseases/ invasive infections and chemical hazards). There were 420,000 deaths reported in 2010 throughout the world. Children under 5 years of age are the most affected age group with foodborne diseases. Virus including Norovirus (NoV), bacteria including *Campylobacter* spp., Enteropathogenic *Escherichia coli*, Enterotoxigenic *E. coli* (ETEC), Shiga toxin-producing *E. coli* (STEC), Non-typhoidal *Salmonella enterica*, *Shigella* spp., *Vibrio cholera*, parasites including *Cryptosporidium* spp., *Entamoeba histolytica*, and *Giardia* spp. are the organisms which lead to the development of diarrheal diseases. Virus including hepatitis A virus (HAV), bacteria including *Brucella* spp. and *Listeria monocytogenes*, *Mycobacterium bovis*, *Salmonella* (*S. typhi*, *S. paratyphi*), protozoal causes including *Toxoplasma gondii*, cestodes including *Echinococcus granulosus*, *Echinococcus multilocularis*, *Taenia solium*, nematodes including *Ascaris* spp., *Trichinella* spp., trematodes including *Fasciola* spp. and *Clonorchis sinensis*. Flukes such as *Paragonimus* spp. and *Opisthorchis* spp., leads to invasive infections. Apart from the organisms' toxins and chemicals namely dioxins, cassava cyanide and aflatoxin, also lead to foodborne hazards. Among the foodborne hazards, the foodborne pathogens (FBP) can be categorized into two major groups of organisms, which cause diarrheal diseases and the organisms which cause invasive infections. NoV and *Campylobacter* spp. are the most prevailed causes for foodborne diseases. Other frequent causes which cause the foodborne diseases are HAV, *Ascaris* spp. (helminth) and *Salmonella typhi*. Foodborne diarrheal diseases are the main cause of death, and it is encountered for 230,000 of the 420,000. Among these deaths, 59,000 was due to the non-typhoidal *S. enterica* (NTS). *E. coli* also contributed significantly to this count having 37,000 deaths due to enteropathogenic *E. coli* (EPEC) whereas 26,000 deaths due to ETEC. NoV was the most common among the viral causes and it accounted for 35,000 deaths. *Salmonella typhi*, *Taenia solium*, hepatitis A, aflatoxin was reported as the non-diarrheal causes of deaths associated with foodborne illnesses. The deaths reported, according to Havelaar et al. (2015) and WHO (2015), was as 52,000, 28,000, 28,000, and 20,000, respectively for causes indicated. Food safety management practices are continuously used by the food industries in the developing world since 1945, i.e., the end of the second world war. Though the good practices have succeeded in controlling foodborne illnesses on a global scale to some extent, yet, these safety standards have pinpointed the weaknesses in the whole global food supply chain. Safety standards themselves prove that single misconduct in food production can lead to a spread of large-scale foodborne diseases throughout the world. Food can be contaminated and can be harboring many pathogens due to various reasons. Food safety standards have been established and materialized. In order to ensure safer food production, the food safety standards have been legalized, keeping the manufactures in line with the safer food production procedures. All these safety protocols have been clearly documented, and industries that are operating in the food business have been strictly advised to keep all their productions in accordance with the provided guidelines. However, foodborne diseases are still playing a vital role in society, having tangible and intangible costs. This highlights the importance of achieving better health education in terms of controlling or minimizing the occurrence of foodborne diseases. FBP and their adverse effects should not be taken lightly into consideration as they have caused devastating scenarios all around the world (Havelaar et al., 2015; WHO, 2015).

8.2 FOODBORNE BACTERIA

8.2.1 CAMPYLOBACTER JEJUNI

Campylobacter enteritis is the most common bacterial gastroenteritis reported worldwide (Nachamkin et al., 2008; Poly and Guerry, 2008). *Campylobacter* was isolated for the first time in 1938 associated with milk-borne outbreak in USA. *Campylobacter* is considered as the most significant or the prevailing foodborne pathogen (bacterial origin) at global level. Transmission of this organism is mainly through the water and food which is contaminated with the organism. Genes of *Campylobacter* is belonging to the Campylobacteriaceae family and Campylobacterales order. *Campylobacter* is a member of class Epsilonproteobacteria and it is composed of 18 species and 6 subspecies (Debruyne et al., 2009). *Campylobacter jejuni* and *C. coli* are the two important species leading to gastrointestinal (GI) illnesses in humans, while *C. jejuni* being the most pathogenic among these two (Dasti et al., 2010). *Campylobacter* is an enteric, fastidious, Gram-negative, oxidase-positive, non-spore-forming bacterium. The width of the bacterium is between 0.2 to 0.8 mm and the length is 0.5 to 8 mm. *Campylobacter* is motile cells with a single flagellum or occasionally multiple. These bacteria have a minimal growth temperature of 30°C, and most *Campylobacter* strains grow at 37°C; *C. jejuni* and *C. coli*, optimum growth temperature ranges from 42 to 45°C. This organism prefers microaerobic environments, where oxygen concentration is between 5% and 10%. *Campylobacter* species are chemoorganotrophs. These organisms are generally found in domestic and wild animals; livestock species, such as pigs, cattle, sheep, and poultry and it can be found as a contaminant of the water. Humans get contact with the organism by the ingestion of contaminated livestock products or the handling of foods of animal origin that were contaminated (Kaakoush et al., 2015). Disease in humans can be a mild, non-inflammatory, self-limiting diarrhea which can develop to severe, inflammatory, bloody diarrhea. It also cared with the abdominal pain which can last for several weeks. Symptoms commonly occur within 1–5 days of *C. jejuni* exposure, and can continue for up to 10 days. Also, it leads to the hemolytic uremic syndrome (HUS), meningitis, pancreatitis, cholecystitis, colitis, and endocarditis. According to the findings by previous studies it has shown that some conditions such as reactive arthritis, peripheral neuropathies, and Miller-Fisher and Guillain-Barrè syndromes can be developed by *C. jejuni* (Nachamkin et al., 2008).

C. jejuni initially colonizes in the small intestine and then moves to the colon, which is the target organ of the organism. Afterward, the organism colonizes in the mucus lining of the gut. *C. jejuni* adheres to epithelial cells by means of different adhesins and their corresponding host cell receptors. *C. jejuni* can disrupt tight junctions of epithelial cells. Then the bacteria enter to the basolateral surface where they reinvade the epithelial cell or be taken up into macrophages. *C. jejuni* are capable of replication inside the macrophages (intracellularly) and induces apoptosis. Subsequently, the activation of the nuclear transcription factor and some signaling pathways are triggered due to the interaction of bacteria with host cells such as macrophages, dendritic cells, and epithelial cells. This activations process leads to the chemokines release where it contributes to inflammatory diarrhea and the infection clearance (Poly and Guerry, 2008). While invading the host cells, *Campylobacter* produces secreted proteins, toxins, and other molecules that can adversely affect the host. These stimulate or inhibit host signal transduction pathways, promote cytokine production, trigger inflammation, change host cell mitogenesis (causing hyperplasia), cause host cell death, or cause water malabsorption/secretion. Ileum and colon are the predilection sites for colonizing the bacteria hence it affects the secretory as well as the absorptive function of the gastrointestinal tract of the host. Thereby it develops the disease called campylobacteriosis with the symptoms of malaise, fever, abdominal pain, and associated with watery diarrhea (Wassenaar and Blaser, 1999). There may be post-infection consequences in some patients following the campylobacteriosis and can affect the nervous system (Guillain–Barré syndrome, GBS; Miller–Fisher syndrome, MFS; Bickerstaff encephalitis) and joints (reactive arthritis, RA). Additionally, the infections with *Campylobacter* spp. leads to inflammatory bowel diseases (IBD) such as ulcerative colitis, Crohn's disease, and irritable bowel syndrome (IBS) (Keithlin et al., 2014).

8.2.2 SALMONELLA SPP.

Salmonella is a widely distributed gram-negative bacterium found in different environments. *Salmonella* is an important foodborne pathogen distributed globally in developing as well as developed countries (Lamas et al., 2018). A second most important or common cause for the gastroenteritis in human is *Salmonella* and it is followed by the *Campylobacter* spp. *Salmonella* was first described by Theobald Smith (1859–1934) and Salmon in 1885 from pigs with hog cholera. It is termed as a "Hog-cholera bacillus," which is now recognized as *Salmonella choleraesuis*. This is transmitted to humans from animals mainly through contaminated food such as eggs, meat, and dairy products. *Salmonella* is belonging to the family Enterobacteriaceae, facultative anaerobe, Gram-negative, oxidase negative, catalase-positive, non-spore-forming bacilli. *S. enterica* and *S. bongori* are the two species of *Salmonella* consists of more than 2,600 serovars (Bhowmick et al., 2011). The *S. enterica* is composed of six different subspecies, *enterica* (I), *salamae* (II), *arizonae* (IIIa), *diarizonae* (IIIb), *houtenae* (IV) and *indica* (VI). *S. bongori*, the other species (V) is consisted with 22 serotypes. This species (*S. bongori*) is less known one and they are mainly associated with cold-blooded animals while human infections are very uncommon (Lamas et al., 2018). *S. enterica* subsp. *enterica* consists of more than 1,531 serovars and some of them (such as *S. typhimurium* and *S. enteritidis*) are clinically significant. Among the cases of human salmonellosis, common organisms involved in heavy clinical burdens globally are *S. typhimurium*, *S. enteritidis*, *S. newport*, *S. derby*, *S. infantis*, *S. schwarzengrund*, *S. heidelberg* and *S. javiana* (Borges et al., 2019). Non-typhoidal salmonellosis (NTS) is most commonly caused by the serovars *Salmonella enterica* serovar Enteritidis and *Salmonella enterica* serovar Typhimurium in human hosts. The majority of human *Salmonella* food poisoning outbreaks are caused by *S. enterica* subsp. *enterica* (Lamas et al., 2018) isolated predominantly from poultry. Two types of disease conditions; typhoid fever and NTS are caused by *Salmonella* spp. Turning to the non-typhoidal *Salmonella* serovars, they cause two main kinds of disease conditions in humans; localized gastroenteritis and invasive bloodstream infection. Localized gastroenteritis saw in individuals with an intact immune system, whereas the second form occurs in individuals with a compromised immune system. Additionally, typhoid fever is the other human disease condition caused by typhoidal *Salmonella* serovars, and it is an invasive infection of the bloodstream (Hiyoshi et al., 2018). *S. typhi* and *S. paratyphi*, cause typhoid fever and clinical manifestations include fever, headache, abdominal pain, and transient diarrhea, which may result in the fatal hepatic, spleen, respiratory, or neurological damage (Dougnon et al., 2017).

Non-typhoidal *Salmonella* (NTS) causes diarrheal disease in humans, which has a mortality rate of around 24% in developing countries (Feasey et al., 2012). There will be a long carrier period subsequently to the infection of *Salmonella* (over 10-week post-infection). These carriers do not show symptoms and can act as reservoirs contributing to the propagation of disease.

The major source of transmission of *Salmonella* to host is through the ingestion of contaminated food or water (Bueno et al., 2012).

Following entry to the body, the organism reaches the stomach and activates acid tolerance response. Upon entry to the small intestine the organism adheres to epithelial cells subsequently it stimulates the signaling pathways which leads to rearrangements of cytoskeletal and disruption of brush border of epithelial cells. This results in *Salmonella*-containing vacuoles (SCV) where this adherence organism can be engulfed, survived, and multiplied (Haraga et al., 2008). There will be a production of several kinds of proinflammatory cytokines (tumor necrosis factor α (TNFα) and Interleukin-8 (IL-8)) in intestinal cells. These factors initiate recruitment and migration of phagocytes into the intestinal lumen (Pontes et al., 2015). Further, in order to prevent the lysosomal enzymes of the host endocytic pathway, *Salmonella* changes the host endocytic trafficking system. *Salmonella* induces *Salmonella*-induced filaments (SIFs), i.e., the formation of a long filamentous membrane and it increase the availability of the nutrient in SCV following the invasion of Salmonella to epithelium of the intestine.

Dendritic cells are taken up through bloodstream to several organs (e.g., spleen, liver) where there is efficient multiplication of the organism. *Salmonella* invade M-cells, which are modified cells of the intestinal wall of the ileum present in Peyer's plaques. M-cells are intimately associated with macrophages, which reside in this tissue. When the epithelial barrier is ruptured, *Salmonella* can invade macrophages and dendritic cells and, from these infected cells, gain systemic circulation. During all stages of this infection, this organism uses specific virulence factors to conduct a more efficient invasion (Herrero-Fresno and Olsen, 2018).

The evading ability of the host environment by the pathogen is due to the virulence of the strain. *Salmonella*, pathogenesis is a kind of complex and it is a multifactorial phenomenon which uses several virulence factors (Skyberg et al., 2006). There are many virulence-associated genes that are encoded for the production of virulence factors, and these genes are located throughout the whole genome, including plasmids, but are mostly clustered on distinct genetic regions, called *Salmonella* pathogenicity islands (SPI) (Wallis and Galyov, 2000). Not only the SPI, there are several virulence factors, virulence regulators (type-3 secretion system (T3SS), quorum sensing) and structural components (fimbriae) that are determining the virulence of the organism (de Jong et al., 2012). Virulence genes of *Salmonella* are clusters in 24 pathogenicity islands, which are crucial in pathogenesis (Espinoza et al., 2017). These horizontally acquired loci encode genes involved in several virulence mechanisms such as the expression of secretion systems, fimbriae, flagella and capsules, serotype conversion and host colonization and subsequent survival within the host (Chen et al., 2019).

8.2.3 ESCHERICHIA COLI

In 1895, Theodor Escherichia first isolated a bacterium described as the *Bacterium coli*, and later, in 1919, the bacterium was renamed as *Escherichia coli* which is highly versatile organisms. *Escherichia coli* belongs to the genus *Escherichia* of the Enterobacteriaceae family, and it is one of the six species in the genus. Most of them are harmless and most abundant in the GI tract of humans and also in warm-blooded animals as commensals having symbiotic relationship with the host, benefiting the host by synthesizing cofactors and competing with the gut invaders to avoid colonization. *E. coli* is a Gram-negative, facultatively anaerobic, motile rod (1–2 μm in length), which grows better at the temperature of 35–37°C. It consists of peritrichous flagella, fimbriae or pili and curli, and some strains have capsules. Despite the beneficial effect on humans and animals, it can also be deadly pathogenic members. Broadly they can be categorized as commensal strains, intestinal pathogenic strains, extraintestinal pathogenic strains (strains leading to extraintestinal sites such as urinary tract, bloodstream, and CNS). Based on the existence of different strains of *E. coli* with variable pathogenic potential they are grouped into different pathovars such as enterohemorrhagic *E. coli* (EHEC), EPEC, enteroinvasive *E. coli* (EIEC; including *Shigella*), ETEC, enteroaggregative *E. coli* (EAEC) and diffusely adherent *E. coli* (DAEC) which are diarrheagenic, and two pathovars uropathogenic *E. coli* (UPEC) and neonatal meningitis *E. coli* (NMEC) which are the most common ExPEC pathovars (Russo and Johnson, 2003; Croxen and Finlay, 2010).

ETEC leads to toxic infection, and it was first reported in 1956 in Calcutta, India. ETEC produces toxins including heat-labile (LT; LT-I and LT-II) and/or heat-stable toxins (ST; STa and STb) following adherence to epithelial cells, but they do not invade epithelial cells. LT produced by ETEC resembles cholera toxin (CT). LT and ST enterotoxin that elicits fluid accumulation and a diarrheal response. Some of the symptoms are vomiting, watery diarrhea, massive dehydration, sunken eyes, and circulatory collapse. Except in some cases, the diarrheic condition is self-limiting that will stay for 3–4 days. But in some cases, there can be fatalities (less than 1%) in young children and infants. EPEC, first pathotype of *E. coli* described, which can cause watery diarrhea, like in ETEC. EPEC adheres to epithelial cells intimately, produces attachment/effacement (A/E) lesion, and is noninvasive. They do not produce any heat-labile (LT), heat-stable (ST), or Shiga toxins (Stx). EPEC adheres to epithelial cells effector molecules are injected into the host cells, thereby activating several cells signaling pathways and promoting a considerable alteration in the epithelial cell's ultrastructure, leading to an attaching and effacing lesion. The integrity of the microvilli structures is disturbed, gradually disappear, and the epithelium loses the ability to absorb nutrients. This group of *E. coli* are reported to be highly invasive. Also, in children and infants, it causes severe inflammation and potentially fatal diarrhea (Bhunia, 2018).

STEC, the group of *E. coli* strains capable of producing Shiga toxin (Stx) are grouped and their most pathogenic subgroup is the enterohemorrhagic *E. coli* (EHEC). EHEC bind strongly to the epithelial cells and this attachment produce attachment/effacement lesions, which stimulates the production of

Stx. EHEC can cause non-bloody and bloody diarrhea and HUS (Karch et al., 2005). Frequently isolated serotype of EHEC *E. coli* is the O157:H7, and the first outbreak was reported in 1982–1983 linked to bloody diarrhea and HUS (Karmali et al., 1983; Riley et al., 1983). Other important non-O157 EHEC serovars are O26, O45, O103, O111, O121, and O145 (Bhunia, 2018).

EIEC causes non-bloody diarrhea and dysentery by invading and multiplying within colonic epithelial cells. The condition is identical to that caused by *Shigella* spp. (Sansonetti, 2002). This group of organisms is incapable of producing toxins such as LT, ST, or Stx.

The bacterial cells will be attached to the colonic epithelial cells and penetration of the enterocytes taken place through the process of endocytosis. Intracellular multiplication of the bacterial cells is taken place subsequently and endocytic vacuole is lysed. This leads to severe inflammation and bloody mucoid diarrhea will be the resulting outcome. The symptoms include profuse diarrhea, abdominal cramp, chills, headache, and fever, though the symptoms diminish after 7–12 days, the carrier stage can be for a long period.

Enteroaggregative *Escherichia coli* (EAEC) was first described in 1983. It leads to acute or persistent watery diarrhea (with or without mucus) in children and is associated with travelers' diarrhea (Wilson et al., 2001). EAEC adhere to epithelial cells and due to the aggregates, appearing like "stacked brick." They do not invade, but there will be a toxin production. EAEC is a pathotype of diarrheagenic *Escherichia coli* (DEC), and the pathogenesis of EAEC involves several steps. Initially, organisms adhere to the mucosal layer of the intestine and it enhances the production of mucus. This adherence forms a thick biofilm on mucosal surface and it will enhance the cytotoxins. The production of cytotoxins enhance the intestinal secretion and damage the intestinal mucosa (Nataro et al., 1995). It produces several types of toxins such as heat-stable, ST-like toxin, EAST (enteroaggregative ST), which responsible for fluid loss, the plasmid-encoded toxin (Pet), which is a serine protease autotransporter that cleaves spectrin protein within the cytoskeleton of the epithelium, resulting in cell elongation and exfoliation. Pic, kind of mucinous toxins which is a protein for intestinal colonization, interferes with the integrity of the mucus membrane. Sat; secreted autotransporter toxin affect the tight junctions of the cells in some organs such as kidney and bladder creating vacuolation in the cells. ShET1; *Shigella* enterotoxin I, stimulates intestinal cAMP- and cGMP-mediated secretion, hemorrhagic necrosis, and shortening of villi, enlarged crypt openings, and formation of crypt abscesses.

DAEC; Diffusely Adhering *E. coli* as implicated by its name, it will adhere to epithelial cells. Specialty of DAEC is that they do not invade or produce a toxin (Bhunia, 2018). The predominant clinical symptom in children is watery diarrhea which is not bloody and sometimes fever, abdominal cramps, and vomiting, sometimes leading to dehydration and shock as well. DAEC produces a diffuse adherence to cells by a fimbrial adhesion, and fimbrial adhesins are the well-studied virulence factors of DAEC. Subsequently to the binding of DAEC to the host cells, the cytoskeleton is altered through signal transduction leading to changes in the structures of the epithelial brush borders and tight junctions. Additionally, an inflammatory reaction also induced (Croxen et al., 2013).

8.2.4 *STAPHYLOCOCCUS AUREUS*

Staphylococci are ubiquitous organisms in the environment (soil, water, and air) on animal and human host (on the skin and mucous membranes) and also it can be found as a contaminant in several foodstuffs such as dairy product (cheese) and in meat products (Kadariya et al., 2014). The name *Staphylococcus* is derived from the Greek words staphylo (i.e., bunch of grapes) and coccus (i.e., grain or berry), come together as *Staphylococcus* (grape-like coccus). *Staphylococci* are Gram-positive (0.5–1.5 μm in diameter) and catalase-positive, and based on the coagulase production, and they are grouped into coagulase-positive (CPS) and coagulase-negative (CNS) staphylococci. In the Bergy's, they arrange themselves as single, in pairs, tetrads, short chains, and grape-like clusters. *Staphylococcus* species are facultative anaerobes or aerobes and have respiratory and fermentative metabolism. *Staphylococci* obtain energy via glycolysis, the hexose monophosphate shunt, and the tricarboxylic acid cycle.

They are catalase-positive, nonmotile, and sporogenous. Young cultures contain capsules but when the cells reach to stationary phase capsules are absent (Martin and Myers, 1994). Staphylococci are mesophilic in nature, where some strains of *S. aureus* have shown the ability to grow at a wide range of temperatures (7–47.8°C). Their optimum growth temperature is 35°C. Whereas the pH range is between 4.5 and 9.3, with the optimum between pH 7.0 and 7.5. Staphylococci can grow at lower A_w, even they can grow at low A_w (0.83 A_w) under ideal conditions. *S. aureus* is highly tolerant to salts and sugars and can grow over an A_w range between 0.83–0.99 (Martin and Myers, 1994).

The genus *Staphylococcus* is a group of Gram-positive bacteria that belong to the family Staphylococcaceae, order Bacillales, class Bacilli, phylum Firmicutes. There are over 40 species of the genus *Staphylococcus* identified (Tong et al., 2015). Many of the species and subspecies in the genus *Staphylococcus* are potentially found in foods, as a result of environmental, human, and animal contamination. Staphylococcal foodborne illness has a long history dating back as early as 1884. This was discovered associated with large food-poisoning outbreak in the United States as spherical organisms from cheese caused. The pathogenic nature of *Staphylococcus* is due to the presence of highly heat-stable enterotoxins by several staphylococcal species, including both coagulase-negative and coagulase-positive. The predominant cause for staphylococcal food poisoning is *S. aureus*, and it is associated with nosocomial and hospital-acquired infections.

S. aureus produces golden-yellow colonies, and they are non-motile. *S. aureus* is catalase-positive, relatively resistant to drying and heat, facultatively anaerobic organism, and this organism grows well in aerobic conditions, but there is an exception having true anaerobic one, i.e., *S. saccharolyticus*.

During the glucose metabolism acetoin is produced as the end product. They ferment mannitol, produces thermonuclease, sensitive to lysostaphin and it coagulate plasma. *S. aureus* can tolerate high concentrations of salt (10–15%) also it is resistant to heat and drying. Some of the *Staphylococcus* spp. are capable to coagulate rabbit plasma quickly (*S. aureus*), some of them are having delayed response (*S. intermedius* and *S. hyicus* subsp. *hyicus*).

S. aureus cell wall is composed of three main components namely, the peptidoglycan, a ribitol teichoic acid, and protein A. Peptidoglycan is composed of repeating units of N-acetylglucosamine β-1,4 linked to N-acetylmuramic acid whereas ribitol teichoic acid bound via N-acetyl mannosaminyl-β-1,4-N-acetylglucosamine to a muramyl-6-phosphate. Protein A covalently linked to the peptidoglycan and it has the characteristics of binding to the Fc component of mammalian immunoglobulin molecules, which results in the autoagglutination of mammalian plasma. The presence of protein A is characteristic of *S. aureus* where most of the other species of staphylococci with no protein A in the cell wall; hence they do not exhibit autoagglutination properties.

Illness due to staphylococcal intoxication is a self-limiting in nature and it will reduce within 24–48 h. But sometimes, it can be severe in some cases with debilitated people, elders, and infants (Murray, 2005). Two to eight hours after the consumption of food, symptoms of staphylococcal intoxication appear. Symptoms of food poisoning include nausea, vomiting, abdominal cramping and retching, with presence or absence of diarrhea, where vomiting is the most frequently observed symptom (Balaban and Rasooly, 2000; Bergdoll and Wong, 2006).

S. aureus produces an array of factors which determine the virulence including some cell wall components; exoenzymes such as proteases, (coagulase), nucleases, like the thermostable deoxyribonuclease known as TNase; and adhesion proteins, staphylococcal enterotoxins (SEs), staphylococcal enterotoxin-like proteins (SEls), superantigens, toxic shock syndrome toxins (TSST), exfoliative toxins (ET), pore-forming hemolysins, ADP-ribosylating toxins. It has been reported that *S. aureus* is having the capacity to produce at least 34 different extracellular proteins (Berube et al., 2013; Rasigade and Vandenesch, 2014).

Among various *S. aureus* toxins and virulence factors known, staphylococcal superantigens (consisting of SEs), toxic shock syndrome toxin 1 (TSST-1) and SEl proteins are most prominent. Staphylococcal food poisoning is due to the ingestion of preformed SEs. Enterotoxins are proteins produced mainly by *S. aureus*, *Staph. hyicus* and *Staph. intermedius* have also been shown to be enterotoxigenic. The *S. aureus* are capable of producing enterotoxins (SEs) throughout the logarithmic phase. These SEs are potent GI exotoxins of *S. aureus*. These toxins are active even in low concentrations and resistant to heat treatment, low pH, proteolytic enzymes, etc. (Larkin et al., 2009). They belong to pyrogenic toxin superantigens (Sags) family (Balaban and Rasooly, 2000; Derzelle et al., 2009). These are potent immunostimulatory toxins of *S. aureus* and some other bacteria (Berube et al., 2013). SEs are toxins of 20–30 kDa that have two separate biological activities: they act on the GI tract and as a superantigen on the immune system that triggers a strong T-cell activation (Otto, 2014). There are 23 serologically distinct staphylococcal superantigens discovered. SAgs which designate as SEs can stimulate the vomiting. Toxins that do not have emetic activity are designated as SEls SAgs.

SAgs are pyrogenic toxin superantigens, which induce food poisoning and TSS. SAgs are powerful non-specific T-cell stimulators that lead to unregulated activation of the immune response. SAgs interact with major histocompatibility complex (MHC) class II molecules on the surface of antigen-presenting cells, and with T-cell receptors (TCR) on specific T-cell subsets and bypass conventional antigen recognition. This leads to the activation of a large number of T-cells (Larkin et al., 2009; Fisher et al., 2018). T cell activation by SAgs leads to proliferation and also a massive release of chemokines and pro-inflammatory cytokines (such as interleukin-1 (IL-1), -2, and -6, γ-interferon, and TNF). Release of this kind of chemicals leads to the TSS.

SEs led to enteric illness by three mechanisms: (i) the release of proinflammatory cytokines as a result of SE-induced superantigenic T-cell proliferation; (ii) the binding of SEs to intestinal mast cells that lead to degranulation; and (iii) a direct effect upon the intestinal epithelium affecting gut transit (Principato and Qian, 2014).

In staphylococcal food poisoning, SAgs control the emetic response through a stimulation of the vagus nerve endings located in the lining of the stomach, causing nausea, cramps, vomiting, and diarrhea. SEs are also capable of penetrating the lining of the gut and it subsequently activate immune responses, i.e., local and systemic. This will stimulate the release of inflammatory mediators (including leukotrienes, histamine, and neurenteric peptide substance P) leading to vomiting (Argudin et al., 2010).

8.2.5 BACILLUS CEREUS

Bacillus cereus is distributed widely in nature, including air, dust, soil, water, plants, animals, and humans. It was firstly isolated by Frankland and Frankland in 1887 from air in a cowshed. Decaying organic matter, fresh, river, and marine waters, fomites, vegetables, and the gut of invertebrates can be the natural reservoir for this organism. These natural reservoirs contaminate the soil and food, leading to colonization of the human gut (Jensen et al., 2003).

They are facultative anaerobic bacteria, Gram-positive spore-forming, large (1.0–1.2 mm by 3.0–5.0 mm) motile rods, and possess two morphologies such as endospore and vegetative cells and they colonize host tissues in a vegetative state (Kotiranta et al., 1998, 2000). The characteristic feature of the spores of *Bacillus cereus*, being resistant to harsh environmental situations, they are well distributed in the environment. The majority of the Bacilli have a significant role in the industry, whereas some are involved in food spoilage producing off-flavor, damaging the structure, and showing unwanted growth in sterile food. Some of the members are virulent/pathogenic with multiple toxins production. Two distinct types of illnesses are reported as diarrheal and emetic. These conditions occurred as a result of ingesting the foods which are contaminated with the organism. Additionally, it can cause fatal systemic infections in immunocompromised patients and neonates. *B. cereus* has variations in virulence, where some of the strains are probiotics in nature, and some of the strains are highly toxic, leading to food-related fatalities. The first diarrheal outbreak of the *Bacillus cereus* was reported in hospitals in Norway (1947–1949). Later on, in the 1950s and it was confirmed as a bacterium causing foodborne disease (Lund et al., 2000; Dierick et al., 2005; Stenfors et al., 2008). The emetic syndrome was first identified in the United Kingdom (early 1970s) after several outbreaks due to the consumption of cooked rice (Stenfors et al., 2008). Among the two illnesses of food-related, the diarrheal type is due to the complex enterotoxins and the enterotoxins are released in the small intestine as the vegetative growth of *B. cereus* takes place. The emetic form of the disease is due to the toxins which are produced by growing cells.

There are three types of toxins identified as the cause for the diarrheal disease which are the pore-forming cytotoxins namely – hemolysin B; Hbl, nonhemolytic enterotoxin; Nhe and cytotoxin K; CytK (Lund and Granum, 1996; Lund et al., 2000). The emetic form of *B. cereus* food poisoning is due to one single toxin, cereulide a small ring formed dodecadepsipeptide. Apart from these toxins, type of cytolysin (cholesterol-binding cytolysin), viz. cereolysin O/hemolysin I, hemolysin II, hemolysin III, and hemolysin IV are also produced as the virulent determinant. Hemolysin III, and hemolysin IV are the cytotoxin K. The other kinds of products by the *B. cereus* include some enzymes namely, lecithinase, phospholipases, collagenase, sphingomyelinase, amylase, protease, and β-lactamase. β-lactamase which produced by this organism involved in the inactivation of penicillin which make them resistant to penicillin.

Cereulide (a sodecadepsipeptide) is another kind of emetic toxin produced by *B. cereus* and this is resistant to acid conditions, proteolysis, and heat. These characteristics of resistance make troublesome as it is not be destroyed by gastric acid as well by the proteolytic enzymes. Even it is not possible to inactivate by reheating foods (Shinagawa et al., 1996). Cereulide causes inhibition of mitochondrial activity (inhibition of fatty acid oxidation) and stimulates the vagus afferent, and diarrheal toxins are presumed to elicit diarrhea. Diarrheal toxins disrupt the integrity of the epithelial cells of the small intestine by acting on the plasma membrane.

B. cereus group of organisms, consisted of eight species and these have shown many similarities with 16S rDNA. These species include, *B. cereus*, *B. pseudomycoides*, *B. mycoides*, *B. weihenstephanensis*, *B. thuringiensis*, *B. cytotoxicus*, *B. anthracis* and *B. toyonensis* (Savini et al., 2013). Preferable temperature for the growth is 35–40°C but there is a possibility to grow at 10–50°C. The pH below 4.5 is not a contusive condition for them to grow. *B. cereus* forms large, gray colonies (3–8 mm in diameter) with a rather flat and "ground-glass" morphology, with irregular borders. In general, they are capable of producing β-hemolysis around colonies except for some strains (Slamti et al., 2004).

B. cereus can be entered to human in many ways; through contaminated foods and also with food ingredients, including dairy products, spices, rice, meat, meat products, vegetables and dried foods (Kramer and Gilbert, 1989; Granum, 2012). There are many other occasions carrying the risk of contamination of materials with *B. cereus* spores and cells; during harvesting of plant materials/products and transporting them to the production/processing places. Contaminations of food processing equipment with spores and cells of these organisms also play a role in spreading the disease. Further, there are some reports on the spreading of these organisms via cow milk. Contamination of the udders of cows can be occurred while grazing, or they can be attached to the udder through bedding material. Also, there is a possibility of getting the organism by the cow through their feedstuff (Kramer and Gilbert, 1989; Andersson et al., 1995).

Also, the *B. cereus* are capable of forming biofilms (Wijman et al., 2007), and it is difficult to remove biofilm cells with general cleaning agents, hence persistence in food industry equipment continuing the contaminations (Ryu and Beuchat, 2005). In addition, lightly heat-treated foods which are stored for a long period under the refrigerated conditions also a favorable environment for *B. cereus*.

8.2.6 CLOSTRIDIUM BOTULINUM

Clostridium spp. is also an organism which is commonly found in the environment and they are found in soil, water, and decomposing plants and animals. *Clostridium* species are Gram-positive, anaerobic, and spore-forming rods. The genus *Clostridium* consists of more than 80 species, and 34 species have been considered pathogenic to man and animals. Among them, *Clostridium botulinum* and *Clostridium perfringens* are usually involved in foodborne outbreaks. This organism was first isolated by Belgian Physician (Emile van Ermengem in 1897) as *Bacillus botulinus* from raw salted ham which is a homemade product and later renamed as *Clostridium botulinum* (Schantz and Johnson, 1997).

There are four discrete groups of bacteria recognized within *C. botulinum* namely, *C. botulinum* Groups I and II (primarily responsible for human botulism), *C. botulinum* Group III is (botulism in various animal species), and *C. botulinum* Group IV (not associated with botulism in humans or animals) (Peck, 2009; Carter and Peck, 2015). Botulism in humans is categorized into three types, namely, infant/intestinal (adult) botulism, wound botulism, and foodborne botulism. The form of infant/intestinal (adult) botulism, is due to the multiplication of cells and formation of neurotoxin in the gut. The wound botulism is an infection associated with cell multiplication and neurotoxin formation in a wound whereas the other form of infection, i.e., foodborne botulism is considered as intoxication due to the ingestion of neurotoxin pre-formed in the food (Carter and Peck, 2015).

Botulism, the disease caused by *C. botulinum* is due to the production of botulinum neurotoxin (BoNT) which is a highly potent toxin. A minute amount of this toxin as little as 30–100 ng has been estimated to be lethal (Peck, 2009). Seven serologically distinct toxins (A, B, C1, D, E, F, and G) are synthesized by different strains of *C. botulinum*, and another toxin designated as H is also reported (Barash and Arnon, 2014; Johnson, 2014). A, B, and E types are reported as the most commonly associated with botulism in humans (Klein, 2003). Moreover, there are more than 30 different subtypes of toxins are reported (Hill and Smith, 2012). Botulinum toxin serotypes, in general, prevent acetylcholine release at the peripheral nerve endings, thus inducing temporary denervation and relaxation of muscles (Dolly and Aoki, 2006). Botulinum neurotoxins have different abilities, such as (i) they target cholinergic nerve endings via binding to ecto-acceptors, (ii) they undergo endocytosis/translocation, and (iii) their light chains act intraneuronal to block acetylcholine release (Dolly and Aoki, 2006). All BoNT serotypes are highly specific in their targeting of cholinergic nerves, therefore the C-terminal region of the toxin binding with high affinity to ecto-acceptors on the presynaptic motor nerve endings (Dolly, 2003). The toxin is rapidly act within 90 minutes and irreversibly bound to the presynaptic neuron (Klein, 2003). Following the binding, via the process of acceptor-mediated endocytosis, the toxin is taken up into the motor neuron (Dolly et al., 1984). Subsequently, the light chain of the toxin is translocated into the cytoplasm of the motor neuron and mediates the toxin's activity. Acetylcholine release at the neuromuscular junction is blocked by BoNT selectively, hence abolishing the motor end-plate potential (Dolly et al., 1987).

Foodborne botulism occurred due to the ingestion of preformed toxins. The common symptoms include in this condition is, having a sudden onset (acute) of descending, symmetrical flaccid paralysis. This flaccid paralysis affects the muscles of the neck and head initially. Difficulty speaking, blurred vision, dry mouth, difficulty in swallowing, slurred speech, and generalized weakness are some of the other signs. When it progresses to severe condition, it develops paralysis in the respiratory system and cardiac system. Also, it is associated with GI symptoms of nausea, vomiting, and diarrhea (Dolly and Aoki, 2006). The time taken for the onset of botulism depends on several factors. Consumption of larger quantities of toxin results in a decreased incubation time. Symptoms generally occur at 12–36 hours, but it can be shorter as (in severe cases, it is around 2 hours), or in mild cases, it can be long as 14 days.

8.2.7 *CLOSTRIDIUM PERFRINGENS*

Clostridium perfringens is an anaerobic, Gram-positive, spore-bearing non-motile rod and is usually distributed in different environments, soil, sewage, and in the intestines of humans and animals (Mueller-Spitz et al., 2010). Due to its ubiquitous nature and its ability to form spores, it has become a greater problem in the food industry. The organism was reported to be the first associate with intestinal disturbances in 1895 (Shandera et al., 1983). The first large-scale outbreak in the world was taken place in 1943, among school children in Leicester, England. *Clostridium perfringens* associated with two very different foodborne diseases. One is relatively mild, classic Type A diarrhea, and it is a more common type in the industrialized world. The second type is severe but rare Type C human necrotic enteritis. Apart from these two gastroenteric diseases, this organism causes human gas gangrene (Brynestad and Granum, 2002). Symptoms of type A foodborne illness include diarrhea and abdominal pain with an incubation period of 8–24 h. Fever is rare, and symptoms subside within 24 h without treatment. Fatalities can occur rarely among elderly or hospitalized individuals. Type C human necrotic enteritis is a severe foodborne illness, often fatal having necrotizing, hemorrhagic jejunitis, commonly called "Pig-Bel" (necrotizing enteritis) and parts of the small intestine can be gangrenous when the disease progress. This can lead to severe toxemia and shock. Persons suffering from enteritis necroticans can survive for a very short time period (less than 48 hours) (Matsuda et al., 2007; Uzal and Mc Clane, 2011).

The pathogenic nature of the *C. perfringens* is due to the production of toxins during sporulation. *C. perfringens* is known to produce over 13 different toxins (Katayama et al., 1996; Petit et al., 1999; Sarker et al., 1999; Smedley et al., 2005). Depending on the type of toxin produced (based upon their production of alpha/CPA, beta /CPB, epsilon/ETX, and iota/ITX), *C. perfringens* can be divided into five toxinotypes: A-E (Omernik and Plusa, 2015). In addition to these four types of major toxins, there are several other toxins also be produced by *C. perfringens* such as enterotoxin (CPE), beta 2 (CPB2), and perfringolysin O (PFO) (Songer and Uzal, 2005).

C. perfringens type C induces human necrotic enteritis (Enteritis necroticans); is the life-threatening illness due to the α-(CPA) and β-toxin (CPB) with β-toxin considered the primary virulence factor responsible for this illness (Sayeed et al., 2008). Self-limiting *C. perfringens* type A food poisoning is due to the type A isolates producing CPE (*Clostridium perfringens* enterotoxins). Apart from the presence of toxins, *C. perfringens* have some other characteristics which help the pathogenicity, such as having very short doubling time and tolerant to some food environment stresses such as radiation, desiccation, freezing, refrigeration, and heat.

During the sporulation of *C. perfringens* produce CPE and toxin is accumulated in the cytoplasm of sporulating cells, localizing in paracrystalline inclusion bodies. At the completion of sporulation, CPE is released into the intestinal lumen and CPE binds to enterocytes through the proteinaceous receptor (Wnek and McClane, 1986). After binding, CPE forms a series of protein complexes in plasma membranes of host cells leading to paracellular permeability changes that may contribute to CPE-induced diarrhea (Singh et al., 2001). Also, it forms pores permeable to water, ions, and small molecules (Hardy et al., 1999).

Type C human necrotic enteritis is due to the β toxin produced by *C. perfringens* type C strains, and it is characterized by hemorrhagic mucosal ulceration or superficial mucosal necrosis of the small intestine in humans, can be acute, deep necrosis in small intestinal mucosa-associated with acute vascular necrosis and massive hemorrhage in the lamina propria and submucosa. (Miclard, 2009a, b). β toxin is known to form pores in the bilayer lipid membranes, which results in membrane permeability leading to cell death or the intestinal necrosis. Host intestine is the first target of C infection, and in the intestine, the toxin is highly active. Beta toxins can damage the intestinal epithelium directly and indirectly. Directly toxins can affect on epithelial cells while indirectly it can lead to endothelial damage. This endothelial damage can progress to thrombosis and intestinal ischemia. Beta toxin and other toxins to be translocated into the circulation to induce its lethal effects (Miclard et al., 2009a, b).

8.2.8 LISTERIA MONOCYTOGENES

Members of the genus *Listeria* include Gram-positive, nonsporulating, rod-shaped aerobic or facultative anaerobic microorganisms. Some of them are pathogenic to humans and animals, whereas some of them are saprophytic in nature. The genus is composed of 10 species viz., *L. innocua*, *L. monocytogenes*, *L. ivanovii*, *L. grayi*, *L. seeligeri*, *L. welshimeri*, *L. rocourtiae* (Leclercq et al., 2010), *L. marthii* (Graves et al., 2010), *L. weihenstephanensis* (Halter et al., 2013), and *L. fleischmanni* (Bertsch et al., 2013). Except for *L. ivanovii* and *L. monocytogenes,* other eight species are saprophytes (Liu, 2013) while both the *L. monocytogenes* and *L. ivanovii* are pathogenic (Liu, 2013). The first description of *L. monocytogenes* is dated back to 1926, and it was by Murray, Webb, and Swann. The first confirmed human case was that of a soldier suffering from meningitis at the end of World War I.

Listeria is incapable of producing spores and capsules. At the temperature of 20 to 25°C, they are motile from its few peritrichous flagella (Galsworthy et al., 1990). It is a small (0.5 μm in diameter and 1 to 2 μm in length), a regular rod with rounded ends. The morphological features can be as single units or they are arranged in short chains, whereas they can be in palisades or in V and Y forms (Jorgensen et al., 1995). Characteristics of colonies on nutrient agar smooth, bluish-gray, translucent, punctiform, and slightly raised with a fine surface texture. They are capable of growing at a wide range of temperatures; 1–2°C to 45°C (Junttila et al., 1988). Optimum pH of *Listeria* is 7 (pH range of 4.5–9.2) (George and Lund, 1992), and they can grow at A_w below 0.93 (Farber and Addison, 1994).

L. monocytogenes mainly cause invasive listeriosis and gastroenteritis. Still, the route of transmission was questionable until the 1980s, and it was identified as the foodborne (Bille, 1990). Human infection may lead to a severe and potentially life-threatening illness known as listeriosis. The organism can be present in some foods, such as ice cream, cheese, unpasteurized milk, raw milk, fermented meats, raw vegetables and sausages, meat, raw and smoked fish, and seafood (Allen et al., 2016; Leong et al., 2016).

L. monocytogenes is considered as the multisystemic invasive pathogen, which can invade multiple host tissues such as the brain, placenta. This pathogen can be considered as an opportunistic pathogen as it affects certain individuals who are debilitated and immunocompromised (Vazquez-Boland et al., 2001) and also the normal healthy individuals. Infection with *Listeria* can be generalized or localized, and there are two main forms of listeriosis defined as fetomaternal/neonatal listeriosis, and listeriosis in adults (Vazquez-Boland et al., 2001; Barbuddhe and Chakraborty, 2009). This can lead to abortion, stillbirth, meningitis, meningoencephalitis, pneumonia, and gastroenteritis. When the organism localizes in the cardiovascular system it leads to endocarditis, myocarditis, arteritis whereas when it is in the respiratory system, pneumonia, pleuritic, and sinusitis can be seen and also when it affects the musculoskeletal system it will develop arthritis, osteomyelitis, etc. (Doganay, 2003).

L. monocytogenes can cause acute, self-limiting, febrile gastroenteritis with the symptoms of fever, watery diarrhea, nausea, headache, joint, and muscle pain. This illness usually occurs 24 hours after the ingestion of bacteria without concomitant invasive infection (Ooi and Lorber, 2005). The GI tract is the route of entry, afterwards strains penetrate the mucosal tissue directly or indirectly. Direct penetration is via invasion of enterocytes and indirect penetration via Peyer's patches (Barbuddhe and Chakraborty, 2009). The central nervous system (CNS) is the other most important predilection site for the organism (Clauss and Lorber, 2008). They invade the CNS by the transportation of bacteria to the CNS within circulating leukocytes in a phagocyte-facilitated mechanism, through the direct invasion of endothelial cells of the blood-brain barrier by bloodborne bacteria via a neural route (cranial nerve invasion) (Drevets et al., 2004).

8.2.9 SHIGELLA

Shigella was isolated and defined by a Japanese Scientist, Dr. Kiyoshi Shiga, in 1897 (Trofa et al., 1999), and it is the causative agent of shigellosis or bacillary dysentery which is an acute, infectious, inflammatory enteritis characterized by scant stools containing blood and mucus. Shigella belongs to the family Enterobacteriaceae, the genus of gammaproteobacteria. *Shigella* is a pathogen confined only to a human pathogen. Epidemics occur most frequently in overcrowded populations with poor sanitary conditions. In endemic areas, shigellosis is common among younger children and adults usually have less susceptibility. Shigella are Gram-negative, non-motile, non-spore-forming, rod-shaped facultative anaerobe with relatively simple nutritional requirements and are very much resembling *Escherichia coli*. Shigella are endemic worldwide, resulting in high morbidity and mortality, particularly in third world countries (Keusch and Acheson, 1996).

There are four species of *Shigella* identified based on the biochemical and serological differences. Those are *S. dysenteriae* (serogroup A) consists of 13 serotypes, *S. flexneri* (serogroup B) consists of 15 serotypes, *S. boydii* (serogroup C) consists of 18 serotypes and *S. sonnei* (serogroup D) consists of single serotype (Niyogi, 2005). *Shigella* is an intracellular pathogen, and the infection occurs following the invasion of cells leading to intense inflammation in the colonic and rectal epithelium. Humans acquired the organisms through ingestion, and the ingestion of a few organisms (10–100 organisms) might lead to the infection. Humans are the natural reservoir and mode of infection is faces of infected people or convalescent carriers. The carriers (Convalescents and subclinical) serve as the main sources (Levine et al., 2007). There is a possibility of spread of the organism through contaminated food. Many food items, including raw vegetables, tossed salads, potato salads, chicken, and shellfish, act as vehicles of the spreading of this organism. Flies serve as vectors (DuPont et al., 1989; Roberts, 1989). The virulence of the organism is due to a large virulence plasmid, and this encodes for the Type III-secretion-system (T3SS), and a set of secreted proteins which are essential for invading the colonic epithelium and other pathogenic sequential. Adhesion, invasion, intracellular replication, and cell-to-cell spread are the main steps involved in the process of colonic infection. The predilection site for the organism or the infection is the large intestine (M cells of the Peyer's patches). As strains of *Shigella* spp. are non-motile, it is still not understood the actual mechanism of reaching the bacteria to M cells and adhere. *Shigella* can tolerate acidic environment in the stomach and it can withstand in the presence of bile.

When the organism reaches to the site of infection, it starts mucosal inflammation and it leads to apoptosis. This facilitates the invading to the M cells, after which the bacteria are phagocytosed. Following the multiplication of organisms within the epithelial cells, it spread to adjacent cell and to lamina propria. Epithelial cells are necrosed and there will be inflammation in lamina propria and submucosa and there will be capillary thrombosis. Due to the sloughed off of dead epithelial cells, ulcers are formed, thereby producing bloody and mucoid stools. It has been found that the invasion of other cells by Dysentery bacilli is very rare (Reller et al., 2006). *Shigella dysenteriae* and certain strains of *Escherichia coli* produce Stx, which are the major virulent determinant. The toxins inhibit protein synthesis by inactivating 60S ribosomes, and they may induce signaling leading to apoptosis. Expression of Shiga toxin has been shown to be iron-regulated, with toxin production increasing under conditions of iron restriction. The A subunit of the toxin possesses the biological activities, whereas the B subunits mediate specific binding and receptor-mediated uptake. Stx produced by the organism is internalized by the host cells. And they are active within the endosomes, later it reaches Golgi apparatus. The A subunit of the toxin has two forms namely A_1 and A_2. A_1 portion inhibit the protein synthesis and leads to cell death. The condition of HUS is due to the action of Shiga toxin on kidney tissues (Chan and Ng, 2016). The common clinical sign of shigellosis is fever and watery diarrhea that progresses to dysentery. Low-grade fever (about 38 to 38.9°C) and watery diarrhea 1–4 days after ingestion of the bacteria can be seen in mild infection. The mild infection may resolve in 4–8 days in adults. The most obvious symptoms are painful abdominal cramps and frequent defecation. These symptoms are aggravated in the severe cases; infected individuals show low-grade or moderate fever and watery diarrhea. This diarrhea progress to dysentery which characterized with bloody, or pus or mucus containing diarrhea. Severe infections last for 3–6 weeks and resolve. In young children, fever, loss of appetite, nausea or vomiting, irritability or drowsiness, diarrhea, and abdominal pain can be seen and they frequently feel an urgent need to pass feces (Gupta et al., 2004; Mattock and Blocker, 2017). According to many published reports, shigellosis may lead to complications, and especially young children are prone to have severe complications. The signs of complications may include high fever, sometimes with seizures or coma, severe diarrhea, severe dehydration, and weight loss, rectal prolapsed, rectal bleeding, and perforation of the large intestine. A HUS where red blood cells are destroyed also has been reported and cause anemia with fatigue. Severe dehydration, mainly in children less than 2 years, can lead to shock and death. Severe conditions can occur in adults as well, in chronically ill, malnourished, or debilitated and in older ages adults can develop. In adults there can be non-diarrheal symptoms such as reactive arthritis, inflammation of the eyes and urethra (Speelman et al., 1984; Anand et al., 1986; Mathan and Mathan, 1986; Bennish et al., 1990; Carter and Hudson, 2009).

8.2.10 VIBRIOS

Vibrios are ubiquitous in the aquatic ecosystem, and the genus *Vibrio* is a member of the *Vibrionaceae* family, and over 80 species have been described, and at least 12 of them are pathogenic to humans. It is mostly associated with seafood, including shellfish and the *Vibrio* species, principally *V. cholera* (O1 and non O1 serogroups), *V. vulnificus* and *V. parahaemolyticus*, are important in human bacterial disease associated with the consumption of seafood. The members of the genus are gram-negative, curved-rod, they are 1.4 to 2.6 μm in length and 0.5–0.8 μm in width, facultative aerobic, non-spore-forming bacteria. The optimum pH range is 8.0–8.8, and the optimum growth temperature range is 20–37°C. They are motile, halophilic, and most of them are oxidase-positive and produce extracellular enzymes such as amylase, gelatinase, chitinase, and DNase (Bhunia, 2018).

8.2.10.1 VIBRIO CHOLERAE

Vibrio cholerae, causative agent of cholera was first described by Filippo Pacini in 1854. It was described as a bacillus of comma-shaped and was isolated from intestinal mucosa of a deceased patient during the Asiatic Cholera Pandemic of 1846–1863. It is a natural member in aquatic environments; where few members of environmental *V. cholerae* is causing cholera. The species has many serogroups (more than 200), and strains O139 and O1 are associated with epidemics of cholera. The non-O1/non-O139 are rarely toxigenic and also seldom involved in an outbreak (Yamai et al., 1997; Weil and Harris, 2015). *V. cholera* O1 consists of two biotypes, namely El Tor and classical. Mode of transmission through the ingestion of contaminated seafood or water. *V. cholerae* often causes asymptomatic infection or mild diarrheal illness, but in severe cases, it causes cholera, infection with profound secretory diarrhea that can be fatal in 6–24 hours. Cholera is one of the most devastating diseases with estimated cases around 2.86 million annually. Deaths due to the cholera is around 95,000, having a range of 21,000 to 143,000 deaths globally (Bhunia, 2018).

Following the ingestion of *V. cholerae*, it passes through the upper GI tract and colonizes in the small intestine. Colonizing factor TCP, involved in bacterial autoaggregation, microcolony formation, and TCP forms matrices to protect the organisms from bile (Jude and Taylor, 2011). Motility is an important factor for colonization in the mucosal surface and for penetrating mucus. The outer-membrane protein OmpU, Several *V. cholerae* protease, also contribute to colonization of *V. cholera* (Richardson, 1991). Haemagglutinin protease (HA/P) is a proteinase, and mucinase disrupts tight intercellular junctions and may facilitate *V. cholerae* penetration of the mucus barrier. Metalloproteinases degrade cell-to-cell junctions and induction of changes in the actin cytoskeleton. The pathogenic strains harbor a cholera toxin prophage (CTXΦ) that carries the genes that encode the CT. CT is the main virulence factor (ADP-ribosylating toxin) and it results in increased cyclic AMP (cAMP) and also chloride secretion. CT is the primary reason of the characteristic profuse watery diarrhea of cholera. In addition to CT, toxin-coregulated pili (TCP) is also important in the virulence and it facilitates the adherence and colonization to mucosa of the small intestine. Other toxins produced by the organisms include ACE (accessory cholera enterotoxin), ZOT (zona occludin toxin), and hemolysin (HlyA), which act on mucosal cells. These toxins affect ion transport pumps for Na^+, Cl^-, HCO^{3-}, and K^+ in the cell thereby altering the ion balance which results in fluid and ion losses (Silva et al., 2006).

8.2.10.2 VIBRIO PARAHAEMOLYTICUS

Vibrio parahaemolyticus is an estuarine marine bacterium causing sporadic outbreaks of gastroenteritis due to the consumption of raw, inadequately cooked, or cross-contaminated seafood. Some strains of *V. parahaemolyticus* are environmental, whereas many strains are pathogenic to humans. It is one of the major foodborne pathogenic vibrios associated with seafood, and *V. parahaemolyticus* was first discovered by a Japanese Scientist, Tsunesaburo Fujino, in 1950 (Wong et al., 2000; Yeung and Boor, 2004). The severity of the disease can be mild and self-limiting but can occasionally become severe in immunocompromised patients. Typical symptoms include diarrhea, abdominal pain, nausea, vomiting, headache, fever, and chills. Infection occurs 4–96 h after consumption of contaminated food and lasts up to three days. Apart from its ability to cause acute gastroenteritis (AGE), it can cause wound infections, septicemia (Yeung and Boor, 2004; Su and Liu, 2007).

V. parahaemolyticus is an Gram-negative, halophilic, motile, facultative anaerobic rod, positive for oxidase, gelatinase, lysine, and ornithine decarboxylase reactions, does not ferment sucrose, and it is able to grow at high pH. Strains of *V. parahaemolyticus* are typically serotyped on the basis of O and K antigens. Several virulence determinants have been associated with the pathogenesis such as the presence of flagella for its motility, multivalent adhesion molecules, siderophores (Vibrioferrin) to scavenge iron, hemolysins (TDH, TRH, and TLH), pathogenicity islands, type three secretion systems (T3SS) (Broberg et al., 2011). TDH has been recognized as the major virulence factor of *V. parahaemolyticus* (Takeda, 1982). TDH act on cellular membranes as a pore-forming toxin that alters ion flux in intestinal cells, thereby leading to a secretory response and diarrhea (Takahashi et al., 2000a, b). TDH-related hemolysin (TRH) is another virulence factor of *V. parahaemolyticus*, which is similar to TDH. The mechanism of TRH also appears to be similar to that of TDH. It induces Ca^{2+}-activated Cl^- channels, which result in altered ion flux (Takahashi et al., 2000b). *V. parahaemolyticus* translocates T3SS1 effectors such as VopQ, VopR, VopS, and VPA0450 into host cells to cause cytotoxicity in different cell types and T3SS2 effectors such as VopA, VopC, VopL, and VopT) are translocated into host cells to cause cytotoxicity of colon epithelial cells or enterotoxicity within the host (Broberg et al., 2011).

8.2.10.3 VIBRIO VULNIFICUS

Vibrio vulnificus, is a halotrophic Gram-negative rod-shaped bacterium that lives in marine and estuarine waters, as free-living or associated with shellfish in natural environments and is a highly lethal opportunistic pathogen of humans (Jones and Oliver, 2009; Hampton et al., 2017). *V. vulnificus* infection was first described in 1976 (Chiang et al., 2003).

Consumption of seafood (primarily raw oysters) contaminated with *V. vulnificus* is the main source of the infection, which results in a severe, fulminant systemic infection. Infection is characterized mainly by primary sepsis, wound infection, and gastroenteritis. There are some other symptoms as well, fever, nausea, hypotensive septic shock, chills, and the formation of secondary lesions on the extremities. More importantly, the primary septicemia caused by *V. vulnificus* is the most lethal infection (Oliver, 2005; Jones and Oliver, 2009).

Strains of the *V. vulnificus* is sub-grouped into three biotypes: biotype 1, 2, and biotype 3. Biotype 1 comprises environmental and clinical isolates, whereas biotype 2 comprises eel pathogens and an opportunistic pathogen of humans (Amaro and Biosca, 1996; Chiang et al., 2003). Biotype 3 was isolated in Israel from persons who had received puncture wounds following contact with Tilapia.

A number of studies have identified many of the virulence factors associated with *V. vulnificus* (Gulig et al., 2005). Some of the virulence factors are external structures of the bacteria such as the capsular polysaccharide (CPS), lipopolysaccharide (LPS) flagellum, factors for acid neutralization, iron toxicity, cytotoxicity, proteins (for attachment and adhesion), outer membrane vesicle (OMV) production (Jones and Oliver, 2009; Vanaja et al., 2016).

8.2.11 AEROMONAS

Aeromonas species are ubiquitous in aquatic environments; they belong to the family *Aeromonadaceae*; which are gram-negative facultative anaerobic rods. First described by Stainer in 1943, since then, at least 32 species have been reported (Beaz-Hidalgo et al., 2015; Marti and Balcazar, 2015; Martínez-Murcia et al., 2016). It is pathogenic of human as well in fish and in human it can produce several kinds of diseases; intestinal and extraintestinal diseases which includes gastroenteritis, septicemia, skin, and soft-tissue infections, biliary tract infection, peritonitis, pneumonia, urinary tract infections, etc. Gastroenteritis is manifested with watery and self-limited diarrhea, fever, abdominal pain, nausea, or vomiting. *Aeromonas hydrophila*, *A. veronii*, *A. dhakensis*, and *Aeromonas caviae* have been identified as the key human pathogens, and it has been associated with various food sources; as fish and seafood, chicken, and red meat, vegetables, raw milk, and dairy products (Wadstrom and Ljungh, 1991). Pathogenesis of the *Aeromonas* gastroenteritis is involved in sequential steps directed locomotion, attachment to GI epithelium, biofilm formation, intestinal colonization, the release of virulence factors, and subsequent development of diseases. Flagella (polar flagellum and multiple lateral flagella), pili, quorum sensing, are the cellular components/mechanisms which aid in the processes of epithelium attachment, cell adherence, biofilm formation, and intestinal colonization. Following the GI tract establishment, *Aeromonas* leads to diarrhea by excreting enterotoxigenic factors or causing invasive gastroenteritis by invading the GI epithelium (Janda and Abbott, 2010). There are a variety of biologically active extracellular products produced by *Aeromonas*, such as hemolysins, cytotoxins, enterotoxins, proteases, leukocidin, phospholipases, endotoxin, outer membrane proteins, and fimbriae or adhesions (Chopra and Houston, 1999).

8.3 FOODBORNE PARASITES

8.3.1 ENTAMOEBA HISTOLYTICA

Entamoeba histolytica is associated with human dysentery and this was described by Schaudinn first in 1903 (Saha et al., 2015). Amoebiasis is the infection caused by *E. histolytica*. The organism exists in two forms; the trophozoite (dividing form) and the cyst which is the dormant form. Infection is due to the ingestion of cysts (infective form) of *E. histolytica* present in food or in water contaminated with feces (Christy and Petri, 2011; Saha et al., 2015). Following the ingestion of cysts, they travel through the digestive tract, and in the ileum-ceca zone, excystation of cysts producing trophozoites is taken place. They colonize the large intestine by adhering to colonic mucins, feed on bacteria of the intestinal flora, and divide. Each cyst forms eight amoebae, and they migrate and multiply in the colon. Amoebae migrate to the intestinal mucosa and develop intestinal amoebiasis (IA); ulcerative colitis (common form) or severe dysentery, bloody diarrhea, abdominal pain. Small-volume mucoid stools and watery diarrhea are common (Christy and Petri, 2011; Faust and Guillen, 2012; Saha et al., 2015). They can damage the extraintestinal sites such as the liver (Amoebic liver abscesses), lungs, and brain (Abscesses). The pathogenesis of amoebiasis is believed to be a multi-step process composed of adhesion, phagocytosis, and proteases secretion. Trophozoites bind to the mucin layer by the Gal/GalNAc lectin, and further amoebic proteases (cysteine proteases) play the main role in mucin degradation, destroying the muco epithelial barrier inducing the overproduction of mucus. Proteolytic enzymes secreted by the organism destroy the brush border microvilli of enteric cells, disrupt the intestinal mucus and epithelial barrier, and facilitate tissue penetration. It also damage (cytotoxic effects) the host epithelial and immune cells by amoebapore, a contact-dependent cytolysin, causing characteristic flask-shaped ulcers, increased intracellular Ca^{2+} levels, ROS production, and apoptosis which destroy the intestinal barrier. Additionally, the phagocytosis is playing a role in virulence, and they tend to phagocytize host cells such as macrophages, RBCs, and lymphocytes (Huston et al., 2003; Mansuri et al., 2014).

8.3.2 *BLASTOCYSTIS*

Blastocystis spp. are unicellular, extracellular, non-motile, intestinal parasitic protists of humans and animals. In humans, *Blastocystis* is a parasite of the colon, and it is shed through feces. It is transmitted by fecal-oral transmission through contaminated food or water or by direct contact. Symptoms of infection with *Blastocystis* are non-specific GI symptoms of diarrhea, nausea, vomiting, and abdominal pain, irritable bowel syndrome, and urticarial. Pathogenesis has not been clearly described, but several pathogenic mechanisms are thought to be involved such as apoptosis, degradation of tight junction proteins resulting in increased permeability, induction of inflammatory cytokines, and downregulation of inducible nitric oxide synthase (Yakoob et al., 2010; Zuel-Fakkar et al., 2011; Jimenez-Gonzalez et al., 2012).

8.3.3 *CRYPTOSPORIDIUM*

Cryptosporidium spp. are obligate protozoan parasites belonging to the phylum Apicomplexa and family Cryptosporidiidae. This organism infects different hosts such as humans, birds, reptiles, and fish. The infection (cryptosporidiosis) is self-limiting in normal healthy individuals, but in immunocompromised individuals, it can lead to protracted and life-threatening. Humans with compromised immune systems can develop chronic diarrhea, cachexia, lack of appetite, nausea, vomiting, and malnutrition. *Cryptosporidium* spp. inhabit the brush borders of the GI, respiratory, and sometimes renal epithelium of various vertebrates. Disease transmission is through direct contact with infected persons or animals, and consumption of contaminated water or food. There are 23 *Cryptosporidium* species reported, and *C. hominis*, *C. parvum* and *C. viatorum* are the species encountered in humans. The infectious stage of *Cryptosporidium* is the oocyst, and this is the stage shed in the environment (Ryan et al., 2016; Pumipuntu and Piratae, 2018). *Cryptosporidium* possesses secretory organelles that secrete exocytose proteins. These proteins facilitate attachment, invasion, and parasitophorous vacuole formation. During the infection, there are alterations in the barrier permeability and changes in ion influx due to the products of damaged epithelial cells or due to the cytokine and/or chemokine responses and also apoptosis. Mononuclear cell infiltration in the lamina propria, mucosal cell apoptosis, mucosal inflammation with villus blunting, and cryptitis are some of the characteristics found histologically, subsequently leading to the loss of barrier function and malabsorption. Voluminous diarrhea associated with the cryptosporidiosis is due to the chloride secretion and impaired glucose absorption (Fayer, 2010; Ryan and Hijjawi, 2015; Ryan et al., 2016).

8.3.4 GIARDIA

Giardia belongs to the phylum Metamonada, class Trepomonadea and to the order Giardiida and family Giardiidae. Among the several species, *Giardia duodenalis* is the cause of giardiasis in mammals, including humans, with a worldwide distribution (Muhsen and Levine, 2012). This is one of the most prevalent GI infections among humans. Two stages are in the life cycle of the organism; trophozoite stages and cyst. The trophozoite is the noninvasive form and these replicate in the intestinal mucosa, whereas the cysts are environmentally resistant stage (Adam, 2001). The infection is transmitted via the fecal-oral route by the ingestion of cysts. Following the ingestion of inert cysts excystation process starts. This process is triggered by acidic conditions of the host stomach and then the cysts will be passed into the small intestine before rupturing. The infection can be asymptomatic, acute, or chronic giardiasis. Generally, the infection is self-limiting or can produce serious effects such as diarrhea and malabsorption. Acute giardiasis is characterized by AGE with anorexia, nausea, vomiting, abdominal pain, watery diarrhea, whereas chronic giardiasis has recurrent chronic diarrhea, postprandial epigastric pain, abdominal distention (Erlandsen and Meyer, 2013).

With the ventral adhesive disk and by the flagella, trophozoites are strongly attached to the enterocytes in the upper small intestine. *Giardia* produces and releases cytopathic substances into the intestinal lumen, and they attack surface glycoproteins and disrupt microvillus membrane integrity. Further, it is reported that *Giardia* induces changes in the gut microbiota, and these changes promote the formation of mucosal B lymphocyte follicles and increases the production of pro-inflammatory cytokines (IL-1ß), which leads to the architectural changes in the intestine (Farthing, 1993).

8.3.5 SARCOCYSTIS

Sarcocystis is a foodborne apicomplexan, intracellular protozoan belongs to the family *Sarcocystidae*, subphylum Apicomplexa, order Eucoccidia. It was first reported in 1843 by Miescher, and this organism is one of the most prevalent parasites in livestock, wildlife, birds, and people. *Sarcocystis* species have intermediate and definitive host life cycle based on a prey-predator relationship. Human acquires intestinal sarcocystosis; *Sarcocystis hominis* from beef and *Sarcocystis suihominis* from pork where the human act as the definitive host. Intestinal infection with *Sarcocystis* spp. is usually considered nonpathogenic for definitive hosts, but there are some reports of intestinal infection with clinical manifestations such as abdominal pain, distension, nausea, and diarrhea. Muscular sarcocystosis in human is sometimes associated with the symptoms of painful edema and erythema, pain in limbs, fever, chronic myositis, and eosinophilia (Chen et al., 1999; Fayer, 2004; Fayer et al., 2015).

8.3.6 TOXOPLASMA GONDII

Toxoplasma gondii is an obligated intracellular unicellular parasite belonging to the Apicomplexan phylum, Coccidia class, Eucoccidiorida order, Sarcocystidae family, *Toxoplasma* genus which was first discovered in 1908, independently, by Splendore and Nicolle and Manceaux. The infection is acquired through: congenital infection (mother-fetus), ingestion of food infected with tissue cysts or oocysts, and drink contaminated with oocysts but sometimes with other routes such as organ transplant and blood transfusion. In immunocompetent people, toxoplasmosis is asymptomatic, but some can develop fever, weakness, sore throat, retinochoroiditis, cervical lymphadenopathy. In immunocompromised patients, it is severe, and toxoplasmic encephalitis can be seen in such patients with the signs of headache, lethargy, loss of memory, dementia, fever, psychomotor slowing, and behavioral changes. *T. gondii* infects food animals, including sheep, goats, pigs, and chickens. Route of infection in humans is ingestion meat or other foods and drinks contaminated with oocysts. Bradyzoites or sporozoites are released from tissue cysts or oocysts by the digestion process, subsequently invade the intestinal epithelial cells and become tachyzoites. Tachyzoites will eventually reach the blood or the lymphatic circulatory system and spread to distant organs. During the invasion and evasion, *T. gondii* is able to modulate the secretion of host cytokines and create a PV that avoids the fusion with the lysosomal system. Further, the expression of inflammatory chemokines in intestinal epithelial cells increases and attracts neutrophils, dendritic cells, and macrophages. Due to the sequential events taken place during the development of the parasite in the host cells, it leads to cell necrosis starting from the intestinal cells (Navia et al., 1986; Contini, 2008; Martina et al., 2011; de Oliveira, 2016; Foroutan-Rad et al., 2016).

8.3.7 TAENIA

Taenia belongs to the phylum Platyhelminthes, class Cestoda, order Cyclophyllidea and family *Taeniidae*, there are 45 *Taenia* spp., and among them, three species are associated with human disease namely: *Taenia saginata* (beef tapeworm), *Taenia solium* (pork tapeworm), and *Taenia asiatica* (Asian tapeworm). Infection with *Taenia* in humans is due to the consumption of improperly cooked meat. Taeniasis and cysticercosis are the two forms of the disease caused by *Taenia*. Taeniasis is the infection caused by adult worm, whereas cysticercosis is caused by metacestodes of *T. solium*, *T. saginata*, *T. asiatica*, *T. crassiceps*, and *T. taeniaeformis*. Individuals with taeniasis are asymptomatic or with mild symptoms such as abdominal pain, nausea, vomiting, changes in the appetite, intermittent diarrhea or constipation, fever, headache, weight loss, anemia, emaciation, dizziness, weakness, anal pruritus, damage, or bleeding in the stomach. Cysticercosis (neurocysticercosis, muscular, and dermatologic cysticercosis, and ocular cysticercosis) is due to the consumption of egg-contaminated vegetables/water or swallowing of egg-containing proglottids. Clinical symptoms of neurocysticercosis include convulsions and/or seizures, intracranial hypertension, headache, nausea, vomiting, vertigo, and papilledema, and psychiatric disturbances. In muscular and dermatologic cysticercosis, acute myositis or muscular pseudohypertrophy and subcutaneous nodules can be seen. Another form of cysticercosis is ophthalmic cysticercosis, which leads to severe pain and blurs or lost vision (Hoberg, 2006; Del Brutto, 2013; Coral-Almeida et al., 2015; Marcin Sierra et al., 2017).

8.3.8 ECHINOCOCCUS

Echinococcus belongs to the family Taeniidae. *Echinococcus granulosus* and *Echinococcus multilocularis* are the two important *Echinococcus* spp. in human infections, which leads to cystic echinococcosis (CE) and alveolar echinococcosis (AE) respectively by *Echinococcus granulosus* and *Echinococcus multilocularis*. Both the forms of the infection are life-threatening with a high fatality rate, but the AE has the most severe consequences. Following the ingestion of the eggs of the parasite by the host, it hatches into the oncosphere. These penetrate the intestinal mucosa and migrate into the organs through the blood circulation. In the tissues such as liver, kidney, spleen, brain, etc., and they develop cysts. CE is a chronic infection, and it takes a long time to show clinical symptoms. The clinical manifestations are variable (Kammerer and Schantz, 1993). It includes hepatic enlargement, epigastric pain, nausea, vomiting, and when the cyst ruptures, there will be an allergic reaction, even fatal anaphylaxis. AE is also a chronic infection, but it can cause a severe and often fatal infection in humans. The liver is the primary location of the infection and its metastases to the lungs, brain, and other tissues. The metacestode invades and destroys tissue like in malignancy. With the rupture of the hepatic cysts, lesions may invade the inferior vena cava and hepatic veins and there will be metastases to the right atrium. There will be a possibility of parasitic pulmonary emboli associated with this, which is fatal. The most common symptoms of AE are cough, chest pain, dyspnea, and hemoptysis. Less frequently, malaise, nausea, and vomiting, and thoracic deformations also can be seen (Ammann, 1996; Morar and Feldman, 2003; Arinc et al., 2009).

8.3.9 PARAGONIMUS

Paragonimiasis is a foodborne zoonotic helminthiasis and the etiology is *Paragonimus* spp. There are more than 50 species identified. The disease paragonimiasis is a subacute to chronic inflammatory granulomatous lung disease. The organism requires three hosts to complete the life cycle, and the definitive host is the human or carnivorous/omnivorous mammals. Human gets the infective larval stage of the life cycle which is the metacercariae by eating contaminated freshwater crabs or crayfish. Adult worms live in the lungs and lead to pulmonary lesions, and they also have extrapulmonary paragonimiasis in other tissues such as the brain, liver, pancreas, and abdominal wall. Metacercariae damage pleural membranes, and it leads to pleuritis and/or pneumothorax. Chest pain, cough, sputum, and dyspnea are significant signs of pulmonary disease. In the intestinal form, abdominal pain and diarrhea can occur. These digestive signs can be seen during the acute phase (period of invasion and the migration of immature flukes). Infective juveniles migrate to extrapulmonary sites, and they form ectopic granulomatous nodules; when it affects the brain, it is called cerebral paragonimiasis, and the condition in the skin is cutaneous paragonimiasis. The severe form of the disease, i.e., cerebral paragonimiasis has symptoms of visual impairment, headache, and epileptic seizures. Several enzymes and biologically active compounds have been characterized in *Paragonimus* as the virulence mediators. Cysteine proteases play important roles in excitement and initial settlement in the hosts, and the biologically active compounds play a role in cell adhesion, migration, and differentiation (Nakamura-Uchiyama et al., 2002; Kim et al., 2004; Song et al., 2010; Jiang et al., 2015; Yang et al., 2015).

8.3.10 FASCIOLA

Fasciola genus belongs to the family *Fasciolidae*, suborder *Echinostomata*, order *Echinostomida* subclass Digenea and class *Trematoda. Fasciola hepatica* and *Fasciola gigantica* are the two important liver flukes known to infect humans. *Fasciola* has a life cycle with a definitive mammalian host and a molluscan intermediate host. The life cycle is composed of several stages, such as eggs, miracidia, sporocysts, rediae, cercariae, metacercariae, and adult flukes. Adult worms live in the bile ducts of the mammalian liver, and they lay eggs. These eggs pass through the bile duct to the GI tract and are discharged with feces. These eggs hatch in the environment and released intermediate stages. The intermediate stage, miracidium penetrates snail tissues and undergo several stages of the life cycle. The stage termed as cercariae are released out of the snail and attach to aquatic vegetation in the water. They form encysted metacercariae, and they are infective to mammalian hosts. When the mammalian host ingests the encysted metacercariae, they exist to juvenile worms. Subsequently, they enter to the intestinal wall, migrate across the peritoneal cavity, afterward move to the liver. In the liver, they burrow into the parenchyma and settle in the bile ducts. Humans are the incidental host, and they get the parasites through the consumption of uncooked, larvae-contaminated vegetables or raw liver of infected animals, or after drinking larvae-containing water. The parenchymal (migratory) phase and the biliary phase are the two phases of fascioliasis in definitive hosts. *Fasciola* spp. produce serine proteases, and it has the ability to digest host macromolecules into peptides, which aid in parasite invasion and development. Some excretory and secretory products such as fatty acid-binding protein; cathepsins suppress the production of pro-inflammatory cytokines. Also, parasites increase the production of bile acids, leading to the reduced production of pro-inflammatory cytokines. Infection in humans can be symptomatic, or it can be acute fascioliasis, chronic fascioliasis, halzoun, or ectopic fascioliasis. Clinical symptoms of acute fascioliasis include fever (40–42°C), nausea, vomiting, in appetite, extreme abdominal pain, flatulence, diarrhea, urticaria (hives), tender hepatomegaly, splenomegaly, jaundice, ascites, anemia. In chronic infection patients may develop the symptoms of nausea, intermittent right upper quadrant abdominal tenderness (pain), biliary cholic, hepatomegaly, fatty food intolerance, pruritus, jaundice, cardiac disorder, pancreatitis, cholangitis, cholecystitis, gallstones, severe anemia (in children), portal cirrhosis, hardening of the liver (fibrosis, also known as pipestem liver), and possible death. If humans consume *Fasciola* infected raw liver, it develops pain, edema, and bleeding as well as respiration difficulty. In ectopic fascioliasis, the organism settles the sites other than the liver such as a peritoneal cavity, intestinal wall, lungs, subcutaneous tissue, lymph nodes (Bestas et al., 2014; Kim et al., 2015; Cwiklinski et al., 2016).

8.3.11 PARAGONIMUS

Paragonimus (lung fluke) is the causative organism of paragonimiasis in animals and humans, which leads to a subacute to chronic inflammatory granulomatous lung disease. Members of the *Paragonimus* genus are distributed in tropical to temperate climate zones. Fifty species of *Paragonimus* have been identified and seven species out of them are known to cause infection in human; *P. uterobilateralis* and *P. africanus* (Africa), *P. kellicotti* and *Paragonimus mexicanus* (America), *P. westermani*, *P. heterotremus* and *P. skrjabini* (Asia) (Blair et al., 2007; Singh et al., 2015). In the life cycle of the parasite, multiple hosts are associated; freshwater snails as the first and crustaceans as the second intermediate hosts. The definitive host can be the carnivorous or omnivorous mammals, including humans. Human acquires the organisms through ingestion of freshwater crabs or crayfish and consuming the raw meat of paratenic hosts, such as wild boar infected with the infective stage larvae live, i.e., metacercariae (Nakamura-Uchiyama et al., 2002; Blair et al., 2007). When the metacercariae reach the duodenum, it penetrates the intestinal wall and penetrates to the peritoneal cavity. Then they penetrate the abdominal muscles till they reach the lungs, i.e., the final destination. From the muscles, they returned to the peritoneal cavity, and these juveniles penetrate through the diaphragm into the pleural cavity where they can burrow into the lungs, damaging pleural membranes and lung tissues. Following the arrival to the lungs, they become adults, and they localize in the bronchiolar lumen and peribronchial tissues and begin laying eggs. Generally, the adult worms in the lungs live as pairs in cysts, causing typical pulmonary lesions. But the extrapulmonary paragonimiasis (ectopic paragonimiasis) also can be taken place where they parasitize the organs other than the lungs such as bran, abdominal wall, liver, pancreas. (Kim et al., 2004; Jiang et al., 2015; Yang et al., 2015). In order to penetrate the host, evade the host immunity, and survive and maturation inside the host, the organism has several virulence-related molecules. Several enzymes and biologically active molecules such as cysteine proteases, leucine amino-peptidase, paramyosin, antioxidant enzymes (glutathione peroxidases) have been identified and characterized. As an example, the cysteine proteases lead to lyse of the host tissue facilitating the migration of worm and also it neutralizes the IgG, thereby it regulates the host immune responses even (Choi et al., 2006; Song et al., 2008). Also, the antioxidant enzymes such as glutathione peroxidase, catalase and glutathione transferase prevent the parasite from the oxidative killing mechanisms of the host cells by the detoxification and scavenging of the potential of antioxidant enzymes (Hong et al., 2000; Kim et al., 2007; Bae et al., 2009). In the early stages of the infection, when the parasites migrate in the peritoneal cavity, human hosts are asymptomatic, but if there is a heavy infection, nonspecific symptoms can be seen, such as easy fatigability, myalgia, and mild fever. Based on the place where the parasite is located, it is reported to have several symptoms, and they can be categorized as intestinal symptoms, pleural symptoms, pulmonary symptoms, and extrapulmonary. Intestinal symptoms (such as abdominal pain and diarrhea) can be seen during the period of invasion and the migration of immature flukes. As the migrating metacercariae affect the pleural membranes, the signs related to pleuritis and pneumothorax also be reported. When the adult organisms are encapsulated in the lung, pulmonary symptoms appear; bronchial mucous production, pathological damage to lung tissues, including blood vessel destruction, remodeling processes, and also granulomatous lesions are formed leading to dry cough with blood-tinged sputum, pleuritic, and chest pain. The chronic symptoms of chest pain, cough, sputum, and dyspnea on exertion are some of the major signs. Though the predilection site in lung tissues for the parasite, the infective juvenile has the ability to migrate to extrapulmonary sites and form granulomatous nodules in several places in the body and it can even lead to severe illnesses associated with cerebral paragonimiasis (Sasaki et al., 2002; Lee et al., 2012; Xia et al., 2015).

8.3.12 CLONORCHIS SINENSIS

Clonorchis sinensis belongs to the genus *Clonorchis* Loss, 19072 known as the Chinese liver fluke is a causative agent of fish-borne clonorchiasis (a disease caused by *C. sinensis*). It is endemic in a number of countries wherein East Asia is the most common human liver fluke. As this organism is reported to cause cholangiocarcinoma, it was reported by the International Agency of Cancer Research and WHO in 2009 as group 1 biocarcinogens (Bouvard et al., 2009). The life cycle includes several hosts, snails as the first intermediate host and fish as the second intermediate host and mammal as the final host (Tang et al., 2016). Freshwater snails (*Parafossarulus* sp., *Alocinma* sp., and *Bithynia* sp.), the first intermediate hosts get the eggs of the parasites through ingestion and the egg hatch to miracidia stage and through several developmental stages they reached to the cercariae. This is the stage released to the water from the snails. Subsequently, it finds the second intermediate host, i.e., fish and certain crustaceans. Cercariae invade the skin of the fish which is mucoid and in the muscle it encysts as metacercariae. Metacercariae get into the final host (cats, dogs, pigs, human, etc.), by ingestion of raw or partially cooked fish containing infective metacercariae. Then the metacercariae encyst in the duodenum and migrate up the bile ducts (where they sexually mature and lay eggs) to the liver. The eggs are passed to the intestine with bile and pass out through feces.

Liver flukes, including *C. sinensis* affect the hepatic system, and pathological consequences of *C. sinensis* infestation occur mainly in the liver, bile ducts, and gallbladder. Pathological changes can be happened due to the parasites themselves, their excretory-secretory products (ESPs; which contain a number of proteases, antioxidants, and metabolic enzymes) and coinfection with pathogenic microorganisms (Saijuntha et al., 2014; Lai et al., 2016; Qian et al., 2016). Mechanical obstruction, inflammation, and subsequently, the bile stagnation can be happened due to the moving and feeding of flukes. Due to the changes in the composition of the bile and action of bacterial enzymes such as glucuronidase stimulate the stone formation in the bile ducts and followed by cholestasis and cholangitis (Sripa et al., 2010; Saijuntha et al., 2014). Though the mechanism of chronic inflammation and cholangiocarcinoma formation is yet to be fully elucidated; it is reported to be associated with different molecular pathways. It is associated with reactive oxygen species (ROS), reactive nitrogen species (RNS), and interleukins and NF-κB of the host (Sripa et al., 2007; Nam et al., 2012). The inflammatory process is induced by NF-κB and HIF-1a like activation factors, and it is due to the interaction of ESPs, toll-like receptor (TLR), pathogen-associated molecular patterns (PAMPs) of the host. Increase of inflammatory cells and proliferation and survival of the malignant cells are the consequences of a series of cellular mechanisms (such as regulating the cell cycle) and factors (inflammatory mediators such as IL-6, IL-1β, IL-8, and TNF-α) (Porta et al., 2011). ESPs involve in several pathological events such as promoting the proliferation of malignant or abnormal cells, suppressing the apoptosis and increase in the metastatic potential (Johnson et al., 2012; Chen et al., 2014).

8.3.13 ANGIOSTRONGYLUS CANTONENSIS

Angiostrongylus cantonensis is a lung nematode (metastrongyloid nematode) widely distributed in the Asia-Pacific region, and it is the causative agent of angiostrongyliasis. The disease has been recognized as one of the major reasons for eosinophilic meningitis (Wang et al., 2008, 2012; Barratt et al., 2016). The life cycle is a simple one, and it propagates through rodent definitive hosts and mollusk intermediate hosts. Adult worms are known to live in pulmonary arteries and also in right ventricle of infected definitive hosts and lay eggs. The first stage of the larvae in the lung migrates through the wall of the respiratory tract to the trachea, afterward down the intestinal tract, and are discharged with the feces. Feces of the definitive host carries the infective stage of the intermediate snail hosts, and they acquire it through ingestion or tissue penetration. In the snails, there are two moltings and develop into third-stage larvae, and it is transmitted to paratenic (transport) hosts (shrimps, land crabs, frog, lizard, and predacious land planarians) through ingestion of infected intermediate host (Wang et al., 2012). Humans do not play any role in the life cycle of the parasite, but they can be the accidental hosts. Humans get the parasite, *A. cantonensis* by ingestion of intermediate or paratenic hosts or vegetables that contain the infective larvae (the third stage). Ingestion of infected raw or undercooked mollusks is the most common route of infection for humans. The infective larvae are digested and released from vectors and invade intestinal tissue, leading to enteritis. The infective larvae penetrate the intestinal tract and enter the blood vessels. Through the circulation, they reach the meninges as they are neurotrophic. During its migration to the predilection site, they do excessive harm to the surrounding tissues as such proteolytic damage, local eosinophilic inflammation, and bleeding. Eosinophilic meningitis and ocular angiostrongyliasis are the most common clinical presentations but sometimes can be fatal encephalitic angiostrongyliasis. When the larvae reach the meninges, they die, and it results in local inflammatory reactions. Surrounding the dead worms, necrosis, and granuloma formation can be seen, and ocular manifestations such as retinal detachment or intraocular hemorrhage, including optic neuritis, also are seen (Shmutzhard et al., 1988; Feng et al., 2013; Barratt et al., 2016).

8.3.14 TRICHINELLA

Trichinella spp. cause trichinellosis, which is a potentially deadly foodborne disease. Humans acquire this infection through ingestion of raw or semi-raw meat and meat-derived products. Most common species of *Trichinella* infections in animals and humans is the *Trichinella spiralis*. The life cycle is composed of two generations in the same host. The new borne larvae, which are released by the gravid female, migrate to the lymphatic and blood vessels; subsequently, they enter to predilection sites such as muscle cells. New-borne larval stages are developed into the infective stages within the muscle cells. When the muscle containing the infective stage is ingested by the new host, larvae are released from the muscle tissues by digestion in the stomach. Subsequently, they reach the duodenum and penetrate into the villa. At the same time, this movement they undergo four molts within and turn into the adult stage. This parasite is transmitted from animal to animal and also to humans, only by the ingestion of skeletal muscle which is infected with larvae of *Trichinella*. Symptoms in humans with the infection of *Trichinella* include some nonspecific symptoms at the start of the infection, including uneasiness, fever, fever-associated chills, headaches, and GI disorders. Generally, the infection can be acute, or chronic in nature and in acute infection, the human will develop pyrexia, eyelid or facial edema and myalgia complicated by myocarditis, thromboembolic disease, and encephalitis. Other complications include encephalitis and secondary infections (bronchopneumonia and sepsis), which usually occur later during infection, that is, in the transitional phase to the chronic stage of disease (Kocięcka, 2000; Neghina et al., 2011a, b; Murrell, 2013; Song et al., 2018).

8.4 FOODBORNE VIRUSES

For the last three decades, foodborne and waterborne viral pathogens are increasingly reporting as one of the leading causes of illness in humans. An increase in the number of cases might be due to the improvement in the diagnostic techniques or increased marketing of fresh and frozen foods. Differing from the bacteria, viruses require a living host to establish the continuation of their life-cycle (Sibley, 2004). Therefore, the viruses cannot be multiplied on food or water, which is not in a living organism. It is recognized that the pathways of foodborne virus transmissions are either food, water or person-to-person contact in most of the cases. These are identified as the major causes of non-bacterial gastroenteritis (Wikswo et al., 2015). As viruses have been recognized as one of the critical sources of foodborne diseases, their role in overall food production and global food supply chain cannot be ignored under any circumstances. Raw or minimally processed foods are the most vulnerable types of foods that can transmit diseases to humans (Molins et al., 2001).

Even though food and water act as transmitting pathways of foodborne viruses, food, and water are futile when it comes to increasing the number of viruses. That being said, the number of viruses does not get expanded when the food and the water is stored or processed (Richards et al., 2010). The transcription and translation of viral nucleic acids require a host or living organisms. By structure, the viruses cannot continue their life-cycle outside of a host (Moreira et al., 2009). Yet, they are capable of surviving outside of a host. Depending on the virus, the time they can survive outside of a living cell varies. But, whenever they are in contact with a susceptible host, they start utilizing the host's metabolism to maintain their cycle. The ultimate outcome of such contact is a form of the disease. Viruses which cause human infections have been grouped in to 22 families and of them 10 are known to be FBP. Viruses from these 10 families can transmit from food and cause diseases ranging from mild diarrhea to severe encephalitis (Velebit et al., 2019).

Viral transmission to humans can happen in various ways, and foodborne viral transmission can take place due to several reasons. Direct contamination of food with the people who are infected, is the most common cause.

In food processing chains, this is known to be the most common cause of contamination (Koopmans et al., 2003), and the viruses transmitted from the fecal-oral route tends to create such contaminations. Once the food is contaminated from a person, the viral particles are ingested, and viruses take refuge in the gut epithelial cells of the host. With the viral invasion of epithelial cells, the virus will continue to spread and replicate themselves at a rapid rate (Harris and Harris, 2001). The replication can happen in the epithelial cells or maybe on some other sites of the host. Wherever the replication takes place, it increases the number of viruses, and ultimately, it will allow the virus to take control of the host's system, which can lead to illnesses. Also, one of the rare cases of foodborne viral transmission is the consumption of animal products derived from infected animals. This has a low probability of happening due to strengthening food safety protocols (Dhama et al., 2013). Even though it is rare, there is still a small chance of success. And that can lead to devastating effects on human health. NoV and also the HAV are the most frequent viruses which are associated with foodborne outbreaks worldwide (WHO, 2008; Li et al., 2015).

8.4.1 NOROVIRUS (NoV)

Human NoV is the major causative agent for AGE globally (Atmar and Estes, 2006; Glass et al., 2009), and the foodborne transmission is a primary route for NoV infection, which spread rapidly through the contaminated food handlers and food products. NoV was first discovered in 1972 associated with a human diarrheal outbreak among the school children in Norwalk, Ohio; it is an RNA virus, belongs to the Caliciviridae family and causes GI infections in the form of AGE (Patel et al., 2008; Glass et al., 2009). Earlier that is prior to the year 2006, rotavirus (RoV) was considered as the main cause for diarrhea in children globally, with the mortality of more than 440,000 children annually (Parashar et al., 2003). But with the introduction of a vaccine against the RoV, the number of cases declined and the NoV took the place of RoV.

NoVs are classified into six gene groups: GI–GVI (Parra et al., 2013). In accordance with the current nomenclature, at least 33 genotypes of these viruses are known. For humans, the most pathogenic are NoVs GI, GII, GIV (Pinto et al., 2012). According to existing literature, the most common genogroup in human infection is the GII gene group. It contains up to 19 genotypes and genotypes GII.3, GII.6, GII.12 are associated with the transmission of the virus through food (Lu et al., 2019). NoV is a small, round, nonenveloped icosahedral virion enclosing a single-stranded, positive-sense polyadenylated RNA. The genome is approximately 7.5-kb, measuring 27–35 nm in diameter (Xi et al., 1990).

Several ways of transmission of NoV infection are known: fecal-oral, person to person contact, and airborne (rarely). The transmission factors of this infection include contaminated food, water; the source of infection is a sick person (Barclay et al., 2014). The widespread infection is contributed by the high resistance of viruses to disinfectants, high viability, and a low infectious dose of pathogens (Glass et al., 2009).

NoV infection is a highly contagious disease and can occur in the form of single cases, group cases, and outbreaks. This pathogen causes over 90% of non-bacterial nature gastroenteritis and about 50% of all cases of gastroenteritis (Karst, 2010). This virus can affect all ages, but cases are high in school children (Shah et al., 2018).

The incubation period for NoV is an average of 24–48 hours. For the development of disease signs such as non-bloody diarrhea, vomiting is from 10 to 51 hours (Cardemil et al., 2017). The onset of the disease is acute; many patients report chills without fever. It should be noted that an increase in body temperature is evident and is characterized mainly by sub febrile conditions (Robilotti et al., 2015). The main clinical manifestations of the disease include nausea and vomiting, accompanied by weakness, dizziness, and headache. Many patients complain of muscle pain. Abdominal pains are observed in all cases of the disease; they can be aching or cramping in nature with localization in the upper abdomen. In immunosuppressed individuals, this virus can lead to persistent infection, and these chronically infected individuals can shed virus for months to years (Chan et al., 2016).

8.4.2 HEPATITIS A VIRUS (HAV)

Viral hepatitis takes a prominent place among all liver diseases. It is one of the most common in modern hepatology. Most lethal outcomes are associated with these infections in patients with acute viral hepatitis, as well as cases of the development of chronic liver diseases, including cirrhosis and hepatocellular carcinoma (Ringehan et al., 2017). The existence of two types of hepatitis was revealed, and one form is known to transmit by the fecal-oral route. This form has a relatively short incubation period, whereas the second form was transmitted parenterally. Later these two forms were named hepatitis A and hepatitis B, respectively. HAV was first visualized by an immune electron microscope (IEM) in 1973 as icosahedral-shaped virions approximately 27 nm in diameter. HAV belongs to the genus Hepatovirus in the Picornaviridae family (Pintó et al., 2007). HAV has initially been assigned to the genus Enterovirus, but it was isolated in a separate genus (Ding and Dabing, 2007). Unlike other members of the family, this virus is more stable in the external environment. Also, it multiplies slowly in cell cultures, requiring more time and virus strains adapted for *in vitro* growth. When transmitting its genome, HAV uses the cellular translation initiation factor eIF4G (Roberts et al., 2009). Therefore, it does not inactivate the synthesis of the cell's own proteins during its cycle, which leads to the absence of a cytopathic effect and complicates the cultivation even more. The HAV genome is a single-stranded (+) RNA 7.5 kb long (Jenkins et al., 2018). The viral genome includes 1 open reading frame (OPC) encoding a polyprotein weighing approximately 250 kDa. There are seven genotypes of HAV; I-VI, where the genotypes I-III and VII, are human and IV-VI genotypes belong to simian (Hussain et al., 2005).

HAV genotype I is the most predominant, followed by genotypes III and II (Genotype II is the least frequent genotype). Besides, each of the human genotypes (I–III) possesses two subgenotypes: A and B (1A is more common than the 1B). Genotype I isolates have been reported from North America, South America, Europe, Asia, Africa, and the Middle East) (Liu et al., 2010). HAV is stable at different environmental conditions making them withstand to different chemical and physical disinfection. Also, it is stable at low pH as low as pH 1 and the infectivity can be retained for 5 hours.

The main route of transmission of this virus to humans is through the fecal-oral route, and this can occur either through direct person-to-person contact or through consumption of food or water contaminated with feces of an infected person. Food can be contaminated with the virus during the cultivation harvesting, processing, packaging, and distribution through unhygienic handling or through the contaminated water or during postharvest processing and distribution through contaminated human sewage either during irrigation or harvesting. Following the entry of the virus through the oral route into a host, it passes to the intestine resisting the gastric acids. In the intestinal cells, there may be limited replication of the viral particles. Subsequently, it enters the portal circulation to end up in the liver. The virus particles enter the hepatocytes, and inside the hepatocyte, there will be significant viral replication. Following the replication, virions are excreted from the hepatocytes to the biliary system or to the bloodstream. Viruses enter the bile are secreted in feces. In contrast, the viruses in blood infect the other hepatocyte continuing the damage.

Hepatitis A is characterized by primary liver damage, which can be explained by the pathogenesis of the disease (Talwani et al., 2011). Liver function suffers from both direct exposure to the virus and the resulting immune response to infection (Walker et al., 2015). The incubation period following HAV infection is 2–6 weeks (average 4 weeks). The disease can be an asymptomatic subclinical infection and mild anicteric hepatitis to classical acute icteric hepatitis (also called acute viral hepatitis), to acute liver failure. This severe form can be fatal.

The incidence of HAV in different countries varies significantly; it is associated with socioeconomic factors affecting the quality of sanitary control and access to drinking water (Mantovani et al., 2015). Primary HAV infection provides long-term immunity, mainly due to specific antibodies produced during the course of the disease (Nainan et al., 2006). In less resource-poor countries with inadequate hygiene (African countries, some countries in Asia, South America, and Eastern Europe), this infection is widespread with nearly universal exposure during early childhood whereas in the countries like Western Europe, Australia, Japan, and North America, with good sanitation and hygiene the disease (HAV infection) occurrence is uncommon. Countries that have the intermediate socioeconomic conditions, the occurrence of HAV infection is comparatively higher (than the countries with well-established sanitary and hygienic conditions).

Often the cause of outbreaks of viral hepatitis A in many countries is the use of insufficiently boiled Clams/Mollusks (the first such outbreak was recorded in 1955 in Sweden) (Croci et al., 2005). This is due to the fact that mollusks filter large amounts of water through their bodies, as a result of which chemical substances, bacteria, and viruses can accumulate in their bodies in high concentrations, especially if they are collected near the place of sewage discharge (Zannella et al., 2017). Food or meat processed or managed by infected people (raw milk, cold meat, and water) can contribute to the transmission of hepatitis A (Petrović and D'Agostino, 2016).

8.4.3 HEPATITIS E VIRUS (HEV)

Hepatitis E virus (HEV) is the leading causative agent of acute (fast-flowing) viral hepatitis worldwide (Clemente-Casares et al., 2016) and reported several outbreaks and also reported as sporadic cases. This is a frequently reported disease in countries with poor hygienic conditions such as limited access to water, sanitation, and health services, affecting thousands of persons (Pérez-Gracia et al., 2015). Annually reported cases of HEV are counted as

20 million infections and 3.3 million symptomatic cases and an estimated 56,600 deaths worldwide. A massive hepatitis E epidemic was recorded in China in 1989, almost 300,000 cases were caused by the consumption of clams harvested from sewage in a polluted area (Fu et al., 2010). This virus is known to be transmitted through contaminated food, water, and through the feces of infected people. However, unlike the HAV, it does not spread by direct personal contact (Kamar et al., 2014). HEV is a member of the Herpes-virus family. Hepatitis E is most often recorded in Central Asia, Africa, and Latin America, i.e., in regions with a hot climate (Khuroo et al., 2016).

In contrast to regions with a hot climate on endemic for Hepatitis E, in countries with a temperate and cold climate, flare incidence of the disease is not recorded (Bouwknegt et al., 2010). The number of reported cases of this hepatitis E infection in recent years not related to traveling to endemic regions has increased among the population living in Europe (Lapa et al., 2015). Such cases of the disease received a special designation called autochthonous Hepatitis E. As per the scientific literature, autochthonous Hepatitis E is caused by the 3rd genotype in Germany, Denmark, France, and the Netherlands (Dreier and Juhl, 2014).

It is a non-enveloped small virus belongs to the genus Hepacivirus in the Hepeviridae family, with a diameter of 27–34 nm of icosahedral symmetry with 7.2-kb single-stranded RNA genome (Drexler et al., 2012). *Hepeviridae* is composed of two genera, namely, *Orthohepevirus* and *Piscihepevirus*, and the genus *Orthohepevirus* is further classified into four species, namely *Orthohepevirus* A, B, C, and D. The *Orthohepevirus* A is made up of four genotypes (i.e., HEV1, HEV2, HEV3, and HEV4). Among them, HEV1 and HEV2 are restricted to human beings and transmitted via contaminated water in developing countries (HEV1 is mainly reported in Asia and HEV2 in Africa and Mexico), and it is associated with large waterborne outbreaks. In contrast, the HEV3 and HEV4 infect mammalian animals, including pigs, wild boar, deer, and seldom human. There are few reports on foodborne outbreaks with HEV3 and HEV4 (Lu et al., 2006; Dalton et al., 2008).

Fecal-oral transmission is the most common route of transmission of infection. Contamination of drinking water supplies with feces is the most common mode of spread of hepatitis E (Said et al., 2009; Brassard et al., 2012). In addition to this primary mode of transmission, zoonotic foodborne transmission, person-to-person transmission (Meng et al., 1998; Teshale et al., 2010), blood transfusion (Colson et al., 2007), solid organ transplantation (Gérolami et al., 2008), nosocomial transmissions (Siddiqui et al., 2005) and vertical transmission (Kumar et al., 2004; Bose et al., 2014) also reported.

Following the entry of the virus through the oral route, infection is initiated in the lining epithelium of the GI tract, and portal vein carries the virus to the liver. In the hepatocytes, the virus multiplication occurs, and from there, it is released to bile and blood, also it can damage the liver.

8.4.4 ROTAVIRUS

Rotavirus infection is a common diarrogenic enteropathogen which represents about 38% of all the diarrhea cases that require hospitalization and about 200,000 deaths each year worldwide. The infection caused by is known as rotavirus gastroenteritis, intestinal flu, and stomach flu (Lin et al., 2014). Even after the introduction of a vaccine against rotaviral infection, still, the annual death reported being >200,000 in low-income countries (Crawford et al., 2017). Rotavirus is a member of the Reoviridae family, and it is a segmented double-stranded RNA (dsRNA) genome encapsulated within a nonenveloped, triple-layered icosahedral capsid. Around 10 different rotavirus species (A–J) have been classified on the basis of sequence and the antigenic properties (Matthijnssens et al., 2012; Bányai et al., 2017) and this species A is the most common infection in children (Gentsch et al., 2005). Rotavirus A is reported to be zoonotic, and there is a possibility of direct animal-to-human transmission (Gentsch et al., 2005; Matthijnssens et al., 2010).

Route of transmission of the rotavirus is through the fecal-oral route, and this virus is highly contagious and stable in the environment. Outside the human body, rotavirus can survive for several months. The virus can also transmit when using household items. That is usually from a sick person or from a recently recovered person from illness. When such a person is using raw water or food, they can get contaminated and transmit the virus to others (Dennehy, 2000). On vegetables and fruits, the virus remains viable for a month, and on various items from 1 to 7 weeks. This virus is most prevalent in infants and young children with the age of <5 years. The disease is characterized by nonspecific symptoms of AGE, including watery, nonbloody diarrhea (generally occurs within 48–72 hours of infection), fever, abdominal cramps, nausea, and headaches (Hagbom et al., 2012). Reinfections can occur throughout life, but with the repeat occurrence of the disease, severity is reduced (Crawford et al., 2017).

The primary site for the rotavirus to be attacked is the enterocytes of small intestinal villi, and they multiply within the enterocytes and induces diarrhea through the destruction of absorptive enterocytes (leading to malabsorption), intestinal secretion stimulated by rotavirus nonstructural protein 4 and activation of the enteric nervous system (Lundgren and Svensson, 2001; Ramig, 2004). Associated with a more severe manifestation of AGE, rotavirus infections can lead to antigenemia and viremia (Candy, 2007). There are some reports of extraintestinal infection affecting the liver, kidney, myocardium, involvement biliary atresia, and neurological involvement (Ramig, 2004).

8.4.5 HUMAN ASTROVIRUS

Astroviruses were first described in 1975 and got their name for a star-shaped structure. Astrovirus is a single-stranded, non-enveloped RNA virus with 6.4–7.7 Kb in size that belongs to the Astroviridae family (Pérot et al., 2017). Viral genome has contained three ORFs designated as ORF1a, 1b, and 2, coding for protease, protease-RdRp fusion, and capsid proteins, respectively. There are two genera in the family Astroviridae: *Avastrovirus* infects avian species, and *Mamastrovirus* infects mammalian species. Three different groups of HAstVs have been reported: the classic (*MAstV 1*), HAstV-MLB (*MAstV 6*), HAstV-VA/HMO (*MAstV 8* and *MAstV 9*) (Bosch et al., 2014). There are eight canonical serotypes (HAstV-1 to HAstV-8) and several noncanonical human genogroups in Human astroviruses (HAstV) (Arias and DuBois, 2017).

Infection occurs through the fecal-oral route, and the infection with AstV is one of the leading causes of pediatric gastroenteritis, second to the rotaviruses and NoVs (Palombo and Bishop, 1996; Walter and Mitchell, 2003). AstV infections are underreported, as the infections are self-resolving and do not require hospitalization. Symptoms occur 3–4 days after infection: diarrhea, nausea, fever, loss of appetite, abdominal pain, and vomiting (Glass et al., 1996). The disease lasts about 5 days, many infections are asymptomatic. However, the virus can be released for another month after the infection. Mostly the infection is asymptomatic, and they act as a carrier. This virus has a specific tropism for epithelial cells of the small intestine and AstV, form a tight barrier, and regulate the passage of nutrients, molecules, ions, other solutes, and water from the mucosal side to the intestinal lumen. As the virus do alterations in the permeability of the intestinal cells, diarrhea, and other clinical manifestations are occurred. The virus is stable in many conditions; it can resist inactivation in the GI tract, can remain intact on dry surfaces up to 90 days, stable in groundwater, and remain at extreme temperatures, resistant to chlorination. Bivalve mollusks (clams, mussels, and oysters) also reported as the common source of infection (Schultz-Cherry et al., 2001;

Kramer et al., 2006). Food preparation and handling can also be a source of contamination (Hansman et al., 2008).

8.4.6 SAPOVIRUS

Human sapovirus was first detected in the stool samples collected from infants with AGE by electron microscopy in the 1970s (Chiba et al., 2000). It can affect humans in all ages, but predominantly it causes sporadic AGE in infants and children (Lee et al., 2012).

Genus *Sapovirus* belongs to the family of caliciviruses (Caliciviridae) combines several genera of viruses, of which two (NoV and Sapovirus) are capable of causing disease in humans (Desselberger, 2019). Sapoviruses are much less common than NoVs, but Sapovirus infection has more acute clinical symptoms than NoV (Oka et al., 2015). SaV virion is about 27–40 nm in diameter with icosahedral symmetry, and the SaV genome is a linear positive-sense single-stranded RNA molecule of about 7.1–7.7 kb with poly-A tail at the 3′-terminus. The SaV genome contains two open-reading frames (ORFs).

SaV is transmitted principally through the fecal-oral route (contaminated food and water, environmental sources, person-to-person contact, and possibly by some other unknown modes). Most often, the disease is self-limiting in nature. The incubation period of the sapovirus gastroenteritis ranges from 1–4 days. The maximum excretion of the virus occurs in the early days of the disease with major clinical signs of include diarrhea and vomiting and there may be additional symptoms such as nausea, chills, stomach/abdominal cramps, myalgia or malaise, headache), but the virus shedding can last up to 2–3 weeks. Diarrhea usually is cured within about a week. It should also be noted that environments that are favorable to NoV are favorable for outbreaks of Sapoviruses (Hansman et al., 2007).

8.4.7 ADENOVIRUS

Adenoviruses were first detected and isolated from adenoid tissue in 1953. Human adenovirus (HAdV) belongs to the family, and it is composed of five genera: *Mastadenovirus*, *Aviadenovirus*, *Atadenovirus*, *Siadenovirus*, and *Icthadenovirus*. The genus *Mastadenovirus* includes all of the species infecting humans as well as simian, murine, bovine, equine, porcine, ovine, canine species. Adenoviruses are nonenveloped double-stranded DNA viruses with a diameter of 70–90 nm. The virion is composed of a complex proteic capsid, which is composed of 252 capsomeres, and a nucleoprotein core that contains the viral genome. There are different ways of transmission of the viruses, including respiratory, fomite, droplets, and fecal-oral transmission through the ingestion of contaminated food or water, and the virus can affect a wide range of issues. Generally, the adenoviral infection is asymptomatic, however it can develop to several diseases; mild upper respiratory tract infections, conjunctivitis, gastroenteritis, hepatitis, myocarditis, and life-threatening pulmonary disease, especially in pediatric and geriatric populations, and in individuals with immune suppression. Adenovirus infections can be epidemic, endemic, or sporadic (Holly and Smith, 2018; Mennechet et al., 2019).

8.4.8 ENTEROVIRUS

The genus Enterovirus belongs to the family *Picornaviridae*, which currently consists of 35 genera. This is a small, nonenveloped, single-stranded positive-sense ribonucleic acid (RNA) virus. The genus Enterovirus is composed of 13 species (i.e., Enterovirus A–J and Rhinovirus A–C).

Among the 13 Enterovirus species, Enterovirus A–D (consisting of a total of 116 serotypes) affect humans, primates, and occasionally dogs (as in the case of Enterovirus C), producing a range of human diseases; Enterovirus E and F occur in cattle, Enterovirus G in pigs, Enterovirus H and J in monkeys, Enterovirus I in camel, and Rhinovirus A–C in humans.

Mode of transmission is through respiratory droplets and fecal-oral and transplacental routes. The incubation period is variable, ranging from 2 to 40 days. The virus can withstand the environmental condition of the stomach which is acidic. The viral binding, translation, replication, assembling that occurs in the mucosal cells of the intestine or oropharyngeal mucosa. Subsequently, the viruses spread to the target tissues and organs through the blood and lymphatics. The clinical manifestations include mild respiratory and GI infections, nonspecific febrile illness, rash, conjunctivitis, and herpangina, to more severe epidemic pleurodynia, poliomyelitis, post poliomyelitis syndrome, meningitis, encephalitis, myopericarditis, paralysis, hepatitis, pancreatitis, and neonatal sepsis (Rao et al., 2012; Chang et al., 2016).

8.5 FOOD ASSOCIATED MICROORGANISMS

The traditional and empirical knowledge of food preservation and preparation of value-added food products has been dated back to the long human history. The use of microorganisms in food preservation and food production is the method of choice since the stone ages. In the last two centuries, there has been a progressive development in the field of food preservation and conservation due to the advances in microbiology, physiology, and biochemistry, and, more recently, genetics, and nutrition. In the food industry, microorganisms are used to produce a number of products, including several enzymes, antimicrobial compounds such as bacteriocins. Due to the wide variety of synthesized enzymes, microorganisms can perform many chemical processes more efficiently and economically than if these processes were carried out by chemical methods (Gurung et al., 2013). The study of the biochemical activity of microorganisms made it possible to select the conditions for their maximum activity as producers of various useful enzymes that are necessary for various chemical reactions and processes (Nigam, 2013).

Bio-preservation of food enhance or improve the shelf life by using microorganisms with antagonistic properties and/or their antimicrobial metabolites. One such kind of predominant and oldest method in the food bio-preservation is the fermentation; the process of transformation of raw materials into the product by the help of microorganisms such as yeasts, molds, and bacteria which apply to dairy products, vegetables, meat, and fish (Voulgari et al., 2010; Kabak and Dobson, 2011). In fact, fermentation has been used for more than 6,000 years as an effective method in food bio-preservation (McGovern et al., 2004).

The main function of the fermentation is the preservation, and it is done through the consumption of nutrients and production of inhibitory compounds, usually ethanol or organic acids, other compounds, such as bacteriocins or short-chain fatty acids. In addition, the fermentation increases the accessibility of nutrients in some foods, such as bread, detoxifies other products, and enhances the flavor of various substrates.

Alcoholic beverages and other products are obtained using yeast (Ciani et al., 2020). Yeast and bacteria are used in the baking industry, while lactic acid bacteria (LAB) are used in the dairy industry (Rizzello et al., 2014). Molds are fungi that form mycelium grow both on liquid and semi-solid, and on solid media and some molds are used for the production of cheeses. However, most of these fungi cause spoilage of the product both due to growth on it or changes in the properties of the product and due to the release of mycotoxins (Hymery et al., 2014). Bakery products are based on the activity of yeast and LAB developing in the dough. The combined action of these microorganisms leads to the fermentation of flour sugars. Yeast causes alcoholic fermentation, while LAB cause lactic acid one (Gänzle et al., 2007). The resulting lactic and other acids acidify the dough, maintaining an optimal pH level for yeast. Carbon dioxide loosens the dough and accelerates its maturation (Komlenić et al., 2012).

Among the bacteria and yeast, molds are the largest and most complex organisms. Favorable conditions for the development of molds are free to access of air and the acidic reaction of the environment (Kuhn et al., 2003). Mold on food products develops in the form of a complex mycelium, and its reproduction occurs through numerous spores. In particular, the most common cause of microbiological spoilage of dairy products is also yeast and mold.

Molds have been used in the food industry, for the manufacture of lactic acid products, in bakery, winemaking, and brewing, in the manufacture of meat and sausages, and the production of citric acid (Dalié et al., 2010). For example, obtaining the famous French cheeses Roquefort and Brie are impossible without mold (Awasti et al., 2020), while in the production of kefir and bread, yeast is used that emit carbon dioxide during fermentation. Among the metabolites produced by microorganisms, food acids (Singh et al., 2017), in particular citric acid, is extensively used. The first reports of organic acid production by fungi date back to the second half of the 19th Century. Currently, the sequence of reactions underlying the metabolism of most acids formed by fungi (gluconic, citric, malic, succinic, and fumaric acids) is relatively well understood (Liaud et al., 2014).

Yeasts represent unicellular immobile microorganisms that do not form mycelium, and it is found in the nature where sugar-containing liquids or media are present (Miura et al., 2003). Yeast plays a significant role in the food industry. The use of yeast in the food industry is mainly associated with the ability to cause alcoholic fermentation: in the baking industry, in the production of beer, wine, alcohol (Loureiro et al., 2003). In the production of wheat bread, yeast *Saccharomyces cerevisiae* is used (Jayaram et al., 2014). Yeast is used for the production of fodder and food protein, vitamins, enzymes, essential amino acids, organic acids, and other biologically active substances (Bekatorou et al., 2006).

Currently, the baking industry is the leading consumer of industrially produced yeast, called baker's yeast. Baker's yeast belongs to the species *Saccharomyces cerevisiae*. This is grown in an environment rich with oxygen, in special containers with sugar beets, nitrogen mixtures, and minerals (Berłowska et al., 2016). These fungi appear in the form of foamy deposits, which are cleaned of impurities using a centrifuge and water. Then the resulting material is dehydrated, compacted, and sold in this form (Gélinas, 2017). Significant amounts of yeast are produced for animal feed and small quantities for human nutritional purposes. The yeast produced directly for such purposes is called primary, and those that are extracted as by-products of the fermentation process, such as brewing, are called secondary (Shurson, 2018). Protein concentrates and isolates, often called unicellular proteins, can be obtained, and used to enrich the proteins of low-protein foods.

One of the great scientists, Louis Pasteur scientific basis of the fermentation processes and modern microbiology, was established, and it was the landmark in the development of fermentation in the food bio-preservation. Whereas Ilja Metchnikoff and his collaborator, Stamen Grigorov, suggested the concept of the functional food products and probiotics for the first time.

The bacteria important in food microbiology in food preservation are clustered into groups according to the fermentation product, for example, LAB, acetic acid bacteria, propionic acid. Proteolytic, saccharolytic, and lipolytic bacteria are distinguished (Lorenzo et al., 2018). LAB are the primary bacteria found in the traditional fermented food, and it combined in one group according to their ability to ferment carbohydrates with the formation of predominantly lactic acid (Rhee et al., 2011). LAB are a group of microaerophilic gram-positive microorganisms that ferment carbohydrates with the formation of lactic acid as one of the main products (De Vuyst et al., 1994). These microorganisms live in a wide range of ecological niches, viz. plant surfaces, environment, GI tract. The genera represent LAB are *Lactobacillus*, *Carnobacterium*, *Lactococcus*, *Weisella*, *Enterococcus*, *Oenococcus*, *Streptococcus*, *Leuconostoc* and *Pediococcus* (Ross et al., 2002). A wide variety of strains are routinely used as starter cultures in several of the food industries; dairy, meat, vegetable, and bakery (Abee et al., 1995). These organisms extend the shelf life of the fermented food by inhibiting the growth of spoilage and pathogenic bacteria due to competition for nutrients. Also, some inhibitors (starter-derived), namely hydrogen peroxide, lactic acid and bacteriocins also can extend the shelf life of the fermented food product. Cottage cheese, sour cream, oil, yogurt is prepared in pure cultures using various starter cultures. Cheese, sauerkraut, pickles, fermented milk products, fermented sausages, and many traditional fermented foods are also associated with the activity of LAB (Ray and Daeschel, 1992; Rezac et al., 2018). The first evidence on milk fermentation was back from approximately 7,000 years ago when the Europeans produced cheese and fermented dairy foods (Salque et al., 2013). Organic acids such as lactic, acetic, propionic, butyric, formic, hydroxyl-phenylacetic acid, and phenylacetic acid are some of the active antagonistic metabolites which are produced by LAB.

Also, they are known to synthesize diverse antagonistic compounds, namely, hydrogen peroxide, ethanol, acetoin, fatty acids, reuterin, diacetyl, reutericyclin, and carbon dioxide. Even they are capable of generating antifungal compounds, i.e., phenyl-lactate, hydroxyphenyl-lactate, propionate, cyclic dipeptides, phenylacetic acid and 3-hydroxy fatty acids. There are several bacteriocins produced by LAB such as nisin, pediocins, lacticins, and enterocins (Muhialdin et al., 2011; Reis et al., 2012; Kaškoniene et al., 2017).

Fermentation is an oxidoreductase process that results in organic acids as (predominantly lactic acid) the main end products of the fermentation. Due to the production of acids, the pH decreases in the surrounding environment, and it affects the non-acidophiles. Moreover, the lactic acid disrupts the cytoplasmic membrane, interference with membrane potential, and reduction in intracellular pH (Shelef, 1994; Axe and Bailey, 1995). Carbon dioxide production due to the fermentation will create an anaerobic environment, and it avoids the growth of aerobic bacteria. Further, the CO_2 dissolves in water, and it generates carbonic acid (Adams and Nicolaides, 1997). In addition to the organic acids and CO_2, LAB produces diacetyl (2,3-butanedione) as a by-product, which exerts antibacterial activity against several notorious bacteria, including *Salmonella*, *Escherichia coli*, *Listeria*, *Aeromonas* and *Yersinia*. In the presence of oxygen, this group of bacteria produces hydrogen peroxide due to the action of flavoprotein oxidases/NADH peroxidase and the H_2O_2 is well known to have an antibacterial effect (Ammor et al., 2006). LAB produces compounds such as reuterin and reutericyclin and these are with antibacterial effect.

Reuterin is known as the 3-hydroxypropionaldehyde (3-HPA), produced by *Lactobacillus reuteri* and by some other LAB inhibit synthesis of DNA. This compound carries the ability of inhibiting many forms of organisms, including fungus, protozoa, bacteria (*Staph. aureus*, *Salmonella choleraesuis*, *E. coli* O157:H7, *Aeromonas hydrophila*, *Yersinia enterocolitica*, and *Campylobacter jejuni*) (Talarico and Dobrogosz 1989: Arqués et al., 2004). Whereas the reutericyclin incapable of inhibiting fungi and yeasts but it has shown the inhibiting effect on gram-positive and gram-negative bacteria (Arqués et al., 2004). Though it is not the primary compounds of LAB as in actinomycete, LAB also produces some kind of antifungal compounds (e.g., benzoic acid, methylhydantoin, mevalonolactone). Some bacteria, including LAB, have the potential of producing other kinds of antimicrobial peptides or protein known as bacteriocins. There are four types of bacteriocins produced by LAB (Klaenhammer, 1993), and these bacteriocins are bactericidal or bacteriostatic. There are several groups of bacteriocins synthesized by LAB: (I) lantibiotics, (II) small heat-stable peptides, (III) large heat-labile proteins, and (IV) complex proteins which require the lipid or carbohydrate moieties. Among the other bacteriocins produced by LAB, the group of antibiotics is well studied. The well-known group within the lantibiotics is Nisin A (a 34-residue antibacterial peptide), which synthesized by several strains of *Lactococcus lactis* and strongly inhibited the growth of a wide range of Gram-positive bacteria (Klaenhammer, 1993; Abee et al., 1995). The cytoplasmic membrane of the bacterial cell is the primary target for nisin activity. It is reported that the use of bacteriocins in the preservation of food, in dairy practice, nitrate is commonly added to cheese milk to prevent outgrowth of clostridia spores. It had shown that the replacement of nitrate by nisin A. It had revealed that the outgrowth spores of *C. tyrobutyricum* in nitrate-free Gouda cheese was inhibited completely following a nisin A producing strain was added to the starter culture (Hugenholtz and de Veer, 1991). It was reported that the Nisin A as a potent inhibitor against the pathogenic microorganisms such as *L. monocytogenes* in Camembert, and also it habited the organism in cottage cheese (Benkerroum and Sandine, 1988; Maisnier-Patin et al., 1992; Ferreira and Lund, 1996). It is also reported that nisin can inhibit the growth of *C. botulinum* spores in cheese spreads (Cleveland et al., 2001). *Bacillus* spp., *E. coli*, and *Salmonella* spp. have been inhibited by nisin exhibiting its broader range of activity against gram-positive and gram-negative bacteria (Plockova et al., 1996; Belfiore et al., 2007). Not only in dairy products, it also showed the inactivation of *L. innocua* and *L. monocytogenes* in caviar sturgeon (*Acipenser transmontanus*) and Pacific salmon (*Oncorhynchus keta*), inhibition of microbial flora of sardine (Al-Holy et al., 2005) and reduce the surface numbers of *L. monocytogenes* (Elotmani and Assobhei, 2004), *Salmonella* and native microflora on vacuum-packed raw shrimps stored at 4°C (Wang et al., 2012).

8.6 CONCLUSION

This chapter recognized several foodborne diseases, caused by vast range of organisms such as bacteria, virus posing severe global burden with negative impact on health, social-economic development, and trade. Among the causes of foodborne diseases, bacterial causes such as *Campylobacter* and *Salmonella* are the most predominant agents. Whereas among the viral causes, NoV is recognized as the leading cause of epidemic of AGE. There are several other causes for foodborne infections discussed namely parasites. This stressed the importance of achieving better health education in terms of controlling or minimizing the occurrence of foodborne diseases.

KEYWORDS

- **bacteria**
- ***Campylobacter***
- ***Entamoeba***
- **foodborne organisms**
- **hepatitis virus**
- **norovirus**
- **parasites**
- ***Salmonella***
- **viruses**

REFERENCES

Abee, T., Krockel, L., & Hill, C., (1995). Bacteriocins: Modes of action and potentials in food preservation and control of food poisoning. *International Journal of Food Microbiology, 28*(2), 169–185.

Adam, R. D., (2001). Biology of *Giardia lamblia*. *Clinical Microbiology Reviews, 14*(3), 447–475.

Adams, M. R., & Nicolaides, L., (1997). Review of the sensitivity of different foodborne pathogens to fermentation. *Food Control, 8*(5, 6), 227–239.

Al-Holy, M., Lin, M., & Rasco, B., (2005). Destruction of *Listeria monocytogenes* in sturgeon (*Acipenser transmontanus*) caviar by a combination of nisin with chemical antimicrobials or moderate heat. *Journal of Food Protection, 68*(3), 512–520.

Ali, M., Nelson, A. R., Lopez, A. L., & Sack, D. A., (2015). Updated global burden of cholera in endemic countries. *PLoS Neglected Tropical Diseases, 9*(6), e0003832.

Allen, K. J., Wałecka-Zacharska, E., Chen, J. C., Katarzyna, K. P., Devlieghere, F., Van, M. E., et al., (2016). *Listeria monocytogenes* – an examination of food chain factors potentially contributing to antimicrobial resistance. *Food Microbiology, 54*, 178–189.

Amaro, C., & Biosca, E. G., (1996). *Vibrio vulnificus* biotype 2, pathogenic for eels, is also an opportunistic pathogen for humans. *Applied and Environmental Microbiology, 62*(4), 1454–1457.

Ammann, R. W., & Eckert, J., (1996). Cestodes: Echinococcus. *Gastroenterology Clinics, 25*(3), 655–689.

Ammor, S., Tauveron, G., Dufour, E., & Chevallier, I., (2006). Antibacterial activity of lactic acid bacteria against spoilage and pathogenic bacteria isolated from the same meat small-scale facility: 1–Screening and characterization of the antibacterial compounds. *Food Control, 17*(6), 454–461.

Anand, B. S., Malhotra, V., Bhattacharya, S. K., Datta, P., Datta, D., Sen, D., et al., (1986). Rectal histology in acute bacillary dysentery. *Gastroenterology, 90*(3), 654–660.

Andersson, A., Ronner, U., & Granum, P. E., (1995). What problems does the food industry have with the spore-forming pathogens *Bacillus cereus* and *Clostridium perfringens*?. *International Journal of Food Microbiology*, *28*(2), 145–155.

Argudín, M. Á., Mendoza, M. C., & Rodicio, M. R., (2010). Food poisoning and *Staphylococcus aureus* enterotoxins. *Toxins*, *2*(7), 1751–1773.

Arias, C. F., & DuBois, R. M., (2017). The astrovirus capsid: A review. *Viruses*, *9*(1), 15.

Arinc, S., Kosif, A., Ertugrul, M., et al., (2009). Evaluation of pulmonary hydatid cyst cases. *International Journal of Surgery*, *7*(3), 192–195.

Arqués, J. L., Fernández, J., Gaya, P., Nuñez, M., Rodríguez, E., & Medina, M., (2004). Antimicrobial activity of reuterin in combination with nisin against foodborne pathogens. *International Journal of Food Microbiology*, *95*(2), 225–229.

Atmar, R. L., & Estes, M. K., (2006). The epidemiologic and clinical importance of norovirus infection. *Gastroenterology Clinics*, *35*(2), 275–290.

Awasti, N., & Anand, S., (2020). The role of yeast and molds in dairy industry: An update. In: Minj, J., Sudhakaran, A., & Kumari, A., (eds.), *Dairy Processing: Advanced Research to Applications* (pp. 243–262). Singapore, Springer.

Axe, D. D., & Bailey, J. E., (1995). Transport of lactate and acetate through the energized cytoplasmic membrane of *Escherichia coli*. *Biotechnology and Bioengineering*, *47*(1), 8–19.

Bae, Y. A., Cai, G. B., Kim, S. H., Zo, Y. G., & Kong, Y., (2009). Modular evolution of glutathione peroxidase genes in association with different biochemical properties of their encoded proteins in invertebrate animals. *BMC Evolutionary Biology*, *9*(1), 1–13.

Balaban, N., & Rasooly, A., (2000). Staphylococcal enterotoxins. *International Journal of Food Microbiology*, *61*(1), 1–10.

Bányai, K., Kemenesi, G., Budinski, I., Földes, F., Zana, B., Marton, S., et al., (2017). Candidate new rotavirus species in Schreiber's bats, Serbia. *Infection, Genetics and Evolution*, *48*, 19–26.

Barash, J. R., & Arnon, S. S., (2014). A novel strain of *Clostridium botulinum* that produces type B and type H botulinum toxins. *The Journal of Infectious Diseases*, *209*(2), 183–191.

Barbuddhe, S. B., & Trinad, C., (2009). *Listeria* as an enteroinvasive gastrointestinal pathogen. In: Sasakawa, C., (ed.), *Molecular Mechanisms of Bacterial Infection Via the Gut* (pp. 173–195). Berlin, Heidelberg: Springer.

Barclay, L., Park, G. W., Vega, E., Hall, A., Parashar, U., et al., (2014). Infection control for norovirus. *Clinical Microbiology and Infection*, *20*(8), 731–740.

Barratt, J., Chan, D., Sandaradura, I., Malik, R., Spielman, D., Lee, R., Marriott, D., et al., (2016). *Angiostrongylus cantonensis*: A review of its distribution, molecular biology and clinical significance as a human pathogen. *Parasitology*, *143*(9), 1087.

Beaz-Hidalgo, R., Latif-Eugenín, F., Hossain, M. J., Berg, K., Niemi, R. M., Rapala, J., et al., (2015). *Aeromonas aquatica* sp. nov., *Aeromonas finlandiensis* sp. nov. and *Aeromonas lacus* sp. nov. isolated from Finnish waters associated with cyanobacterial blooms. *Systematic and Applied Microbiology*, *38*(3), 161–168.

Bekatorou, A., Psarianos, C., & Koutinas, A. A., (2006). Production of food grade yeasts. *Food Technology and Biotechnology*, *44*(3), 407–415.

Belfiore, C., Castellano, P., & Vignolo, G., (2007). Reduction of *Escherichia coli* population following treatment with bacteriocins from lactic acid bacteria and chelators. *Food Microbiology*, *24*(3), 223–229.

Benkerroum, N., & Sandine, W. E., (1988). Inhibitory action of nisin against *Listeria monocytogenes*. *Journal of Dairy Science*, *71*(12), 3237–3245.

Bennish, M. L., Harris, J. R., Wojtyniak, B. J., & Struelens, M., (1990). Death in shigellosis: Incidence and risk factors in hospitalized patients. *Journal of Infectious Diseases*, *161*(3), 500–506.

Bergdoll, M. S., & Lee, W. A. C., (2006). Staphylococcal intoxications. In: Cliver, D. O., Potter, M., & Riemann, H. P., (eds.), *Foodborne Infections and Intoxications* (pp. 523–562). San Diego, California: Academic Press, Elsevier Inc.

Berłowska, J., Pielech-Przybylska, K., Balcerek, M., Dziekońska-Kubczak, U., Patelski, P., Dziugan, P., & Kręgiel, D., (2016). Simultaneous saccharification and fermentation of sugar beet pulp for efficient bioethanol production. *BioMed Research International*, *2016*, 1–10.

Bertsch, D., Rau, J., Eugster, M. R., Haug, M. C., Lawson, P. A., Lacroix, C., & Meile, L., (2013). *Listeria fleischmannii* sp. nov., isolated from cheese. *International Journal of Systematic and Evolutionary Microbiology*, *63*(2), 526–532.

Berube, B. J., & Wardenburg, J. B., (2013). *Staphylococcus aureus* α-toxin: Nearly a century of intrigue. *Toxins*, *5*(6), 1140–1166.

Bestas, R., Yalçin, K., & Çiçek, M., (2014). Cholestasis caused by *Fasciola gigantica*. *Türkiye Parazitolojii Dergisi*, *38*(3), 201.

Bhowmick, P. P., Devegowda, D., Ruwandeepika, H. D., Fuchs, T. M., Srikumar, S., Karunasagar, I., et al., (2011). gcpA (stm1987) is critical for cellulose production and biofilm formation on polystyrene surface by *Salmonella enterica* serovar weltevreden in both high and low nutrient medium. *Microbial Pathogenesis*, *50*(2), 114–122.

Bhunia, A. K., (2018). *Foodborne Microbial Pathogens: Mechanisms and Pathogenesis* (pp. 1–165). New York: Springer.

Bille, J., (1990). Epidemiology of human listeriosis in Europe, with special reference to the Swiss outbreak. *Foodborne listeriosis*, *71*, 74.

Blair, D., Takeshi, A., & Wenlin, W., (2007). Paragonimiasis. In: Murrell, K. D., & Fried, B., (eds.), *Foodborne Parasitic Zoonoses: World Class Parasites* (pp. 117–150). Boston, MA: Springer.

Borges, K. A., Furian, T. Q., Souza, S. N. D., Salle, C. T. P., Moraes, H. L. D. S., & Nascimento, V. P. D., (2019). Antimicrobial resistance and molecular characterization of *Salmonella enterica* serotypes isolated from poultry sources in Brazil. *Brazilian Journal of Poultry Science*, *21*(1).

Bosch, A., Pintó, R. M., & Guix, S., (2014). Human astroviruses. *Clinical Microbiology Reviews*, *27*(4), 1048–1074.

Bose, P. D., Das, B. C., Hazam, R. K., Kumar, A., Medhi, S., & Kar, P., (2014). Evidence of extrahepatic replication of hepatitis E virus in human placenta. *Journal of General Virology*, *95*(6), 1266–1271.

Bouvard, V., Baan, R., Straif, K., Grosse, Y., Secretan, B., El Ghissassi, F., et al., (2009). A review of human carcinogens – part B: Biological agents. *The Lancet Oncology*, *10*(4), 321–322.

Bouwknegt, M., Rutjes, S. A., & De Roda, H. A. M., (2010). Hepatitis E virus risk profile. *Identifying Potential Animal, Food and Water Sources for Human Infection* (pp. 1–60). RIVM report, Bilthoven, the Netherlands.

Brassard, J., Gagné, M. J., Généreux, M., & Côté, C., (2012). Detection of human foodborne and zoonotic viruses on irrigated, field-grown strawberries. *Applied and Environmental Microbiology*, *78*(10), 3763–3766.

Broberg, C. A., Calder, T. J., & Orth, K., (2011). *Vibrio parahaemolyticus* cell biology and pathogenicity determinants. *Microbes and Infection*, *13*(12, 13), 992–1001.

Brynestad, S., & Granum, P. E., (2002). *Clostridium perfringens* and foodborne infections. *International Journal of Food Microbiology*, *74*(3), 195–202.

Bueno, S. M., Riquelme, S., Riedel, C. A., & Kalergis, A. M., (2012). Mechanisms used by virulent *Salmonella* to impair dendritic cell function and evade adaptive immunity. *Immunology*, *137*(1), 28–36.

Candy, D. C., (2007). Rotavirus infection: A systemic illness?. *PLoS Med.*, *4*(4), e117.

Cardemil, C. V., Parashar, U. D., & Hall, A. J., (2017). Norovirus infection in older adults: Epidemiology, risk factors, and opportunities for prevention and control. *Infectious Disease Clinics*, *31*(4), 839–870.

Carter, A. T., & Peck, M. W., (2015). Genomes, neurotoxins and biology of *Clostridium botulinum* group I and group II. *Research in Microbiology*, *166*(4), 303–317.

Carter, J. D., & Hudson, A. P., (2009). Reactive arthritis: Clinical aspects and medical management. *Rheumatic Disease Clinics*, *35*(1), 21–44.

Chan, Y. S., & Ng, T. B., (2016). Shiga toxins: From structure and mechanism to applications. *Applied Microbiology and Biotechnology*, *100*(4), 1597–1610.

Chang, P. C., Chen, S. C., & Chen, K. T., (2016). The current status of the disease caused by enterovirus 71 infections: Epidemiology, pathogenesis, molecular epidemiology, and vaccine development. *International Journal of Environmental Research and Public Health*, *13*(9), 890.

Chen, K., Dong, N., Chan, E. W. C., & Chen, S., (2019). Transmission of ciprofloxacin resistance in *Salmonella* mediated by a novel type of conjugative helper plasmids. *Emerging Microbes and Infections*, *8*(1), 857–865.

Chen, W., Wang, X., Lv, X., Tian, Y., Xu, Y., Mao, Q., et al., (2014). Characterization of the secreted cathepsin B cysteine proteases family of the carcinogenic liver fluke *Clonorchis sinensis*. *Parasitology Research*, *113*(9), 3409–3418.

Chen, X., Zuo, Y., & Zuo, W., (1999). Observation on the clinical symptoms and sporocyst excretion in human volunteers experimentally infected with *Sarcocystis hominis*. *Chinese Journal of Parasitology and Parasitic Diseases*, *17*(1), 25–27.

Chiang, S. R., & Chuang, Y. C., (2003). *Vibrio vulnificus* infection: Clinical manifestations, pathogenesis, and antimicrobial therapy. *Journal of Microbiology, Immunology, and Infection*, *36*(2), 81–88.

Chiba, S., Nakata, S., Numata-Kinoshita, K., & Honma, S., (2000). Sapporo virus: History and recent findings. *The Journal of Infectious Diseases*, *181*(S2), S303–S308.

Choi, J. H., Lee, J. H., Yu, H. S., Jeong, H. J., Kim, J., Hong, Y. C., et al., (2006). Molecular and biochemical characterization of hemoglobinase, a cysteine proteinase, in *Paragonimus westermani*. *The Korean Journal of Parasitology*, *44*(3), 187.

Chopra, A. K., & Houston, C. W., (1999). Enterotoxins in *Aeromonas*-associated gastroenteritis. *Microbes and Infection*, *1*(13), 1129–1137.

Christy, N. C., & Petri, W. A., (2011). Mechanisms of adherence, cytotoxicity and phagocytosis modulate the pathogenesis of *Entamoeba histolytica*. *Future Microbiology*, *6*(12), 1501–1519.

Ciani, M., Canonico, L., Oro, L., & Comitini, F., (2020). Footprint of nonconventional yeasts and their contribution in alcoholic fermentations. In: Grumezescu, A. M., & Holban, A. M., (eds.), *Biotechnological Progress and Beverage Consumption* (pp. 435–465). United Kingdom, Elsevier.

Clauss, H. E., & Lorber, B., (2008). Central nervous system infection with *Listeria monocytogenes*. *Current Infectious Disease Reports*, *10*(4), 300–306.

Clemente-Casares, P., Ramos-Romero, C., Ramirez-Gonzalez, E., & Mas, A., (2016). *Hepatitis E* virus in industrialized countries: The silent threat. *BioMed Research International, 2016*, 1–18.

Cleveland, J., Montville, T. J., Nes, I. F., & Chikindas, M. L., (2001). Bacteriocins: Safe, natural antimicrobials for food preservation. *International Journal of Food Microbiology, 71*(1), 1–20.

Colson, P., Coze, C., Gallian, P., Henry, M., De Micco, P., & Tamalet, C., (2007). Transfusion-associated hepatitis E, France. *Emerging Infectious Diseases, 13*(4), 648.

Contini, C., (2008). Clinical and diagnostic management of toxoplasmosis in the immunocompromised patient. *Parasitologia, 50*(1, 2), 45.

Coral-Almeida, M., Gabriël, S., Abatih, E. N., Praet, N., Benitez, W., & Dorny, P., (2015). *Taenia solium* human cysticercosis: A systematic review of sero-epidemiological data from endemic zones around the world. *PLoS Neglected Tropical Diseases, 9*(7), e0003919.

Crawford, S. E., Ramani, S., Tate, J. E., Parashar, U. D., Svensson, L., Hagbom, M., et al., (2017). Rotavirus infection. *Nature Reviews Disease Primers, 3*(1), 1–16.

Croci, L., De Medici, D., Di Pasquale, S., & Toti, L., (2005). Resistance of hepatitis A virus in mussels subjected to different domestic cookings. *International Journal of Food Microbiology, 105*(2), 139–144.

Croxen, M. A., & Finlay, B. B., (2010). Molecular mechanisms of *Escherichia coli* pathogenicity. *Nature Reviews Microbiology, 8*(1), 26–38.

Croxen, M. A., Law, R. J., Scholz, R., Keeney, K. M., Wlodarska, M., & Finlay, B. B., (2013). Recent advances in understanding enteric pathogenic *Escherichia coli*. *Clinical Microbiology Reviews, 26*(4), 822–880.

Cwiklinski, K., O'neill, S. M., Donnelly, S., & Dalton, J. P., (2016). A prospective view of animal and human Fasciolosis. *Parasite Immunology, 38*(9), 558–568.

Dalié, D. K. D., Deschamps, A. M., & Richard-Forget, F., (2010). Lactic acid bacteria – Potential for control of mould growth and mycotoxins: A review. *Food Control, 21*(4), 370–380.

Dalton, H. R., Bendall, R., Ijaz, S., & Banks, M., (2008). Hepatitis E: An emerging infection in developed countries. *The Lancet Infectious Diseases, 8*(11), 698–709.

Dasti, J. I., Tareen, A. M., Lugert, R., Zautner, A. E., & Groß, U., (2010). *Campylobacter jejuni*: A brief overview on pathogenicity-associated factors and disease-mediating mechanisms. *International Journal of Medical Microbiology, 300*(4), 205–211.

De Jong, H. K., Parry, C. M., Van, D. P. T., & Wiersinga, W. J., (2012). Host–pathogen interaction in invasive salmonellosis. *PLOS Pathog., 8*(10), e1002933.

De Oliveira, G. B., da Silva, M. A. L., Wanderley, L. B., Da Cunha, C. C., Ferreira, E. C. B., De Medeiros, Z. M., et al., (2016). Cerebral toxoplasmosis in patients with acquired immune deficiency syndrome in the neurological emergency department of a tertiary hospital. *Clinical Neurology and Neurosurgery, 150*, 23–26.

De Vuyst, L., & Vandamme, E. J., (1994). Lactic acid bacteria and bacteriocins: Their practical importance. In: De Vuyst, L., & Vandamme, E. J., (eds.), *Bacteriocins of Lactic acid Bacteria* (pp. 1–11). Boston, MA: Springer.

Debruyne, L., On, S. L., De Brandt, E., & Vandamme, P., (2009). Novel *Campylobacter lari*–like bacteria from humans and molluscs: Description of *Campylobacter peloridis* sp. nov., *Campylobacter lari* subsp. *concheus* subsp. nov. and *Campylobacter lari* subsp. *lari* subsp. nov. *International Journal of Systematic and Evolutionary Microbiology, 59*(5), 1126–1132.

Del, B. O. H., Wadia, N. H., Dumas, M., Cruz, M., Tsang, V. C., & Schantz, P. M., (1996). Proposal of diagnostic criteria for human cysticercosis and neurocysticercosis. *Journal of the Neurological Sciences*, *142*(1, 2), 1–6.

Dennehy, P. H., (2000). Transmission of rotavirus and other enteric pathogens in the home. *The Pediatric Infectious Disease Journal*, *19*(10), S103–S105.

Derzelle, S., Dilasser, F., Duquenne, M., & Deperrois, V., (2009). Differential temporal expression of the staphylococcal enterotoxins genes during cell growth. *Food Microbiology*, *26*(8), 896–904.

Desselberger, U., (2019). Caliciviridae other than noroviruses. *Viruses*, *11*(3), 286.

Dhama, K., Rajagunalan, S., Chakraborty, S., Verma, A. K., Kumar, A., Tiwari, R., et al., (2013). Foodborne pathogens of animal origin - diagnosis, prevention, control and their zoonotic significance: A review. *Pakistan Journal of Biological Sciences*, *16*(20), 1076.

Dierick, K., Van, C. E., Swiecicka, I., Meyfroidt, G., Devlieger, H., Meulemans, A., et al., (2005). Fatal family outbreak of *Bacillus cereus*-associated food poisoning. *Journal of Clinical Microbiology*, *43*(8), 4277–4279.

Ding, C., & Zhang, D., (2007). Molecular analysis of duck hepatitis virus type 1. *Virology*, *361*(1), 9–17.

Doganay, M., (2003). Listeriosis: Clinical presentation. *FEMS Immunology and Medical Microbiology*, *35*(3), 173–175.

Dolly, J. O., & Aoki, K. R., (2006). The structure and mode of action of different botulinum toxins. *European Journal of Neurology*, *13*, 1–9.

Dolly, J. O., Black, J., Williams, R. S., & Melling, J., (1984). Acceptors for botulinum neurotoxin reside on motor nerve terminals and mediate its internalization. *Nature*, *307*(5950), 457–460.

Dolly, J. O., Lande, S., & Wray, D. W., (1987). The effects of *in vitro* application of purified botulinum neurotoxin at mouse motor nerve terminals. *The Journal of Physiology*, *386*(1), 475–484.

Dolly, O., (2003). Synaptic transmission: Inhibition of neurotransmitter release by botulinum toxins. *Headache: The Journal of Head and Face Pain*, *43*, 16–24.

Dougnon, T. V., Legba, B., Deguenon, E., Hounmanou, G., Agbankpe, J., Amadou, A., et al., (2017). Pathogenicity, epidemiology and virulence factors of *Salmonella* species: A review. *Notulae Scientia Biologicae*, *9*(4), 460–466.

Dreier, J., & Juhl, D., (2014). Autochthonous hepatitis E virus infections: A new transfusion-associated risk?. *Transfusion Medicine and Hemotherapy*, *41*(1), 29–39.

Drevets, D. A., Leenen, P. J., & Greenfield, R. A., (2004). Invasion of the central nervous system by intracellular bacteria. *Clinical Microbiology Reviews*, *17*(2), 323–347.

Drexler, J. F., Seelen, A., Corman, V. M., Tateno, A. F., Cottontail, V., Zerbinati, R. M., et al., (2012). Bats worldwide carry hepatitis E virus-related viruses that form a putative novel genus within the family hepeviridae. *Journal of Virology*, *86*(17), 9134–9147.

DuPont, H. L., Levine, M. M., Hornick, R. B., & Formal, S. B., (1989). Inoculum size in shigellosis and implications for expected mode of transmission. *The Journal of Infectious Diseases*, *159*(6), 1126–1128.

Elotmani, F., & Assobhei, O., (2004). *In vitro* inhibition of microbial flora of fish by nisin and lactoperoxidase system. *Letters in Applied Microbiology*, *38*(1), 60–65.

Erlandsen, S. L., & Meyer, E. A., (2013). Trophozoite motility and the mechanism of attachment. In: Erlandsen, S. L., & Meyer, E. A., (eds.), *Giardia and Giardiasis: Biology, Pathogenesis, and Epidemiology* (pp. 33–60). New York, USA: Springer Science.

Espinoza, R. A., Silva-Valenzuela, C. A., Amaya, F. A., Urrutia, Í. M., Contreras, I., & Santiviago, C. A., (2017). Differential roles for pathogenicity islands SPI-13 and SPI-8 in the interaction of *Salmonella enteritidis* and *Salmonella typhi* with murine and human macrophages. *Biological Research, 50.*

Farber, J. M., & Addison, C. J., (1994). RAPD typing for distinguishing species and strains in the genus *Listeria. Journal of Applied Bacteriology, 77*(3), 242–250.

Farthing, M. J. G., (1993). Pathogenesis of giardiasis. *Transactions of the Royal Society of Tropical Medicine and Hygiene, 87*, 17–21.

Faust, D. M., & Guillen, N., (2012). Virulence and virulence factors in *Entamoeba histolytica*, the agent of human amoebiasis. *Microbes and Infection, 14*(15), 1428–1441.

Fayer, R., (2004). *Sarcocystis* spp. in human infections. *Clinical Microbiology Reviews, 17*(4), 894–902.

Fayer, R., (2010). Taxonomy and species delimitation in *Cryptosporidium. Experimental Parasitology, 124*(1), 90–97.

Fayer, R., Esposito, D. H., & Dubey, J. P., (2015). Human infections with *Sarcocystis* species. *Clinical Microbiology Reviews, 28*(2), 295–311.

Feasey, N. A., Dougan, G., Kingsley, R. A., Heyderman, R. S., & Gordon, M. A., (2012). Invasive non-typhoidal *Salmonella* disease: An emerging and neglected tropical disease in Africa. *The Lancet, 379*(9835), 2489–2499.

Feng, Y., Nawa, Y., Sawanyavisuth, K., Lv, Z., & Wu, Z. D., (2013). Comprehensive review of ocular angiostrongyliasis with special reference to optic neuritis. *The Korean Journal of Parasitology, 51*(6), 613.

Ferreira, M. A. S. S., & Lund, B. M., (1996). The effect of nisin on *Listeria monocytogenes* in culture medium and long-life cottage cheese. *Letters in Applied Microbiology, 22*(6), 433–438.

Fisher, E. L., Otto, M., & Cheung, G. Y., (2018). Basis of virulence in enterotoxin-mediated staphylococcal food poisoning. *Frontiers in Microbiology, 9*, 436.

Foroutan-Rad, M., Majidiani, H., Dalvand, S., Daryani, A., Kooti, W., Saki, J., et al., (2016). Toxoplasmosis in blood donors: A systematic review and meta-analysis. *Transfusion Medicine Reviews, 30*(3), 116–122.

Fu, H., Li, L., Zhu, Y., Wang, L., Geng, J., Chang, Y., et al., (2010). Hepatitis E virus infection among animals and humans in Xinjiang, China: Possibility of swine to human transmission of sporadic hepatitis E in an endemic area. *The American Journal of Tropical Medicine and Hygiene, 82*(5), 961–966.

Galsworthy, S. B., Girdler, S., & Koval, S. F., (1990). Chemotaxis in *Listeria monocytogenes. Acta Microbiologica Hungarica, 37*(1), 81–85.

Gänzle, M. G., Vermeulen, N., & Vogel, R. F., (2007). Carbohydrate, peptide and lipid metabolism of lactic acid bacteria in sourdough. *Food Microbiology, 24*(2), 128–138.

Gélinas, P., (2017). Compressed baker's yeast: Mapping patents on post-fermentation processes. *Comprehensive Reviews in Food Science and Food Safety, 16*(3), 456–476.

Gentsch, J. R., Laird, A. R., Bielfelt, B., Griffin, D. D., Bányai, K., Ramachandran, et al., (2005). Serotype diversity and reassortment between human and animal rotavirus strains: Implications for rotavirus vaccine programs. *Journal of Infectious Diseases, 192*(S1), S146–S159.

George, S. M., & Lund, B. M., (1992). The effect of culture medium and aeration on growth of *Listeria monocytogenes* at pH 4.5. *Letters in Applied Microbiology, 15*(2), 49–52.

Gérolami, R., Moal, V., & Colson, P., (2008). Chronic hepatitis E with cirrhosis in a kidney-transplant recipient. *New England Journal of Medicine*, *358*(8), 859–860.

Glass, R. I., Noel, J., Mitchell, D., Herrmann, J. E., Blacklow, N. R., Pickering, L. K., et al., (1996). The changing epidemiology of astrovirus-associated gastroenteritis: A review. *Viral Gastroenteritis*, 287–300.

Glass, R. I., Parashar, U. D., & Estes, M. K., (2009). Norovirus gastroenteritis. *New England Journal of Medicine*, *361*(18), 1776–1785.

Granum, P. E., & Lindbäck, T., (2012). *Bacillus cereus*. In: Doyle, M. P., & Buchanan, R. L., (eds.), *Food Microbiology: Fundamentals and Frontiers* (pp. 491–502). New York, USA: Wiley.

Graves, L. M., Helsel, L. O., Steigerwalt, A. G., Morey, R. E., Daneshvar, M. I., Roof, S. E., et al., (2010). *Listeria marthii* sp. nov., isolated from the natural environment, Finger Lakes National Forest. *International Journal of Systematic and Evolutionary Microbiology*, *60*(6), 1280–1288.

Gulig, P. A., Bourdage, K. L., & Starks, A. M., (2005). Molecular pathogenesis of *Vibrio vulnificus*. *Journal of Microbiology*, *43*(S1), 118–131.

Gupta, A., Polyak, C. S., Bishop, R. D., Sobel, J., & Mintz, E. D., (2004). Laboratory-confirmed shigellosis in the United States, 1989–2002: Epidemiologic trends and patterns. *Clinical Infectious Diseases*, *38*(10), 1372–1377.

Gurung, N., Ray, S., Bose, S., & Rai, V., (2013). A broader view: Microbial enzymes and their relevance in industries, medicine, and beyond. *BioMed Research International*, 1372–1377.

Hagbom, M., Sharma, S., Lundgren, O., & Svensson, L., (2012). Towards a human rotavirus disease model. *Current Opinion in Virology*, *2*(4), 408–418.

Halter, E. L., Neuhaus, K., & Scherer, S., (2013). *Listeria weihenstephanensis* sp. nov., isolated from the water plant *Lemna trisulca* taken from a freshwater pond. *International Journal of Systematic and Evolutionary Microbiology*, *63*(2), 641–647.

Hampton, C. M., Guerrero-Ferreira, R. C., Storms, R. E., Taylor, J. V., Yi, H., Gulig, P. A., et al., (2017). The opportunistic pathogen *Vibrio vulnificus* produces outer membrane vesicles in a spatially distinct manner related to capsular polysaccharide. *Frontiers in Microbiology*, *8*, 2177.

Hansman, G. S., Oka, T., Li, T. C., Nishio, O., Noda, M., & Takeda, N., (2008). Detection of human enteric viruses in Japanese clams. *Journal of Food Protection*, *71*(8), 1689–1695.

Hansman, G. S., Saito, H., Shibata, C., Ishizuka, S., Oseto, M., Oka, T., et al., (2007). Outbreak of gastroenteritis due to sapovirus. *Journal of Clinical Microbiology*, *45*(4), 1347–1349.

Haraga, A., Ohlson, M. B., & Miller, S. I., (2008). *Salmonellae* interplay with host cells. *Nature Reviews Microbiology*, *6*(1), 53–66.

Hardy, S. P., Denmead, M., Parekh, N., & Granum, P. E., (1999). Cationic currents induced by *Clostridium perfringens* type A enterotoxin in human intestinal CaCO-2 cells. *Journal of Medical Microbiology*, *48*(3), 235–243.

Harris, K. F., & Harris, L. J., (2001). Ingestion-egestion theory of cuticula-borne virus transmission. In: Harris, K. F., Smith, O. P., & Duffus, J. E., (eds.), *Virus-Insect-Plant Interactions* (pp. 111–132). USA: Academic Press.

Havelaar, A. H., Kirk, M. D., Torgerson, P. R., Gibb, H. J., Hald, T., Lake, R. J., et al., (2015). World Health Organization global estimates and regional comparisons of the burden of foodborne disease in 2010. *PLoS Medicine*, *12*(12), e1001923.

Herrero-Fresno, A., & Olsen, J. E., (2018). *Salmonella typhimurium* metabolism affects virulence in the host – A mini-review. *Food Microbiology*, *71*, 98–110.

Hill, K. K., & Smith, T. J., (2012). Genetic diversity within *Clostridium botulinum* serotypes, botulinum neurotoxin gene clusters and toxin subtypes. In: Rummel, A., & Binz, T., (eds.), *Botulinum Neurotoxins* (pp. 1–20). Verlag Berlin, Heidelberg: Springer.

Hiyoshi, H., Tiffany, C. R., Bronner, D. N., & Bäumler, A. J., (2018). Typhoidal *Salmonella* serovars: Ecological opportunity and the evolution of a new pathovar. *FEMS Microbiology Reviews*, *42*(4), 527–541.

Hoberg, E. P., (2006). Phylogeny of *Taenia*: Species definitions and origins of human parasites. *Parasitology International*, *55*, S23–S30.

Holly, M. K., & Smith, J. G., (2018). Adenovirus infection of human enteroids reveals interferon sensitivity and preferential infection of goblet cells. *Journal of Virology*, *92*(9).

Hong, S. J., Kang, S. Y., Chung, Y. B., Chung, M. H., Oh, Y. J., Kang, I., et al., (2000). *Paragonimus westermani*: A cytosolic glutathione S-transferase of a σ-class in adult stage. *Experimental Parasitology*, *94*(3), 180–189.

Hugenholtz, J., & De Veer, G. J. C. M., (1991). Application of nisin A and nisin Z in dairy technology. *Nisin and Novel Lantibiotics*, 440–447.

Hussain, Z., Das, B. C., Husain, S. A., Asim, M., Chattopadhyay, S., Malik, A., et al., (2005). Hepatitis A viral genotypes and clinical relevance: Clinical and molecular characterization of hepatitis A virus isolates from northern India. *Hepatology Research*, *32*(1), 16–24.

Huston, C. D., Boettner, D. R., Miller-Sims, V., & Petri, Jr. W. A., (2003). Apoptotic killing and phagocytosis of host cells by the parasite *Entamoeba histolytica*. *Infection and Immunity*, *71*(2), 964–972.

Hymery, N., Vasseur, V., Coton, M., Mounier, J., Jany, J. L., Barbier, G., et al., (2014). Filamentous fungi and mycotoxins in cheese: A review. *Comprehensive Reviews in Food Science and Food Safety*, *13*(4), 437–456.

Janda, J. M., & Abbott, S. L., (2010). The genus *Aeromonas*: Taxonomy, pathogenicity, and infection. *Clinical Microbiology Reviews*, *23*(1), 35–73.

Jayaram, V. B., Cuyvers, S., Verstrepen, K. J., Delcour, J. A., & Courtin, C. M., (2014). Succinic acid in levels produced by yeast (*Saccharomyces cerevisiae*) during fermentation strongly impacts wheat bread dough properties. *Food Chemistry*, *151*, 421–428.

Jenkins, A., Minhas, R., Morris, C., & Berry, N., (2018). Genomic sequence of the WHO international standard for hepatitis A virus RNA. *Genome Announcements*, *6*(19).

Jensen, G. B., Hansen, B. M., Eilenberg, J., & Mahillon, J., (2003). The hidden lifestyles of *Bacillus cereus* and relatives. *Environmental Microbiology*, *5*(8), 631–640.

Jiang, A., Ge, L., You, Y., & Shen, Z., (2015). The diagnosis of cutaneous paragonimiasis after the external migration of the parasite from a punch biopsy site. *JAAD Case Reports*, *1*(5), 239–240.

Jimenez-Gonzalez, D. E., Martinez-Flores, W. A., Reyes-Gordillo, J., Ramirez-Miranda, M. E., Arroyo-Escalante, S., Romero-Valdovinos, M., et al., (2012). Blastocystis infection is associated with irritable bowel syndrome in a Mexican patient population. *Parasitology Research*, *110*(3), 1269–1275.

Johnson, C., Han, Y., Hughart, N., McCarra, J., Alpini, G., & Meng, F., (2012). Interleukin-6 and its receptor, key players in hepatobiliary inflammation and cancer. *Translational Gastrointestinal Cancer*, *1*(1), 58.

Johnson, E. A., (2014). Validity of botulinum neurotoxin serotype H. *The Journal of Infectious Diseases*, *210*(6), 992–993.

Jones, M. K., & Oliver, J. D., (2009). *Vibrio vulnificus*: Disease and pathogenesis. *Infection and Immunity*, *77*(5), 1723–1733.

Jørgensen, F., Stephens, P. J., & Knøchel, S., (1995). The effect of osmotic shock and subsequent adaptation on the thermotolerance and cell morphology of *Listeria monocytogenes*. *Journal of Applied Bacteriology*, *79*(3), 274–281.

Jude, B. A., & Taylor, R. K., (2011). The physical basis of type 4 pilus-mediated microcolony formation by *Vibrio cholerae* O1. *Journal of Structural Biology*, *175*(1), 1–9.

Junttila, J. R., Niemelä, S. I., & Hirn, J., (1988). Minimum growth temperatures of *Listeria monocytogenes* and non-haemolytic listeria. *Journal of Applied Bacteriology*, *65*(4), 321–327.

Kaakoush, N. O., Castaño-Rodríguez, N., Mitchell, H. M., & Man, S. M., (2015). Global epidemiology of *Campylobacter* infection. *Clinical Microbiology Reviews*, *28*(3), 687–720.

Kabak, B., & Dobson, A. D., (2011). An introduction to the traditional fermented foods and beverages of Turkey. *Critical Reviews in Food Science and Nutrition*, *51*(3), 248–260.

Kadariya, J., Smith, T. C., & Thapaliya, D., (2014). *Staphylococcus aureus* and staphylococcal foodborne disease: An ongoing challenge in public health. *BioMed Research International*, *2014*.

Kamar, N., Dalton, H. R., Abravanel, F., & Izopet, J., (2014). *Hepatitis E* virus infection. *Clinical Microbiology Reviews*, *27*(1), 116–138.

Kammerer, W. S., & Schantz, P. M., (1993). Echinococcal disease. *Infectious Disease Clinics of North America*, *7*(3), 605–618.

Karch, H., Tarr, P. I., & Bielaszewska, M., (2005). Enterohaemorrhagic *Escherichia coli* in human medicine. *International Journal of Medical Microbiology*, *295*(6, 7), 405–418.

Karmali, M., Petric, M., Steele, B., & Lim, C., (1983). Sporadic cases of haemolytic-uraemic syndrome associated with faecal cytotoxin and cytotoxin-producing *Escherichia coli* in stools. *The Lancet*, *321*(8325), 619–620.

Karst, S. M., (2010). Pathogenesis of noroviruses, emerging RNA viruses. *Viruses*, *2*(3), 748–781.

Kaškonienė, V., Stankevičius, M., Bimbiraitė-Survilienė, K., Naujokaitytė, G., Šernienė, L., Mulkytė, K., et al., (2017). Current state of purification, isolation and analysis of bacteriocins produced by lactic acid bacteria. *Applied Microbiology and Biotechnology*, *101*(4), 1323–1335.

Katayama, S., Dupuy, B., Cole, S. T., Daube, G., & China, B., (1996). Genome mapping of *Clostridium perfringens* strains with I-CeuI shows many virulence genes to be plasmid-borne. *Molecular and General Genetics*, *251*(6), 720–726.

Keithlin, J., Sargeant, J., Thomas, M. K., & Fazil, A., (2014). Systematic review and meta-analysis of the proportion of *Campylobacter* cases that develop chronic sequelae. *BMC Public Health*, *14*(1), 1–19.

Keusch, G. T., & Acheson, D. W., (1996). *Shigella* infection. In: Paradise, L. J., Bendinelli, M., & Friedman, H., (eds.), *Enteric Infections and Immunity* (pp. 79–100). Boston, MA: Springer.

Khuroo, M. S., Khuroo, M. S., & Khuroo, N. S., (2016). Transmission of hepatitis E virus in developing countries. *Viruses*, *8*(9), 253.

Kim, A. J., Choi, C. H., Choi, S. K., Shin, Y. W., Park, Y. K., Kim, L., et al., (2015). Ectopic human *Fasciola hepatica* infection by an adult worm in the mesocolon. *The Korean Journal of Parasitology*, *53*(6), 725.

Kim, E. A., Juhng, S. K., Kim, H. W., Kim, G. D., Lee, Y. W., Cho, H. J., et al., (2004). Imaging findings of hepatic paragonimiasis: A case report. *Journal of Korean Medical Science*, *19*(5), p.759.

Kim, T. Y., Lee, J. Y., Im Kim, T., Moon, K. H., Kang, S. Y., & Hong, S. J., (2007). Molecular cloning and enzymatic characterization of a class μ-glutathione S-transferase of *Paragonimus westermani*. *Parasitology Research*, *101*(5), 1225–1231.

Klaenhammer, T. R., (1993). Genetics of bacteriocins produced by lactic acid bacteria. *FEMS Microbiology Reviews*, *12*(1–3), 39–85.

Klein, A. W., (2003). Complications, adverse reactions, and insights with the use of botulinum toxin. *Dermatologic Surgery*, *29*(5), 549–556.

Kocięcka, W., (2000). Trichinellosis: Human disease, diagnosis and treatment. *Veterinary Parasitology*, *93*(3, 4), 365–383.

Komlenić, D. K., Slačanac, V., & Jukić, M., (2012). Influence of acidification on dough rheological properties. In: De Vicente, J., (ed.), *Rheology* (pp. 265–292).

Koopmans, M., Vennema, H., Heersma, H., Van, S. E., Van, D. Y., Brown, D., et al., (2003). Early identification of common-source foodborne virus outbreaks in Europe. *Emerging Infectious Diseases*, *9*(9), 1136.

Kotiranta, A., Haapasalo, M., Kari, K., Kerosuo, E., Olsen, I., Sorsa, T., et al., (1998). Surface structure, hydrophobicity, phagocytosis, and adherence to matrix proteins of *Bacillus cereus* cells with and without the crystalline surface protein layer. *Infection and Immunity*, *66*(10), 4895–4902.

Kotiranta, A., Lounatmaa, K., & Haapasalo, M., (2000). Epidemiology and pathogenesis of *Bacillus cereus* infections. *Microbes and Infection*, *2*(2), 189–198.

Kramer, A., Schwebke, I., & Kampf, G., (2006). How long do nosocomial pathogens persist on inanimate surfaces? A systematic review. *BMC Infectious Diseases*, *6*(1), 1–8.

Kramer, J. M., Gilbert, R. J., & Doyle, M. P., (1989). Foodborne bacterial pathogens. In: Doyle (ed.), *Foodborne Bacterial Pathogens* (pp. 22–70). Marcel Dekker Inc., New York and Basel.

Kuhn, D. M., & Ghannoum, M. A., (2003). Indoor mold, toxigenic fungi, and *Stachybotrys chartarum*: Infectious disease perspective. *Clinical Microbiology Reviews*, *16*(1), 144–172.

Kumar, A., Beniwal, M., Kar, P., Sharma, J. B., & Murthy, N. S., (2004). *Hepatitis E* in pregnancy. *International Journal of Gynecology and Obstetrics*, *85*(3), 240–244.

Lai, D. H., Hong, X. K., Su, B. X., Liang, C., Hide, G., Zhang, X., et al., (2016). Current status of *Clonorchis sinensis* and clonorchiasis in China. *Transactions of the Royal Society of Tropical Medicine and Hygiene*, *110*(1), 21–27.

Lamas, A., Miranda, J. M., Regal, P., Vázquez, B., Franco, C. M., & Cepeda, A., (2018). A comprehensive review of non-enterica subspecies of *Salmonella enterica*. *Microbiological research*, *206*, 60–73.

Lapa, D., Capobianchi, M. R., & Garbuglia, A. R., (2015). Epidemiology of hepatitis E virus in European countries. *International Journal of Molecular Sciences*, *16*(10), 25711–25743.

Larkin, E. A., Carman, R. J., Krakauer, T., & Stiles, B. G., (2009). *Staphylococcus aureus*: The toxic presence of a pathogen extraordinaire. *Current Medicinal Chemistry*, *16*(30), 4003–4019.

Le Cann, P., Ranarijaona, S., Monpoeho, S., Le Guyader, F., & Ferré, V., (2004). Quantification of human astroviruses in sewage using real-time RT-PCR. *Research in Microbiology*, *155*(1), 11–15.

Leclercq, A., Clermont, D., Bizet, C., Grimont, P. A., Le Fleche-Mateos, A., Roche, S. M., et al., (2010). *Listeria rocourtiae* sp. nov. *International Journal of Systematic and Evolutionary Microbiology*, *60*(9), 2210–2214.

Lee, L. E., Cebelinski, E. A., Fuller, C., Keene, W. E., Smith, K., Vinjé, J., & Besser, J. M., (2012). Sapovirus outbreaks in long-term care facilities, Oregon and Minnesota, USA, 2002–2009. *Emerging Infectious Diseases*, *18*(5), 873.

Leong, D., Alvarez-Ordóñez, A., Jooste, P., & Jordan, K., (2016). *Listeria monocytogenes* in food: Control by monitoring the food processing environment. *African Journal of Microbiol Research*, *10*(1), 1–14.

Levine, M. M., Kotloff, K. L., Barry, E. M., Pasetti, M. F., & Sztein, M. B., (2007). Clinical trials of *Shigella* vaccines: Two steps forward and one step back on a long, hard road. *Nature Reviews Microbiology*, *5*(7), 540–553.

Li, D., De Keuckelaere, A., & Uyttendaele, M., (2015). Fate of foodborne viruses in the "farm to fork" chain of fresh produce. *Comprehensive Reviews in Food Science and Food Safety*, *14*(6), 755–770.

Liaud, N., Giniés, C., Navarro, D., Fabre, N., Crapart, S., Herpoël-Gimbert, I., Levasseur, A., Raouche, S., & Sigoillot, J. C., (2014). Exploring fungal biodiversity: Organic acid production by 66 strains of filamentous fungi. *Fungal Biology and Biotechnology*, *1*(1), 1–10.

Lin, C. L., Chen, S. C., Liu, S. Y., & Chen, K. T., (2014). Disease caused by rotavirus infection. *The open Virology Journal*, *8*, 14.

Liu, D., (2013). Molecular approaches to the identification of pathogenic and nonpathogenic *listeriae*. *Microbiology Insights*, *6*, 59–69.

Liu, W., Zhai, J., Liu, J., & Xie, Y., (2010). Identification of recombination between subgenotypes IA and IB of hepatitis A virus. *Virus Genes*, *40*(2), 222–224.

Liu, W., Zhang, L., Yi, H., Shi, J., Xue, C., Li, H., et al., (2014). Qualitative detection of class IIa bacteriocinogenic lactic acid bacteria from traditional Chinese fermented food using a YGNGV-motif-based assay. *Journal of Microbiological Methods*, *100*, 121–127.

Lorenzo, J. M., Munekata, P. E., Dominguez, R., Pateiro, M., Saraiva, J. A., & Franco, D., (2018). Main groups of microorganisms of relevance for food safety and stability: General aspects and overall description. In: Barba, F. J., Santana, A. S., Orlien, V., & Koubaa, M., (eds.), *Innovative Technologies for Food Preservation* (pp. 53–107). USA: Academic Press, Elsevier.

Loureiro, V., & Malfeito-Ferreira, M., (2003). Spoilage yeasts in the wine industry. *International Journal of Food Microbiology*, *86*(1, 2), 23–50.

Lu, L., Li, C., & Hagedorn, C. H., (2006). Phylogenetic analysis of global hepatitis E virus sequences: Genetic diversity, subtypes and zoonosis. *Reviews in Medical Virology*, *16*(1), 5–36.

Lu, L., Zhong, H., Xu, M., Su, L., Cao, L., Jia, R., & Xu, J., (2019). Genetic diversity and epidemiology of Genogroup II noroviruses in children with acute sporadic gastroenteritis in Shanghai, China, 2012–2017. *BMC Infectious Diseases*, *19*(1), 1–10.

Lund, T., & Granum, P. E., (1996). Characterization of a non-haemolytic enterotoxin complex from *Bacillus cereus* isolated after a foodborne outbreak. *FEMS Microbiology Letters*, *141*(2, 3), 151–156.

Lund, T., De Buyser, M. L., & Granum, P. E., (2000). A new cytotoxin from *Bacillus cereus* that may cause necrotic enteritis. *Molecular Microbiology*, *38*(2), 254–261.

Lundgren, O., & Svensson, L., (2001). Pathogenesis of rotavirus diarrhea. *Microbes and Infection*, *3*(13), 1145–1156.

Macé, S., Haddad, N., Zagorec, M., & Tresse, O., (2015). Influence of measurement and control of microaerobic gaseous atmospheres in methods for *Campylobacter* growth studies. *Food Microbiology*, *52*, 169–176.

Maisnier-Patin, S., Deschamps, N., Tatini, S. R., & Richard, J., (1992). Inhibition of *Listeria monocytogenes* in Camembert cheese made with a nisin-producing starter. *Le Lait*, *72*(3), 249–263.

Mansuri, M. S., Bhattacharya, S., & Bhattacharya, A., (2014). A novel alpha kinase EhAK1 phosphorylates actin and regulates phagocytosis in *Entamoeba histolytica. PLoS Pathog, 10*(10), e1004411.

Mantovani, S. A., Delfino, B. M., Martins, A. C., Oliart-Guzmán, H., Pereira, T. M., Branco, F. L., et al., (2015). Socioeconomic inequities and hepatitis A virus infection in Western Brazilian Amazonian children: Spatial distribution and associated factors. *BMC Infectious Diseases, 15*(1), 1–12.

Marcin, S. M., Arroyo, M., Cadena, T. M., Ramírez, C. N., García, H. F., Taboada, D., et al., (2017). Extraparenchymal neurocysticercosis: Demographic, clinicoradiological, and inflammatory features. *PLoS Neglected Tropical Diseases, 11*(6), p.e0005646.

Marti, E., & Balcázar, J. L., (2015). *Aeromonas rivipollensis* sp. nov., a novel species isolated from aquatic samples. *Journal of Basic Microbiology, 55*(12), 1435–1439.

Martin, S. E., & Myers, E. R., (1994). *Staphylococcus aureus*. In: Hui, Y. H., Gorhan, J. R., Murrell, K. D., & Cliver, D. O., (eds.), *Foodborne Disease Handbook: Diseases Caused by Bacteria* (pp. 345–394). New York: Marcel Dekker, Inc.

Martina, M. N., Cervera, C., Esforzado, N., Linares, L., Torregrosa, V., Sanclemente, G., Hoyo, I., et al., (2011). *Toxoplasma gondii* primary infection in renal transplant recipients. Two case reports and literature review. *Transplant International, 24*(1), e6–e12.

Martínez-Murcia, A., Beaz-Hidalgo, R., Navarro, A., Carvalho, M. J., Aravena-Román, M., Correia, A., et al., (2016). *Aeromonas lusitana* sp. nov., isolated from untreated water and vegetables. *Current Microbiology, 72*(6), 795–803.

Mathan, M. M., & Mathan, V. I., (1986). Ultrastructural pathology of the rectal mucosa in *Shigella* dysentery. *The American Journal of Pathology, 123*(1), 25.

Matsuda, T., Okada, Y., Inagi, E., Tanabe, Y., Shimizu, Y., Nagashima, K., et al., (2007). Enteritis necroticans 'pigbel' in a Japanese diabetic adult. *Pathology International, 57*(9), 622–626.

Matthijnssens, J., Heylen, E., Zeller, M., Rahman, M., Lemey, P., & Van, R. M., (2010). Phylodynamic analyses of rotavirus genotypes G9 and G12 underscore their potential for swift global spread. *Molecular Biology and Evolution, 27*(10), 2431–2436.

Matthijnssens, J., Otto, P. H., Ciarlet, M., Desselberger, U., Van, R. M., & Johne, R., (2012). VP6-sequence-based cutoff values as a criterion for rotavirus species demarcation. *Archives of Virology, 157*(6), 1177–1182.

Mattock, E., & Blocker, A. J., (2017). How do the virulence factors of *Shigella* work together to cause disease?. *Frontiers in Cellular and Infection Microbiology, 7*, 64.

McGovern, P. E., Zhang, J., Tang, J., Zhang, Z., Hall, G. R., Moreau, R. A., et al., (2004). Fermented beverages of pre-and proto-historic China. *Pro. Natl. Acad. Sci. USA, 101*(51), 17593–17598.

Meng, X. J., Halbur, P. G., Shapiro, M. S., Govindarajan, S., Bruna, J. D., Mushahwar, I. K., Purcell, R. H., & Emerson, S. U., (1998). Genetic and experimental evidence for cross-species infection by swine hepatitis E virus. *Journal of Virology, 72*(12), 9714–9721.

Mennechet, F. J., Paris, O., Ouoba, A. R., Salazar, A. S., Sirima, S. B., Takoudjou, D. G. R., Diarra, A., et al., (2019). A review of 65 years of human adenovirus seroprevalence. *Expert Review of Vaccines, 18*(6), 597–613.

Miclard, J., Jäggi, M., Sutter, E., Wyder, M., Grabscheid, B., & Posthaus, H., (2009a). *Clostridium perfringens* beta-toxin targets endothelial cells in necrotizing enteritis in piglets. *Veterinary Microbiology, 137*(3, 4), 320–325.

Miclard, J., Van, B. J., Wyder, M., Grabscheid, B., & Posthaus, H., (2009b). *Clostridium perfringens* β-toxin binding to vascular endothelial cells in a human case of enteritis necroticans. *Journal of Medical Microbiology*, *58*(6), 826–828.

Mihalov-Kovács, E., Gellért, Á., Marton, S., Farkas, S. L., Fehér, E., Oldal, M., Jakab, F., Martella, V., & Bányai, K., (2015). Candidate new rotavirus species in sheltered dogs, Hungary. *Emerging Infectious Diseases*, *21*(4), 660.

Miura, N. N., Adachi, Y., Yadomae, T., Tamura, H., Tanaka, S., & Ohno, N., (2003). Structure and biological activities of β-glucans from yeast and mycelial forms of *Candida albicans*. *Microbiology and Immunology*, *47*(3), 173–182.

Molins, R. A., Motarjemi, Y., & Käferstein, F. K., (2001). Irradiation: A critical control point in ensuring the microbiological safety of raw foods. *Food Control*, *12*(6), 347–356.

Morar, R., & Feldman, C., (2003). Pulmonary echinococcosis. *European Respiratory Journal*, *21*(6), 1069–1077.

Moreira, D., & López-García, P., (2009). Ten reasons to exclude viruses from the tree of life. *Nature Reviews Microbiology*, *7*(4), 306–311.

Moser, L. A., & Schultz-Cherry, S., (2005). Pathogenesis of astrovirus infection. *Viral Immunology*, *18*(1), 4–10.

Mueller-Spitz, S. R., Stewart, L. B., Klump, J. V., & McLellan, S. L., (2010). Freshwater suspended sediments and sewage are reservoirs for enterotoxin-positive *Clostridium perfringens*. *Applied and Environmental Microbiology*, *76*(16), 5556–5562.

Muhialdin, B. J., Hassan, Z., & Sadon, S. K., (2011). Antifungal Activity of *Lactobacillus fermentum* Te007, *Pediococcus pentosaceus* Te010, *Lactobacillus pentosus* G004, and *L. paracasi* D5 on selected foods. *Journal of Food Science*, *76*(7), M493–M499.

Muhsen, K., & Levine, M. M., (2012). A systematic review and meta-analysis of the association between *Giardia lamblia* and endemic pediatric diarrhea in developing countries. *Clinical Infectious Diseases*, *55*(S4), S271–S293.

Murray, R. J., (2005). Recognition and management of *Staphylococcus aureus* toxin-mediated disease. *Internal Medicine Journal*, *35*, S106–119.

Murrell, K. D., (2013). Zoonotic foodborne parasites and their surveillance. *Revue Scientifique et Technique (International Office of Epizootics)*, *32*(2), 559–569.

Nachamkin, I., Shadomy, S. V., Moran, A. P., Cox, N., Fitzgerald, C., Ung, H., et al., (2008). Anti-ganglioside antibody induction by swine (A/NJ/1976/H1N1) and other influenza vaccines: Insights into vaccine-associated Guillain-Barré syndrome. *The Journal of Infectious Diseases*, *198*(2), 226–233.

Nainan, O. V., Xia, G., Vaughan, G., & Margolis, H. S., (2006). Diagnosis of hepatitis A virus infection: A molecular approach. *Clinical Microbiology Reviews*, *19*(1), 63–79.

Nakamura-Uchiyama, F., Mukae, H., & Nawa, Y., (2002). Paragonimiasis: A Japanese perspective. *Clinics in Chest Medicine*, *23*(2), 409–420.

Nam, J. H., Moon, J. H., Kim, I. K., Lee, M. R., Hong, S. J., Ahn, J. H., et al., (2012). Free radicals enzymatically triggered by *Clonorchis sinensis* excretory–secretory products cause NF-κB-mediated inflammation in human cholangiocarcinoma cells. *International Journal for Parasitology*, *42*(1), 103–113.

Nataro, J. P., Yikang, D., Cookson, S., Cravioto, A., Savarino, S. J., Guers, L. D., Levine, M. M., & Tacket, C. O., (1995). Heterogeneity of enteroaggregative *Escherichia coli* virulence demonstrated. *Journal of Infectious Diseases*, *171*(2), 465–468.

Navia, B. A., Petito, C. K., Gold, J. W., Cho, E. S., Jordan, B. D., & Price, R. W., (1986). Cerebral toxoplasmosis complicating the acquired immune deficiency syndrome: Clinical and neuropathological findings in 27 patients. *Annals of Neurology*, *19*(3), 224–238.

Neghina, R., & Neghina, A. M., (2011b). Reviews on trichinellosis (IV): Hepatic involvement. *Foodborne Pathogens and Disease*, *8*(9), 943–948.

Neghina, R., Neghina, A. M., & Marincu, I., (2011a). Reviews on trichinellosis (III): Cardiovascular involvement. *Foodborne Pathogens and Disease*, *8*(8), 853–860.

Nigam, P. S., (2013). Microbial enzymes with special characteristics for biotechnological applications. *Biomolecules*, *3*(3), 597–611.

Niyogi, S. K., (2005). Shigellosis. *Journal of Microbiology*, *43*(2), 133–143.

Norhana, M. W., Poole, S. E., Deeth, H. C., & Dykes, G. A., (2012). Effects of nisin, EDTA and salts of organic acids on *Listeria monocytogenes*, *Salmonella* and native microflora on fresh vacuum packaged shrimps stored at 4°C. *Food Microbiology*, *31*(1), 43–50.

Oka, T., Wang, Q., Katayama, K., & Saif, L. J., (2015). Comprehensive review of human sapoviruses. *Clinical Microbiology Reviews*, *28*(1), 32–53.

Oliver, J. D., (2005). Wound infections caused by *Vibrio vulnificus* and other marine bacteria. *Epidemiology and Infection*, *133*(3), 383–391.

Omernik, A., & Płusa, T., (2015). Toxins of *Clostridium perfringens* as a natural and bioterroristic threats. *Polski merkuriusz lekarski: Organ Polskiego Towarzystwa Lekarskiego*, *39*(231), 149–152.

Ooi, S. T., & Lorber, B., (2005). Gastroenteritis due to *Listeria monocytogenes*. *Clinical Infectious Diseases*, *40*(9), 1327–1332.

Otto, M., (2014). *Staphylococcus aureus* toxins. *Current Opinion in Microbiology*, *17*, 32–37.

Palombo, E. A., & Bishop, R. F., (1996). Annual incidence, serotype distribution, and genetic diversity of human astrovirus isolates from hospitalized children in Melbourne, Australia. *Journal of Clinical Microbiology*, *34*(7), 1750–1753.

Pang, X. L., Lee, B. E., Tyrrell, G. J., & Preiksaitis, J. K., (2009). Epidemiology and genotype analysis of sapovirus associated with gastroenteritis outbreaks in Alberta, Canada: 2004–2007. *The Journal of Infectious Diseases*, *199*(4), 547–551.

Parashar, U. D., Hummelman, E. G., Bresee, J. S., Miller, M. A., & Glass, R. I., (2003). Global illness and deaths caused by rotavirus disease in children. *Emerging Infectious Diseases*, *9*(5), 565.

Parish, M. E., & Higgins, D. P., (1989). Survival of *Listeria monocytogenes* in low pH model broth systems. *Journal of Food Protection*, *52*(3), 144–147.

Parra, G. I., Azure, J., Fischer, R., Bok, K., Sandoval-Jaime, C., Sosnovtsev, S. V., et al., (2013). Identification of a broadly cross-reactive epitope in the inner shell of the norovirus capsid. *PLoS One*, *8*(6), e67592.

Patel, M. M., Widdowson, M. A., Glass, R. I., Akazawa, K., Vinjé, J., & Parashar, U. D., (2008). Systematic literature review of role of noroviruses in sporadic gastroenteritis. *Emerging Infectious Diseases*, *14*(8), 1224.

Peck, M. W., (2009). Biology and genomic analysis of *Clostridium botulinum*. *Advances in Microbial Physiology*, *55*, 183–320.

Pérez-Gracia, M. T., García, M., Suay, B., & Mateos-Lindemann, M. L., (2015). Current knowledge on hepatitis E. *Journal of Clinical and Translational Hepatology*, *3*(2), 117.

Pérot, P., Lecuit, M., & Eloit, M., (2017). Astrovirus diagnostics. *Viruses*, *9*(1), 10.

Petit, L., Gibert, M., & Popoff, M. R., (1999). *Clostridium perfringens*: Toxinotype and genotype. *Trends in Microbiology*, *7*(3), 104–110.

Petrović, T., & D'Agostino, M., (2016). Viral contamination of food. In: Barros-Velazquez, J., (ed.), *Antimicrobial Food Packaging* (pp. 65–79). USA: Elsevier.

Pinto, P., Wang, Q., Chen, N., Dubovi, E. J., Daniels, J. B., Millward, L. M., et al., (2012). Discovery and genomic characterization of noroviruses from a gastroenteritis outbreak in domestic cats in the US. *PloS One*, *7*(2), e32739.

Pintó, R. M., Aragonès, L., Costafreda, M. I., Ribes, E., & Bosch, A., (2007). Codon usage and replicative strategies of hepatitis A virus. *Virus Research*, *127*(2), 158–163.

Plockova, M., Stepanek, M., Demnerova, K., Curda, L., & Svirakova, E., (1996). Effect of nisin for improvement in shelf-life and quality of processed cheese. *Advances in Food Sciences*, *18*(3, 4), 78–83.

Poly, F., & Guerry, P., (2008). Pathogenesis of *Campylobacter*. *Current Opinion in Gastroenterology*, *24*(1), 27–31.

Pontes, M. H., Lee, E. J., Choi, J., & Groisman, E. A., (2015). *Salmonella* promotes virulence by repressing cellulose production. *Proc. Natl. Acad. Sci. USA.*, *112*(16), 5183–5188.

Porta, C., Riboldi, E., & Sica, A., (2011). Mechanisms linking pathogens-associated inflammation and cancer. *Cancer Letters*, *305*(2), 250–262.

Principato, M., & Qian, B. F., (2014). Staphylococcal enterotoxins in the etiopathogenesis of mucosal autoimmunity within the gastrointestinal tract. *Toxins*, *6*(5), 1471–1489.

Pumipuntu, N., & Piratae, S., (2018). Cryptosporidiosis: A zoonotic disease concern. *Veterinary World*, *11*(5), 681.

Qian, M. B., Utzinger, J., Keiser, J., & Zhou, X. N., (2016). Clonorchiasis. *The Lancet*, *387*(10020), 800–810.

Ramig, R. F., (2004). Pathogenesis of intestinal and systemic rotavirus infection. *Journal of Virology*, *78*(19), 10213–10220.

Rao, C. D., Yergolkar, P., & Shankarappa, K. S., (2012). Antigenic diversity of enteroviruses associated with nonpolio acute flaccid paralysis, India, 2007–2009. *Emerging Infectious Diseases*, *18*(11), 1833.

Rasigade, J. P., & Vandenesch, F., (2014). *Staphylococcus aureus*: A pathogen with still unresolved issues. *Infection, Genetics and Evolution*, *21*, 510–514.

Ray, R. C., & Didier, M., (2014). *Microorganisms and Fermentation of Traditional Foods* (p. 380). Florida, USA: CRC Press.

Reis, J. A., Paula, A. T., Casarotti, S. N., & Penna, A. L. B., (2012). Lactic acid bacteria antimicrobial compounds: Characteristics and applications. *Food Engineering Reviews*, *4*(2), 124–140.

Reller, M. E., Nelson, J. M., Mølbak, K., Ackman, D. M., Schoonmaker-Bopp, D. J., Root, T. P., & Mintz, E. D., (2006). A large, multiple-restaurant outbreak of infection with *Shigella flexneri* serotype 2a traced to tomatoes. *Clinical Infectious Diseases*, *42*(2), 163–169.

Rezac, S., Kok, C. R., Heermann, M., & Hutkins, R., (2018). Fermented foods as a dietary source of live organisms. *Frontiers in Microbiology*, *9*, 1785.

Rhee, S. J., Lee, J. E., & Lee, C. H., (2011). Importance of lactic acid bacteria in Asian fermented foods. *Microbial Cell Factories*, *10*(1), 1–13.

Richards, G. P., McLeod, C., & Le Guyader, F. S., (2010). Processing strategies to inactivate enteric viruses in shellfish. *Food and Environmental Virology*, *2*(3), 183–193.

Richardson, K., (1991). Roles of motility and flagellar structure in pathogenicity of *Vibrio cholerae*: Analysis of motility mutants in three animal models. *Infection and Immunity*, *59*(8), 2727–2736.

Riley, L. W., Remis, R. S., Helgerson, S. D., McGee, H. B., Wells, J. G., Davis, B. R., et al., (1983). Hemorrhagic colitis associated with a rare *Escherichia coli* serotype. *New England Journal of Medicine*, *308*(12), 681–685.

Ringehan, M., McKeating, J. A., & Protzer, U., (2017). Viral hepatitis and liver cancer. *Philosophical Transactions of the Royal Society B: Biological Sciences*, *372*(1732), 20160274.

Rizzello, C. G., Curiel, J. A., Nionelli, L., Vincentini, O., Di Cagno, R., Silano, M., Gobbetti, M., & Coda, R., (2014). Use of fungal proteases and selected sourdough lactic acid bacteria for making wheat bread with an intermediate content of gluten. *Food Microbiology*, *37*, 59–68.

Roberts, L. O., Jopling, C. L., Jackson, R. J., & Willis, A. E., (2009). Viral strategies to subvert the mammalian translation machinery. *Progress in Molecular Biology and Translational Science*, *90*, 313–367.

Roberts, T., (1989). Human illness costs of foodborne bacteria. *American Journal of Agricultural Economics*, *71*(2), 468–474.

Robilotti, E., Deresinski, S., & Pinsky, B. A., (2015). Norovirus. *Clinical Microbiology Reviews*, *28*(1), 134–164.

Ross, R. P., Morgan, S., & Hill, C., (2002). Preservation and fermentation: Past, present and future. *International Journal of Food Microbiology*, *79*(1, 2), 3–16.

Russo, T. A., & Johnson, J. R., (2003). Medical and economic impact of extraintestinal infections due to *Escherichia coli*: Focus on an increasingly important endemic problem. *Microbes and Infection*, *5*(5), 449–456.

Ryan, U., & Hijjawi, N., (2015). New developments in *Cryptosporidium* research. *International Journal for Parasitology*, *45*(6), 367–373.

Ryan, U., Zahedi, A., & Paparini, A., (2016). *Cryptosporidium* in humans and animals – a one health approach to prophylaxis. *Parasite Immunology*, *38*(9), 535–547.

Ryser, E. T., & Marth, E. H., (2007). *Listeria, Listeriosis, and Food Safety* (p. 873). Florida, USA: CRC Press.

Ryu, J. H., & Beuchat, L. R., (2005). Biofilm formation and sporulation by *Bacillus cereus* on a stainless steel surface and subsequent resistance of vegetative cells and spores to chlorine, chlorine dioxide, and a peroxyacetic acid–based sanitizer. *Journal of Food Protection*, *68*(12), 2614–2622.

Saha, A., Gaurav, A. K., Bhattacharya, S., & Bhattacharya, A., (2015). Molecular basis of pathogenesis in amoebiasis. *Current Clinical Microbiology Reports*, *2*(4), 143–154.

Said, B., Ijaz, S., Kafatos, G., Booth, L., Thomas, H. L., Walsh, A., Ramsay, M., & Morgan, D., (2009). Hepatitis E outbreak on cruise ship. *Emerging Infectious Diseases*, *15*(11), 1738.

Saijuntha, W., Sithithaworn, P., Kaitsopit, N., Andrews, R. H., & Petney, T. N., (2014). Liver flukes: *Clonorchis* and *Opisthorchis*. In: Toledo, R., & Fried, B., (eds.), *Digenetic Trematodes* (pp. 153–199). New York: Springer.

Salque, M., Bogucki, P. I., Pyzel, J., Sobkowiak-Tabaka, I., Grygiel, R., Szmyt, M., & Evershed, R. P., (2013). Earliest evidence for cheese making in the sixth millennium BC in Northern Europe. *Nature*, *493*(7433), 522–525.

Sansonetti, P., (2002). Host–pathogen interactions: The seduction of molecular cross talk. *Gut*, *50*(S3), iii2–iii8.

Sarker, M. R., Carman, R. J., & McClane, B. A., (1999). Inactivation of the gene (cpe) encoding *Clostridium perfringens* enterotoxin eliminates the ability of two cpe-positive *C. perfringens* type A human gastrointestinal disease isolates to affect rabbit ileal loops. *Molecular Microbiology*, *33*(5), 946–958.

Sasaki, M., Kamiyama, T., Yano, T., Nakamura-Uchiyama, F., & Nawa, Y., (2002). Active hepatic capsulitis caused by *Paragonimus westermani* infection. *Internal Medicine*, *41*(8), 661–663.

Savini, V., Polilli, E., Marrollo, R., Astolfi, D., Fazii, P., & D'Antonio, D., (2013). About the *Bacillus cereus* group. *Internal Medicine*, *52*(5), 649.

Sayeed, S., Uzal, F. A., Fisher, D. J., Saputo, J., Vidal, J. E., Chen, Y., Gupta, P., Rood, J. I., & McClane, B. A., (2008). Beta toxin is essential for the intestinal virulence of *Clostridium perfringens* type C disease isolate CN3685 in a rabbit ileal loop model. *Molecular Microbiology*, *67*(1), 15–30.

Schantz, E. J., & Johnson, E. A., (1997). Botulinum toxin: The story of its development for the treatment of human disease. *Perspectives in Biology and Medicine*, *40*(3), 317–327.

Schmutzhard, E., Boongird, P., & Vejjajiva, A., (1988). Eosinophilic meningitis and radiculomyelitis in Thailand, caused by CNS invasion of *Gnathostoma spinigerum* and *Angiostrongylus cantonensis*. *Journal of Neurology, Neurosurgery and Psychiatry*, *51*(1), 80–87.

Schultz-Cherry, S., King, D. J., & Koci, M. D., (2001). Inactivation of an astrovirus associated with poult enteritis mortality syndrome. *Avian Diseases*, 76–82.

Shah, M. P., & Hall, A. J., (2018). Norovirus illnesses in children and adolescents. *Infectious Disease Clinics*, *32*(1), 103–118.

Shandera, W. X., Tacket, C. O., & Blake, P. A., (1983). Food poisoning due to *Clostridium perfringens* in the United States. *The Journal of Infectious Diseases*, *147*(1), 167–170.

Shelef, L. A., (1994). Antimicrobial effects of lactates: A review. *Journal of Food Protection*, *57*(5), 445–450.

Shinagawa, K., Ueno, Y., Hu, D., Ueda, S., & Sugii, S., (1996). Mouse lethal activity of a HEp-2 vacuolation factor, cereulide, produced by *Bacillus cereus* isolated from vomiting-type food poisoning. *Journal of Veterinary Medical Science*, *58*(10), 1027–1029.

Shurson, G. C., (2018). Yeast and yeast derivatives in feed additives and ingredients: Sources, characteristics, animal responses, and quantification methods. *Animal Feed Science and Technology*, *235*, 60–76.

Sibley, L. D., (2004). Intracellular parasite invasion strategies. *Science*, *304*(5668), 248–253.

Siddiqui, A. R., Jooma, R. A., & Smego, Jr. R. A., (2005). Nosocomial outbreak of hepatitis E infection in Pakistan with possible parenteral transmission. *Clinical Infectious Diseases*, *40*(6), 908–909.

Silva, A. J., Leitch, G. J., Camilli, A., & Benitez, J. A., (2006). Contribution of hemagglutinin/protease and motility to the pathogenesis of El Tor biotype cholera. *Infection and Immunity*, *74*(4), 2072–2079.

Singh, R., Kumar, M., Mittal, A., & Mehta, P. K., (2017). Microbial metabolites in nutrition, healthcare and agriculture. *3 Biotech*, *7*(1), 15.

Singh, T., Hiromu, S., Devi, K., & Singh, W., (2015). First case of *Paragonimus westermani* infection in a female patient in India. *Indian Journal of Medical Microbiology*, *33*, 156.

Singh, U., Mitic, L. L., Wieckowski, E. U., Anderson, J. M., & McClane, B. A., (2001). Comparative biochemical and immunocytochemical studies reveal differences in the effects of *Clostridium perfringens* enterotoxin on polarized caco-2 cells versus vero cells. *Journal of Biological Chemistry*, *276*(36), 33402–33412.

Skyberg, J. A., Logue, C. M., & Nolan, L. K., (2006). Virulence genotyping of *Salmonella* spp. with multiplex PCR. *Avian Diseases*, *50*(1), 77–81.

Slamti, L., Perchat, S., Gominet, M., Vilas-Bôas, G., Fouet, A., Mock, M., Sanchis, V., et al., (2004). Distinct mutations in PlcR explain why some strains of the *Bacillus cereus* group are nonhemolytic. *Journal of Bacteriology*, *186*(11), 3531–3538.

Smedley, J. G., Fisher, D. J., Sayeed, S., Chakrabarti, G., & Mc Clane, B. A., (2005). The enteric toxins of *Clostridium perfringens*. In: Aktories, K., & Just, I., (eds.), *Reviews of Physiology, Biochemistry and Pharmacology* (pp. 183–204). Berlin, Heidelberg: Springer.

Song, S. M., Park, J. H., Kim, J., Kim, S. I., Hong, Y. C., Kong, H. H., et al., (2008). Identification and characterization of *Paragonimus westermani* leucine aminopeptidase. *Parasitology International*, *57*(3), 334–341.

Song, S. M., Shin, J. W., de Guzman, J. V., Kim, J., Yu, H. S., Jha, B. K., et al., (2010). *Paragonimus westermani*: Identification and characterization of the fasciclin I domain-containing protein. *Experimental Parasitology*, *125*(2), 76–83.

Song, Y. Y., Zhang, Y., Yang, D., Ren, H. N., Sun, G. G., Jiang, P., et al., (2018). The immune protection induced by a serine protease inhibitor from the foodborne parasite *Trichinella spiralis*. *Frontiers in Microbiology*, *9*, 1544.

Songer, J. G., & Uzal, F. A., (2005). Clostridial enteric infections in pigs. *Journal of Veterinary Diagnostic Investigation*, *17*(6), 528–536.

Speelman, P., Kabir, I., & Islam, M., (1984). Distribution and spread of colonic lesions in shigellosis: A colonoscopic study. *Journal of Infectious Diseases*, *150*(6), 899–903.

Sripa, B., Kaewkes, S., Intapan, P. M., Maleewong, W., & Brindley, P. J., (2010). Foodborne trematodiases in Southeast Asia: Epidemiology, pathology, clinical manifestation and control. *Advances in Parasitology*, *72*, 305–350.

Sripa, B., Kaewkes, S., Sithithaworn, P., Mairiang, E., Laha, T., Smout, M., et al., (2007). Liver fluke induces cholangiocarcinoma. *PLoS Med.*, *4*(7), e201.

Stenfors, A. L. P., Fagerlund, A., & Granum, P. E., (2008). From soil to gut: *Bacillus cereus* and its food poisoning toxins. *FEMS Microbiology Reviews*, *32*(4), 579–606.

Stiles, M. E., & Holzapfel, W. H., (1997). Lactic acid bacteria of foods and their current taxonomy. *International Journal of Food Microbiology*, *36*(1), 1–29.

Su, Y. C., & Liu, C., (2007). *Vibrio parahaemolyticus*: A concern of seafood safety. *Food Microbiology*, *24*(6), 549–558.

Takahashi, A., Kenjyo, N., Imura, K., Myonsun, Y., & Honda, T., (2000b). Cl^- secretion in colonic epithelial cells induced by the *Vibrio parahaemolyticus* hemolytic toxin related to thermostable direct hemolysin. *Infection and Immunity*, *68*(9), 5435–5438.

Takahashi, A., Sato, Y., Shiomi, Y., Cantarelli, V. V., Iida, T., Lee, M., et al., (2000a). Mechanisms of chloride secretion induced by thermostable direct haemolysin of *Vibrio parahaemolyticus* in human colonic tissue and a human intestinal epithelial cell line. *Journal of Medical Microbiology*, *49*(9), 801–810.

Takeda, Y., (1982). Thermostable direct hemolysin of *Vibrio parahaemolyticus*. *Pharmacology and Therapeutics*, *19*(1), 123–146.

Talarico, T. L., & Dobrogosz, W. J., (1989). Chemical characterization of an antimicrobial substance produced by *Lactobacillus reuteri*. *Antimicrobial Agents and Chemotherapy*, *33*(5), 674–679.

Talwani, R., Gilliam, B. L., & Howell, C., (2011). Infectious diseases and the liver. *Clinics in Liver Disease*, *15*(1), 111–130.

Tang, Z. L., Huang, Y., & Yu, X. B., (2016). Current status and perspectives of *Clonorchis sinensis* and clonorchiasis: Epidemiology, pathogenesis, omics, prevention and control. *Infectious Diseases of Poverty*, *5*(1), 1–12.

Teshale, E. H., Grytdal, S. P., Howard, C., Barry, V., Kamili, S., Drobeniuc, J., Hill, V. R., et al., (2010). Evidence of person-to-person transmission of hepatitis E virus during a large outbreak in Northern Uganda. *Clinical Infectious Diseases*, *50*(7), 1006–1010.

Tong, S. Y., Schaumburg, F., Ellington, M. J., Corander, J., Pichon, B., Leendertz, F., et al., (2015). Novel staphylococcal species that form part of a Staphylococcus aureus-related complex: The non-pigmented *Staphylococcus argenteus* sp. nov. and the non-human primate-associated *Staphylococcus schweitzeri* sp. nov. *International Journal of Systematic and Evolutionary Microbiology*, *65*(Pt 1), 15.

Trofa, A. F., Ueno-Olsen, H., Oiwa, R., & Yoshikawa, M., (1999). Dr. Kiyoshi Shiga: Discoverer of the dysentery bacillus. *Clinical Infectious Diseases*, *29*(5), 1303–1306.

Uzal, F. A., & McClane, B. A., (2011). Recent progress in understanding the pathogenesis of *Clostridium perfringens* type C infections. *Veterinary Microbiology*, *153*(1, 2), 37–43.

Vanaja, S. K., Russo, A. J., Behl, B., Banerjee, I., Yankova, M., Deshmukh, S. D., et al., (2016). Bacterial outer membrane vesicles mediate cytosolic localization of LPS and caspase-11 activation. *Cell*, *165*(5), 1106–1119.

Vázquez-Boland, J. A., Kuhn, M., Berche, P., Chakraborty, T., Domínguez-Bernal, G., Goebel, W., et al., (2001). *Listeria* pathogenesis and molecular virulence determinants. *Clinical Microbiology Reviews*, *14*(3), 584–640.

Velebit, B., Djordjevic, V., Milojevic, L., Babic, M., Grkovic, N., Jankovic, V., et al., (2019). The common foodborne viruses: A review. In: *IOP Conference Series: Earth and Environmental Science* (Vol. 333, No. 1, pp. 1–8).

Voulgari, K., Hatzikamari, M., Delepoglou, A., Georgakopoulos, P., Litopoulou-Tzanetaki, E., & Tzanetakis, N., (2010). Antifungal activity of non-starter lactic acid bacteria isolates from dairy products. *Food Control*, *21*(2), 136–142.

Wadström, T., & Ljungh, Å., (1991). *Aeromonas* and *Plesiomonas* as food- and waterborne pathogens. *International Journal of Food Microbiology*, *12*(4), 303–311.

Walker, C. M., Feng, Z., & Lemon, S. M., (2015). Reassessing immune control of hepatitis A virus. *Current Opinion in Virology*, *11*, 7–13.

Wallis, T. S., & Galyov, E. E., (2000). Molecular basis of *Salmonella*-induced enteritis: Microreview. *Molecular Microbiology*, *36*(5), 997–1005.

Walter, J. E., & Mitchell, D. K., (2003). Astrovirus infection in children. *Current Opinion in Infectious Diseases*, *16*(3), 247–253.

Wang, Q. P., Wu, Z. D., Wei, J., Owen, R. L., & Lun, Z. R., (2012). Human *Angiostrongylus cantonensis*: An update. *European Journal of Clinical Microbiology and Infectious Diseases*, *31*(4), 389–395.

Wassenaar, T. M., & Blaser, M. J., (1999). Pathophysiology of *Campylobacter jejuni* infections of humans. *Microbes and Infection*, *1*(12), 1023–1033.

Weil, A. A., & Harris, J. B., (2015). *Vibrio cholerae*. In: Tang, Y., Sussman, M., Liu, D., Poxton, I., & Schwartzman, J., (eds.), *Molecular Medical Microbiology* (pp. 1079–1098). London, UK: Academic Press.

Wijman, J. G., De Leeuw, P. P., Moezelaar, R., Zwietering, M. H., & Abee, T., (2007). Air-liquid interface biofilms of *Bacillus cereus*: Formation, sporulation, and dispersion. *Applied and Environmental Microbiology*, *73*(5), 1481–1488.

Wikswo, M. E., Kambhampati, A., Shioda, K., Walsh, K. A., Bowen, A., & Hall, A. J., (2015). Outbreaks of acute gastroenteritis transmitted by person-to-person contact, environmental contamination, and unknown modes of transmission – United States, 2009–2013. *Morbidity and Mortality Weekly Report: Surveillance Summaries*, *64*(12), 1–16.

Wilson, A., Evans, J., Chart, H., Cheasty, T., Wheeler, J. G., Tompkins, D., & Smith, H. R., (2001). Characterization of strains of enteroaggregative *Escherichia coli* isolated during the infectious intestinal disease study in England. *European Journal of Epidemiology, 17*(12), 1125–1130.

Wnek, A. P., & McClane, B. A., (1986). Comparison of receptors for *Clostridium perfringens* type A and cholera enterotoxins in isolated rabbit intestinal brush border membranes. *Microbial Pathogenesis, 1*(1), 89–100.

Wong, H. C., Liu, S. H., Ku, L. W., Lee, I. Y., Wang, T. K., Lee, Y. S., Lee, C. L., Kuo, L. P., & Shih, D. Y. C., (2000). Characterization of *Vibrio parahaemolyticus* isolates obtained from foodborne illness outbreaks during 1992 through 1995 in Taiwan. *Journal of Food Protection, 63*(7), 900–906.

World Health Organization, (2008). *Foodborne Disease Outbreaks: Guidelines for Investigation and Control* (p. 145). World Health Organization, France, WHO.

World Health Organization, (2015). *WHO Estimates of the Global Burden of Foodborne Diseases: Foodborne Disease Burden Epidemiology Reference Group 2007–2015* (p. 253). World Health Organization, Switzerland.

Xi, J. N., Graham, D. Y., Wang, K. N., & Estes, M. K., (1990). Norwalk virus genome cloning and characterization. *Science, 250*(4987), 1580–1583.

Xia, Y., Ju, Y., Chen, J., & You, C., (2015). Cerebral paragonimiasis: A retrospective analysis of 27 cases. *Journal of Neurosurgery: Pediatrics, 15*(1), 101–106.

Yakoob, J., Jafri, W., Beg, M. A., Abbas, Z., Naz, S., Islam, M., & Khan, R., (2010). Irritable bowel syndrome: Is it associated with genotypes of *Blastocystis hominis*. *Parasitology Research, 106*(5), 1033–1038.

Yamai, S., Okitsu, T., Shimada, T., & Katsube, Y., (1997). Distribution of serogroups of *Vibrio cholerae* non-O1 non-O139 with specific reference to their ability to produce cholera toxin, and addition of novel serogroups. *Kansenshogaku zasshi. The Journal of the Japanese Association for Infectious Diseases, 71*(10), 1037–1045.

Yang, X., Xu, M., Wu, Y., & Xiang, B., (2015). Pancreatic paragonimiasis mimics pancreatic cystic-solid tumor – A case report. *Pancreatology, 15*(5), 576–578.

Yeung, P. M., & Boor, K. J., (2004). Epidemiology, pathogenesis, and prevention of foodborne *Vibrio parahaemolyticus* infections. *Foodborne Pathogens and Disease, 1*(2), 74–88.

Zannella, C., Mosca, F., Mariani, F., Franci, G., Folliero, V., Galdiero, M., Tiscar, P. G., & Galdiero, M., (2017). Microbial diseases of bivalve mollusks: Infections, immunology and antimicrobial defense. *Marine Drugs, 15*(6), 182.

Zuel-Fakkar, N. M., Abdel, H. D. M., &Hassanin, O. M., (2011). Study of zuel isolates in urticaria: A case–control study. *Clinical and Experimental Dermatology: Experimental Dermatology, 36*(8), 908–910.

CHAPTER 9

Pathogenic Microbes in the Food Systems

ATIF LIAQAT, SAMREEN AHSAN, RABIA IQBAL,
MUHAMMAD FARHAN JAHANGIR CHUGHTAI, and ADNAN KHALIQ

ABSTRACT

Food safety is becoming an imperative public health issue in all over the world. Foodborne pathogens include bacteria, viruses, fungi, and parasites, infecting humans by contaminated food or water. Foodborne bacteria include various species from the genera *Bacillus*, *Brucella*, *Campylobacter*, *Clostridium*, *Cronobacter*, *Helicobacter*, *Listeria*, *Mycobacterium*, *Escherichia*, *Staphylococcus*, *Salmonella*, *Shigella*, *Streptococcus*, *Vibrio*, *Yersinia*, etc. Foodborne viruses, comprising mostly of hepatitis A, E viruses, rotavirus, astrovirus, and norovirus (NoV), can transfer from infected living organism to food and water. Foodborne bacteria and viruses truly are a main cause of concern, but it is difficult to escape the conclusion that mycotoxins in foods are responsible for much higher numbers of human deaths than are foodborne bacteria and viruses. Foodborne fungi produce mycotoxins [aflatoxin, ochratoxin A, fumonisins, and zearalenone (ZEN)] that are cancerogenic metabolites in infected foods. Foodborne parasitic diseases are becoming common and can transmit directly or indirectly in humans from meat and water sources. These parasites include protozoa and helminths. Control of contamination, proper cooking and adopting good hygienic practices during handling/preparation of food can result in the provision of safe food. Overall, there is an urgent need for better monitoring and control of foodborne pathogens using new technologies.

Food Microbial and Molecular Biology: From Fundamentals to Applications. Saher Islam, Devarajan Thangadurai, Jeyabalan Sangeetha, Zaira Zaman Chowdhury (Eds.)

9.1 INTRODUCTION

Since ancient era, human beings have been using various products from different sources for health benefits (Liaqat et al., 2019). A strong correlation between food consumption and human illnesses was also documented in very old times as quoted by Hippocrates in 460 BC. Foodborne pathogens significantly affect human health and economy. Every year, a large number of people are affected through foodborne zoonotic diseases principally in developing countries. These illnesses are caused significantly by bacteria, fungi, parasites, and viruses carried by food so commonly known as foodborne pathogens. About 200 different types of illnesses have been reported and classified as: (I) food intoxication and (II) food infection. Food poisoning is an illness caused by consumption of food that is contaminated with foodborne pathogens. It can be food infection if someone has consumed food containing live microbes or it may be food intoxication if someone has consumed toxins produced by them (Bintsis, 2017; Zahoor et al., 2018).

9.2 FOODBORNE BACTERIA

The characteristics of significantly important foodborne bacterial agents have been described below, along with the illnesses and reported outbreaks these have resulted in. Bacterial pathogens mainly include *B. cereus*, *C. jejuni*, *C. botulinum*, *C. perfringens*, *C. sakazakii*, *L. monocytogenes*, *E. coli*, *S. aureus*, *Salmonella* spp., *Shigella* spp., *Vibrio* spp., and *Y. enterocolitica*. All these pathogenic bacteria exhibit different morphological, cultural, and physiological characteristics, which are important to know for their risk of illnesses.

9.2.1 *BACILLUS*

Various species of *Bacillus* are ubiquitous and found in soil, marine, and freshwater environments. *Bacillus* spores often contaminate the raw milk on farm. *Bacillus cereus* (*B. cereus*), family Bacillaceae, is a spore-forming, Gram-positive, motile rod. It is one of the chief foodborne pathogenic bacteria and frequently found in food production surroundings (Rajkowski and Smith, 2001; Bacon and Sofos, 2003; Arnesen et al., 2008; Zahoor et al., 2018). The spores of this bacterium have appendages and may also contain pili which, in comparison to any other bacillus spores, are more hydrophobic.

These spores are adhesive endospores and such characteristic allows them to adhere to various diverse kinds of surfaces, making them resistant to be removed through sanitation and cleaning processes (Rajkowski and Smith, 2001; Arnesen et al., 2008). Depending upon the strain, the temperatures range for the growth of vegetative form of *B. cereus* is 4–15 to 35–55°C, however, prefer to grow at 30–40°C (Andersson et al., 1995). The pH requirement for the growth of this organism 4.9 to 9.3 and a partial growth is observed on meat at a pH of 4.35 (Bacon and Sofos, 2003; FDA, 2012). The organism grows at a minimum water activity of 0.93 which is also recommended as 0.912 as the fried rice have water activity in the range of 0.912 to 0.961, readily supporting the growth of *B. cereus* (ICMSF 1996; Rajkowski and Smith, 2001).

There are two types of toxins produced by the *B. cereus*, one the diarrheal and other being the emetic (vomiting) one which cause two kinds of diseases. Growth of this organism in the small intestine results in the production of diarrheal toxins which causes diarrhea and abdominal pain. While the growth in food produces emetic toxins which results in vomiting and nausea. The endospores of *B. cereus*, surviving during cooking, germinate, and subsequently proliferate during storage. Vegetables, soups, sauces, puddings, milk, meat, and their products are often associated with *B. cereus* diarrheal food poisoning which occurs due to the action of thermo-labile enterotoxin complex. The late onset of this type is characterized by the symptoms of nausea, abdominal pain, and diarrhea following about 8–16 hours of incubation period. The disease symptoms usually persist for 12–24 hours. Pastry, noodles, pasta, cooked, and fried rice are the foods associated with the *B. cereus* emetic food poisoning which involves the action of thermo-stable toxins. The rapid onset of this type comprises the symptoms principally of vomiting and nausea appearing after 1–5 hours of incubation time and persist for 6–24 hours (Rajkowski and Smith, 2001; Arnesen et al., 2008).

9.2.2 BRUCELLA

Brucella was named on the scientist Sir David Bruce, who found these responsible for brucellosis. The species of this genus are Gram-negative, oxidase +ve, catalase +ve, oval rod-shaped bacteria. Optimal temperature for their growth is 37°C (Adams and Moss, 2008). Three important species of this genus (1) *Brucella suis*, (2) *Brucella abortus*, and (3) *Brucella melitensis* have been detected in the milk of cattle. These can cause disease in humans, known as brucellosis, which is an example of milk borne zoonotic disease. All over the world, more than 0.5 million cases (among humans, wild life, domestic animals across Africa) of this disease are documented every year. The disease may be transferred to humans through contact with diseased animals or consuming their products (Githui et al., 2000; Godfroid et al., 2011; Karshima, 2013). Symptoms of brucellosis may comprise physical weakness, joint, and back pain, fatigue, headache, sweats, etc. Central nervous system (CNS) and cardiac system can also be affected as a result of infection. Common sources of *Brucella* contamination include buffaloes, sheep, and goat. These bacteria can be inactivated by pasteurization treatment (Zahoor et al., 2018).

9.2.3 CLOSTRIDIUM

The bacteria of the genus *Clostridium* are Gram-positive, motile, aerotolerant/obligate anaerobes, spore formers, rod-shaped and belong to the Bacillaceae family. These bacteria are ubiquitous in nature however, mainly present in the intestines of animals and soil. It exhibits the potential to produce endospores under unfavorable conditions which germinate in the presence of organic substrates for vegetative growth (Bacon and Sofos, 2003).

9.2.4 CLOSTRIDIUM BOTULINUM (CL. BOTULINUM)

Cl. botulinum is one of the well-known species of genus *Clostridium* which is capable to produce lethal poison named as botulinum neurotoxin. This toxin is produced during cellular growth and released upon cell lysis. The probability for botulinum toxin has been reported in canned foods (corn, lobster, beets, green beans, spinach, ripe olives, peppers, soups, tuna fish, asparagus, mushrooms, chicken, ham, stuffed eggplant, sausage, smoked, and salted fish) which can cause the botulism in humans and animals (Bacon and Sofos, 2003; FDA, 2012; Bintsis, 2017).

On the basis of antigen specificity of toxin generated by each strain, botulinum neurotoxins have been classified into seven different types: A, B, C, D, E, F, and G neurotoxins. In humans, botulism is caused by the types A, B, E, and F, while C and D type neurotoxins are responsible for this disease in mammals and birds. After ingestion of neurotoxin, symptoms appear in 12–36 hours which result in vomiting and nausea initially which further led to visual impairment, acute flaccid paralysis and possibly death. Many foods borne botulism outbreaks have been associated with Group I *Cl. botulinum*, proteolytic in nature, which can be controlled through thermal processing of canned or bottled foods at 121°C for 3 minutes (Bacon and Sofos, 2003; FDA, 2012; Carter and Peck, 2015).

9.2.5 CLOSTRIDIUM PERFRINGENS (CL. PERFRINGENS)

Previously known as *Clostridium welchii*, *Cl. perfringens* belongs to the genus *Clostridium* and the family Bacillaceae. Morphologically, *Cl. perfringens* is encapsulated rod which is non-motile in nature and produce toxin made up of protein. It produces spores under unfavorable conditions which can resist to heat, desiccation, and radiation. *Cl. perfringens* is the highest widespread *Clostridium* species observed in the human clinical samples, excluding feces. The temperature range for the vegetative growth is 6.0–50°C with an optimal of 43–47°C. The organism prefers to grow on a pH of 6.0–7.2 while can grow on 5.0–9.0 pH. Depending upon the strain, it requires a minimum water activity of 0.93 (Bacon and Sofos, 2003; Bintsis, 2017).

It is a ubiquitous pathogen, often linked with the dust contamination and found in the shellfish as well as meat products. There have been five toxin producing types of *Cl. perfringens* identified as A, B, C, D, and E. The organism is recognized as the second most common cause of foodborne bacterial illnesses as reported in the US, mainly associated with the type A. The infection results when a highly contaminated food (viable vegetative cells $>10^{6-7}$) is consumed. It is associated with simple wound infections to myonecrosis, intra-abdominal sepsis, intravascular hemolysis, pneumonia, gangrenous cholecystitis, etc. Outbreaks caused by *Cl. perfringens* are preventable if contamination is controlled on farms or slaughterhouses and through proper preparation in restaurants (Bacon and Sofos, 2003; Grass et al., 2013; Bintsis, 2017).

9.2.6 CAMPYLOBACTER

Campylobacter jejuni (*C. jejuni*) belongs to the family Campylobacteriaceae and frequently results in diarrhea. The genus *Campylobacter* includes 18 species, 6 subspecies and 2 biovars which are Gram-negative, spiral shaped (0.2–5.0 μm long and 0.2–0.9 μm wide), microaerophilic bacteria requiring about 5% O_2 and 10% CO_2 in the surroundings. The pathogenic species for humans grow at a narrow temperature range of almost 30–46°C and categorized as thermophiles (Humphrey et al., 2007; Scallan et al., 2011).

C. jejuni is broadly distributed in nature. It can make colonies in the intestine of birds and mammals. Contaminated foods transmit this organism to human beings. In host, this organism first attaches with the epithelial cells and then invades the epithelial layer. Damage to the epithelial cells result in diarrheal illness and in case of systemic infection, it leads to more severe illness. *C. coli* and *C. jejuni* are the chief basis of acute foodborne gastroenteritis. These bacteria can be transmitted from animals to humans. Children are at more risk to disease. A number of wild and healthy household animals contain *Campylobacter* species as their normal intestinal microflora. These animals include wild birds, cats, dogs, rodents, goats, sheep, geese, ducks, pigs, sheep, marine mammals, and rodents. Water bodies such as streams are also often associated with these species. Contaminated water, unpasteurized milk, undercooked or raw poultry meat and cross-contamination are often associated with campylobacteriosis. In US, *C. jejuni* alone causes about 0.85 million illnesses, 8,500 hospitalizations, and 76 deaths every year (Schaffner et al., 2004; Scallan et al., 2011; Bintsis, 2017; Zahoor et al., 2018).

9.2.7 *CRONOBACTER*

The genus *Coronobacter* comprises seven species as *Cr. muytjensii*, *Cr. malonaticus*, *Cr. universalis*, *Cr. dublinensis*, *Cr. condimenti*, *Cr. turicensis* and *Cr. sakazakii*. These bacteria are Gram negative rods (Jaradat et al., 2014).

Cronobacter sakazakii (*Cr. sakazakii*), formerly recognized as *Enterobacter sakazakii*, produces yellow pigment at incubation temperature of 30°C. Primarily isolated from powdered formula, it is related with meningitis, necrotizing enterocolitis, and septicemia in the infants through infections in immune-compromised adults and in elderly people have also been documented (Bintsis, 2017).

The potential sources of contamination may be retail foods (dried flours, dried milk powder, dried meats, nuts, spices, legumes) and hygienically poor home settings. However, epidemiological studies have indicated that most of the *Cr. sakazakii* outbreaks and infant infections are associated with the powdered formula milk contaminated with this pathogen (Kandhai et al., 2004; Hochel et al., 2012; Bintsis, 2017).

9.2.8 ESCHERICHIA

Escherichia coli (*E. coli*) is a non-spore former, facultative anaerobe, rod shaped bacterium which may be with or without flagella, so it may be motile or non-motile in nature. It has the ability to ferment glucose and other simple sugars into acetic, lactic, and formic acids. It can grow at a wide range of pH from 4.3–10 with an optimum growth at 6.0–8.0. The majority of *E. coli* strains are harmless while others are pathogenic which are responsible for mortality and morbidity cases all over the world at a low dose. *E. coli* spreads by consumption of food and water contaminated with the waste material of infected animals and humans. Animal products are contaminated during slaughtering and processing while the plant products are contaminated in the field through cattle manure, used as fertilizer (Garcia et al., 2010; Croxen et al., 2013; Bintsis, 2017).

Based on the pathogenicity mechanism, pathogenic *E. coli* is classified into six groups: 1. EHEC (Enterohemorrhagic *E. coli*), 2. ETEC (enterotoxigenic *E. coli*), 3. A/EEC (Attaching and Effacing *E. coli*), 4. EIEC (enteroinvasive *E. coli*), 5. EAggEC (Enteroaggregative *E. coli*), and 6. EPEC (Enteropathogenic *E. coli*). EHEC which was previously known as VTEC (Verotoxin-producing *E. coli*) is also referred as STEC (Shiga-toxin producing *E. coli*) (Garcia et al., 2010; Croxen et al., 2013). Infection due to STEC results in mild to severe diarrhea. The infections may develop in to hemolytic uremic syndrome (HUS), characterized by bleeding, which can cause kidney failure and even prove fatal. STEC strain O157:H7 is a zoonotic pathogen having intestines of cattle as a major reservoir. Several outbreaks have been documented due to the ingestion of undercooked ground meat and fresh produce which were contaminated with STEC O157:H7. Due to this virulent strain every year 63,000 illnesses, 2,100 hospitalizations, and 20 deaths are recorded (Scallan et al., 2011).

9.2.9 HELICOBACTER

Helicobacter pylori, previously known as *Campylobacter pylori*, are the member of the genus *Helicobacter*. Its infection is linked with gastroenteritis, stomach, and duodenal ulcer in humans. It is transmitted through milk and milk products and pasteurization treatment is recommended for its destruction. For an accurate diagnosis in humans, gastric biopsy (expensive treatment) is needed (Jayarao and Henning, 2001; Adams and Moss, 2008; Zahoor et al., 2018).

9.2.10 MYCOBACTERIUM

The species of this genus are Gram positive and non-spore forming bacteria. Majority of non-dangerous species of this genus can be found in surroundings; however, two ancient and potentially dangerous diseases (leprosy and tuberculosis (TB)) are associated with them. TB is considered as the chronic and contagious bacterial disease which can be transferred from animals to humans. It is associated with wild species, humans, and domestic animals. Granulomas in the lungs, kidney, intestines, lymph nodes, and liver are formed as a result of the occurrence of this disease. Most of the developed countries have got rid of this disease but still documented in the developing countries. The bacterium responsible for TB in humans, is recognized as *Mycobacterium tuberculosis* (Admas and Moss, 2008; Zahoor et al., 2018).

Mycobacterium bovis is another member of this genus. This bacterium has been found in the feces, milk, respiratory, and vaginal secretions of animals. It is transmitted through aerosols when cattle are close to each other. Human infection occurs primarily through unpasteurized milk products, however, undercooked or raw meat has also been found linked with transmission of this bacterium. Inactivation of this bacterium can be achieved through pasteurization. It is a zoonotic disease and reported cases are significantly greater in people dealing with animals. Cough, chest pain and fever may be observed in human's respiratory infection. *Mycobacterium bovis* has been found resistant against medicines which are used to treat *Mycobacterium tuberculosis* (Girma et al., 2014; Zahoor et al., 2018).

9.2.11 LISTERIA

Listeria monocytogenes (*L. monocytogenes*) is comprised of 13 serotypes which include 1/2a, 1/2b, 1/2c, 3a, 3b, 3c, 4a, 4ab, 4b, 4c, 4d, 4e, and 7. Majority of the foodborne infections result from the 1/2a, 1/2b, and 4b serotypes. This bacterium is one of the dominant causes of death in an immuno-compromised person, the elderly, newborns, and pregnant women. Its infection during pregnancy may lead to birth defects, miscarriage, and stillbirths. It can be found in soil, water, raw foods, undercooked meat, and decaying vegetable contents (FDA, 2012; Buchanan et al., 2017). This bacterium can resist extreme temperatures ranging 1–45°C and can grow on refrigeration temperature. However, it can be destroyed through cooking and pasteurization. *L. monocytogenes* infection results in listeriosis, a severe infection generally caused by eating the contaminated food. With a high mortality rate (20–30%), listeriosis is considered to be the deadliest foodborne risk. Infection due to this bacterium can result in gastroenteritis, meningitis, and septicemia. Once this bacterium gets entry to the food processing area, it can persist there for years, thereby contaminating the foods (Bintsis, 2017; Buchanan et al., 2017).

9.2.12 *SALMONELLA*

Salmonella was named on the scientist (Dr. Daniel Salmon) who isolated the *S. choleraesuis* first time from pig's intestine. These bacteria belong to the family Enterobacteriaceae. This genus is divided into two species (*S. bongori* and *S. enterica*) and further subdivided into serotypes based on antigenic properties of flagella and surface. *Salmonella* species are generally known by their serotypes, e.g., *S. enterica* subsp. *enterica* is further subdivided into various serotypes such as *S. typhimurium* and *S. enteritidis* (Bacon and Sofos, 2003; FDA, 2012). *S. enterica* causes typhoid fever. Colonization and host infection by *Salmonella* species result due to the presence of numerous pathogenicity islands, encoding several virulence factors. Every year more than 1000,000 people in the US come in contact with *Salmonella* resulting in 19,000 hospitalizations and 380 deaths on an average, as per estimations of CDC. *Salmonella* species are the intestinal residents of many wild animals and livestock. Foods commonly associated with *Salmonella* outbreaks involve meat and eggs; however, fruits and vegetables may also be found contaminated with these bacteria (Bintsis, 2017; Zahoor et al., 2018).

9.2.13 *SHIGELLA*

Bacteria of this genus are Gram-negative, facultative anaerobes, non-spore-formers, non-motile rods. *Shigella* belongs to the family Enterobacteriaceae and comprises four serogroups which were previously regarded as species. These include *S. dysenteriae*, *S. flexneri*, *S. boydii* and *S. sonnei* belonging to the serogroups A, B, C, and D, respectively. The serogroups A, B, and C further contain 38 serotypes and serogroup D has only 1. There is a close resemblance between DNA of *Shigella* species and *E. coli*. Some biochemical properties are also common between these two types of bacteria. *Shigella* species can grow at 6–48°C temperature with an optimal of 37°C. The pH for the growth of these organisms ranges 4.8–9.3 with an optimum 6.0–8.0 (ICMSF, 1996; Bacon and Sofos, 2003).

Environments with poor hygiene and sanitation are frequently associated with these types of bacteria. Primarily it is transmitted through person-to-person contact. Shigellosis is associated with consumption of foods and water contaminated with these bacteria. Previously, foods linked with shigellosis outbreak included fresh produce, milk, chicken, salads, shellfish, etc., served at schools, cruise ships, airlines, restaurants, and military mess halls. Following an incubation period of 12–50 hours, gastrointestinal (GI) infections occur including watery diarrhea along with abdominal cramps, fatigue, fever, and malaise which can involve any of the *Shigella* serogroup (ICMSF, 1996; Bacon and Sofos, 2003; Bintsis, 2017).

9.2.14 *STAPHYLOCOCCUS*

These are non-spore-forming, Gram-positive, non-motile, facultative anaerobes. However, *Staph. aureus* subsp. *anaerobius* and *Staph. saccharolyticus* proliferate more quickly in the presence of oxygen. Morphologically, these cocci bacteria exist as single, in a pair, short chains, tetrads or clusters which resemble with the bunch of grapes. It is considered as one of the most resistant pathogens which can survive for long times in dry state and can be isolated from sewage, water, dust, and air. Its species are prevalent in nature and can be isolated from blood, mouth, mammary glands, skin, and skin glands of birds and mammals, upper respiratory, genitourinary (GU), and intestinal tract of infected persons. Depending on the strain type, *Staph. aureus* can grow at a temperature ranging 7.0 to 47.8°C, however, the optimal temperature for growth is 40–45°C. At 10–46°C temperature it produces enterotoxins. It can grow at pH ranging 4.5–9.3 but optimal for growth is 7.0–7.5. It can also tolerate very high levels of salt (more than 10% NaCl). This bacterium can grow at a 0.83 water activity; however, production of enterotoxins needs a minimum of 0.86 water activity (Bacon and Sofos, 2003; FDA, 2012; Bintsis, 2017).

Staph. aureus can produce nine different types of enterotoxins, known as pyrogenic toxins, and represented as A, B, C1, C2, C3, D, E, F, and G. Among them type A and D cause most of the outbreaks. Foods may be contaminated by direct contact, through skin fragments and respiratory tract droplets. The symptoms of infection may include nausea, vomiting, dizziness, chills, headache, abdominal cramps, weakness, perspiration, diarrhea (may or may not contain blood) and muscular cramping. Every year food poisoning due to staphylococci results in 241,188 illnesses, 1,064 hospitalizations and 6 deaths in the US according to the CDC estimates. Foods associated with staphylococci include pork sausages, salmon steaks, cream pies, shrimps, ground turkey, oysters, ground beef, and milk which are considered as carriers of this pathogen (Bacon and Sofos, 2003; FDA, 2012; Bintsis, 2017).

9.2.15 *STREPTOCOCCUS*

Three diverse groups based on the genome studies of streptococci make them worthy of genus status. Dairy or lactic streptococci now belong to the *Lactococcus* genus. Numerous bacteria, previously thought to be different *Streptococcus* species, are currently categorized as subspecies of *Lactococcus lactis*. However, the absence of an antigen (Group N) in *Streptococcus thermophilus* (yogurt starter) makes the reason to classify it under *Streptococcus* genus. Streptococci (Group A), carried by humans, may infect animal udder. These can result in mastitis, scarlet fever, and sore throat. There may be a rapid multiplication of these bacteria in unprocessed (unpasteurized) or improperly cooled milk. However, pasteurization can control these bacteria (Adams and Moss, 2008; Zahoor et al., 2018).

9.2.16 *VIBRIO*

The members of this group are Gram-negative, motile, curved, or straight rods, non-spore formers and facultative anaerobes in nature. There are > than 35 species in the genus *Vibrio*, belonging to the family Vibrionaceae. More than $1/3^{rd}$ of these species is harmful for human beings, including *V. parahaemolyticus*, *V. cholera* and *V. vulnificus*. All pathogenic species of this genus are reported to grow mainly in marine environments of temperate or tropical regions and need sodium for optimum growth, however, *V. cholera* has also been found in freshwater. *V. cholera* can grow at temperature ranging 10–43°C with an optimal of 30–37°C. The preferable pH requirement for growth is 7.6, however, can grow at a range of 5.0–9.6 pH. Least water activity for growth of this bacterium is 0.97, preferably growing at 0.984. The raw and contaminated foods such as oysters and crustaceans may be a vehicle for entry of *V. cholera* in human body. The symptoms appear after incubation time of several hours to 5 days and include vomiting, diarrhea, and hypoglycemia. Severe dehydration occurs as a result of vomiting which leads to muscle cramping (ICMSF, 1996; Bacon and Sofos, 2003; Wu et al., 2014).

Other *Vibrio* species, *V. parahaemolyticus*, has been found very often in clinical samples collected in the US. It grows at a temperature ranging 5.0–44°C with an optimal of 30–37°C. Optimum pH for the growth of this bacterium is 7.6–8.6 while it can grow on 4.8–11.0 pH. Minimum water activity for its growth is 0.94 while preferring 0.981. Transmission of this bacterium has been associated with foods like prawns, seaweeds, oysters, clams, and crabs. Food poisoning due to this bacterium results in gastroenteritis which is related with eating raw, undercooked, or re-contaminated seafood. After 4–96 hours of incubation symptoms are demonstrated comprising vomiting, nausea, slight fever, headache, abdominal cramps, and watery or sometimes bloody diarrhea. Four different kinds of hemolytic components have been reported in *V. parahaemolyticus* such as lysophospholipase, phospholipase A, a thermolabile direct hemolysin, a thermostable direct hemolysin (TDH). This bacterium is invasive in nature. It can cross lamina propria and has been detected in liver, spleen, heart, and pancreas (Bacon and Sofos, 2003; Bintsis, 2017).

9.2.17 *YERSINIA*

These are Gram-negative, non-spore formers, facultative anaerobes, rod shaped bacteria, belonging to the Enterobacteriaceae family. Only three species, including *Y. pestis*, *Y. enterocolitica* and *Y. pseudotuberculosis* are considered to be pathogenic for animals and humans. *Yersinia* includes 10 species of which *Y. enterocolitica* can grow at temperature ranging 0-45°C with an optimal of 25–30°C. The optimum pH for growth of this bacterium is 7.6 and can grow at 4.0–10 pH. It is broadly present in the surroundings and has been detected in seafood, water contaminated with sewage, raw milk, soil, pigs, and poultry. As it can grow on refrigeration temperature, so contamination of refrigerated foods may become a health risk. Infection of this bacterium in human beings results in foodborne gastroenteritis (Bacon and Sofos, 2003; Adams and Moss, 2008; Bintsis, 2017).

After consuming contaminated food and following incubation time of few to 7 days symptoms of infection may appear. Mostly raw or undercook pork is associated with the yersiniosis infection. Children may suffer from intestinal yersiniosis for 4 weeks, while in the case of adults, it can persist for 1–2 weeks. Symptoms of intestinal yersiniosis comprise watery or bloody diarrhea along with vomiting, abdominal pain, and fever. Children below

15 years of age and immuno-compromised people are more susceptible to infection. Extraintestinal infections due to yersiniosis involve meningitis, septicemia, glomerulonephritis, myocarditis, thyroiditis, and Reiter syndrome. Toxins produced by this bacterium are stable against enzyme degradation, heat, and also during long storage times. pH stability of these toxins is comparable to that of heat-stable toxins produced by ETEC (Bacon and Sofos, 2003; FDA, 2012; Bintsis, 2017).

9.3 FOODBORNE VIRUSES

A virus needs a live host to replicate and it uses food only as a carrier. The food becomes contaminated during processing by infected food handler. Viruses are resistant to temperature, pH, and disinfection (Koopmans et al., 2004). These transfer genetic material that may be RNA/DNA enclosed in protecting covering of protein.

All enteric viruses are resistant to low pH and can survive for several weeks and months. Unlike bacteria, the viruses are resistant to high temperature, mostly used for cooking, and oxidizing agents, e.g., ozone, chlorine, however, UV light can inactivate viruses in water or on surfaces. The food and water-borne viruses mainly transmit from fecal-oral route, e.g., hepatitis A virus (HAV) and hepatitis E virus (HEV) mostly transmit to human bodies from those crops that have been irrigated from organic manure (Bozkurt et al., 2021; Treagus et al., 2021) (Table 9.1).

TABLE 9.1 Characteristics of Foodborne Viruses

Foodborne Viruses	Genus	Source	Incubation Time	References
Hepatitis A virus (HAV)	*Hepatovirus*	Shellfish, soft fruits (strawberries and raspberries), lettuce	3–6 weeks	Appleton (2000)
Hepatitis E virus (HEV)	*Herpesvirus*	Shellfish and produce, as well as animal meats	2 weeks to 2 months	Yugo and Meng (2013)
Rotavirus	*Rotavirus*	Fruits, vegetables, water, and other produce	48 h	Baert et al. (2009)
Astrovirus	*Mammoastrovirus*	Meat and other	3–5 days	Lee et al. (2013)
Noroviruses	*Norovirus*	Oysters to romaine lettuce, raspberries, and other produce	48 h	Robilotti et al. (2015)

9.3.1 HEPATITIS A VIRUS (HAV)

HAV is a member of the family Picornaviridae. It is highly resistant and stable in extreme environmental conditions such as heat and drying, low pH (2.0), UV light, gamma rays and low levels of chlorine and ozone than other picornaviruses (Grohmann and Lee, 2003). HAV starts replication as it is ingested, commonly via contaminated water or food. After penetration and replication in the intestinal epithelial cells, HAV enters blood and then blood carries the virus to its target site, the liver where it replicates in hepatocytes and liver macrophages. When an immune response is induced, cytotoxic T cells destroy infected liver cells, severely disrupting regular body functions controlled by the liver (Cliver and Matsui, 2002). HAV is excreted from liver to bile and released from the body in feces. Consequently, HAV may enter from feces to water and inadequately treated sewage water becomes a significant risk for shellfish, water for recreational use, water used in agriculture and may accidentally mix in drinking water. The route of its transmission is fecal-oral. HAV endures to be shed in low numbers for up to 2 weeks following the onset of symptoms, which include fever, loss of appetite, nausea, and abdominal discomfort, often followed by jaundice lasting several days. In infants and children younger than five years of age, infection is often mild or asymptomatic. Immunity to hepatitis A is usually lifetime and death is rare (Grohmann and Lee, 2003; Grove et al., 2006).

9.3.2 HEPATITIS E VIRUS (HEV)

HEV is mainly associated with waterborne sources of contamination, however; foodborne sources have also been reported. Consumption of raw and undercooked animal meat is mainly responsible for acute hepatitis in humans. Infectious HEV has been known to be found in sewage water, untreated water, animal meat, shellfish, and their products (Yugo and Meng, 2013).

9.3.3 ROTAVIRUS AND ASTROVIRUS

Rotaviruses and Astroviruses are less commonly associated with outbreaks of foodborne gastroenteritis, and these are also difficult to quantify because of mild symptoms appearance so not reported routinely. Both of these viruses mostly affect young children, usually under the age of 1 year. The

Astroviruses are morphologically divergent viruses because of their five- or six-points star seen by electron microscope on the exterior of certain particles (Appleton, 2000).

9.3.4 NOROVIRUSES (NoVs)

The NoV, genus classified as Norwalk-like viruses include the species of Norwalk virus (NLV). These viruses cause gastroenteritis in humans, a generally mild disease featuring symptoms such as vomiting, nausea, abdominal pain, diarrhea, malaise, muscle pain, anorexia, headache, and low-grade fever in infected persons. Symptoms normally appear from 1–2 days following contaminated food consumption and last for 1–8 days (Grove et al., 2006).

Now a day's genome-based detection methods are available for viral detection in solid food and water in which amplification is done for fragments of the viral RNA. Nested (qualitative) reverse transcriptase-polymerase chain reaction-based (RT-PCR) and real-time (quantitative) RT-PCR have been used for detection and quantification of virus in concentrated water samples (Villar et al., 2006). These molecular techniques, particularly PCR based methods, provide sensitive, rapid, and quantitative analytical tools to assess the microbial quality of food and water and to evaluate the efficacy of virus removed from drinking source (Girones et al., 2010). Recently, ELISA based assays have been developed for detection of astroviruses and NLV (Koopmans et al., 2002).

9.4 FOODBORNE FUNGI

9.4.1 TOXIGENIC FUNGI AND MYCOTOXINS

Foodborne molds are of various types and cause 25% of agricultural food crop deterioration due to the release of certain enzymes, e.g., lipase, protease, cellulase, and some volatile compounds. The growth of fungi on food items release toxins known as "mycotoxins." These are secondary metabolites and major potential danger of their presence in the human diet is the inability to detect them biologically due to their less than 1,000 Da molecular weight. These mycotoxins cause human diseases and qualitative loss of food items. There are several natural storage fungi that have been reported on food commodities, e.g., *Aspergillus* spp., *Penicillium* spp., *Fusarium* spp., *Alternaria* spp., etc., but *Aspergillus* spp. and mainly *Aspergillus flavus* is dominant, which is a matter of great concern due to secretion of high level of aflatoxins (AFs) in affected food items. The other mycotoxins are ochratoxin A, fumonisins, and zearalenone (ZEN). *Fusarium* species are destructive pathogens on cereal crops and other commodities which produce mycotoxins during pre- and post-harvesting. Certain species of *Aspergillus* and *Penicillium* are also plant pathogens or commensals, but these genera are more commonly associated with commodities during the process of drying and storage. The most significant toxigenic species and mycotoxins are described below. Toxicity from mycotoxins may be acute, chronic, mutagenic, and teratogenic. The acute toxicity causes kidney or liver damage and cancer that lead to death. Sometimes it causes skin necrosis, leucopenia, or immunodeficiency. The chronic effects are low but can cause tumor or cancer. The DNA replication has showed mutagenic and teratogenicity effects caused by mycotoxins (Pitt, 2000) (Table 9.2).

9.4.1.1 AFLATOXINS (AFs)

Aflatoxins (AFs) are highly important mycotoxins due to their significant burden on agriculture sector (Alshannaq and Yu, 2017). The three most important crops affected by these toxins are peanuts, maize, and cottonseed. These toxins are mostly produced by *Aspergillus flavus* and *Aspergillus parasiticus*. These toxins may cause acute liver cirrhosis, tumors, teratogenicity, mutation, and immune suppression. Mainly produced AFs include B_1, B_2, G_1 and G_2. Where the letter 'B' denotes blue and 'G' is for green bright color produced by these AFs under UV light on thin-layer chromatographic plates while the subscript numbers 1 and 2 show major and minor compounds. The aflatoxin B_1 can also increase oxidative stress that can damage the cells by lipid peroxidation because of formation of reactive oxygen species (ROS). The AFs B1 and B2 are ingested from feed-in lactating mammals where it is hydroxylated and converted to aflatoxin M, which can be transferred to infants or others by feeding milk (Pitt, 2000; Wu et al., 2014).

9.4.1.2 OCHRATOXIN A

Ochratoxin A was mainly designated as a metabolite of *Aspergillus ochraceus*. *A. ochraceus* is species mainly found in seeds, nuts, green coffee beans and

TABLE 9.2 Characteristics of Mycotoxin

Mycotoxin	Fungus	Food Source	International Agency for Research on Cancer (IARC)	References
Aflatoxins B1, B2, G1, G2	*Aspergillus flavus* *A. parasiticus*	Maize, peanuts, tree nuts, spices, cottonseed	Group 1 (The agent is carcinogenic to humans)	Dawei et al. (2017); Lien et al. (2019)
Aflatoxin M1 (B1 metabolite)	Indirectly produced by *Aspergillus flavus* *A. parasiticus*	Milk and milk products	Group 1 (The agent is carcinogenic to humans)	Moosavy et al. (2019)
Ochratoxin A	*Aspergillus ochraceus* *A. carbonarius* *A. niger*	Nuts, green coffee beans, fruits, milk, meat, and in human milk	Group 2B (The agent is probably carcinogenic to humans)	Eygue et al. (2020)
Fumonisins	*Fusarium verticillioides* *F. proliferatum* *A. niger*	Maize and its products and can also occur in black tea, soybean, sorghum, wheat, barley, figs, asparagus, and medicinal plants	Group 2B (The agent is probably carcinogenic to humans)	Wangia et al. (2019)
Zearalenone	*F. graminearum* and *F. semitectum*	Corn, wheat, barley, sorghum, and rye	Group 3 (The agent is not classifiable as to its carcinogenicity to humans)	Zhang et al. (2020)

fruits. Ochratoxin A has also been reported in animal milk, meat as well as human milk. These species are widespread in tropical foods and survive sun drying. Ochratoxin A is a plausible carcinogen, that can cause urinary tract cancer and kidney damage (Wu et al., 2014; Alshannaq and Yu, 2017).

9.4.1.3 FUMONISINS

Currently, more than 28 fumonisins have been identified and divided into four groups (A, B, C, and P), among which 70–80% is only FB1. Fumonisins are mainly produced by *F. verticillioides* and *F. proliferatum*. These mainly affect the crop of maize and its products but can also be present in black tea, soybean, sorghum, wheat, barley, figs, asparagus, and medicinal plants. Fumonisins appear to be the cause of esophageal cancer in southern Africa, parts of China and elsewhere. The toxic action of fumonisins appears to result from competition with sphingosine in sphingolipid metabolism (Wu et al., 2014; Alshannaq and Yu, 2017).

9.4.1.4 ZEARALENONE (ZEN)

Zearalenone (ZEN) is produced by *Fusarium* species, mainly *F. graminearum* and *F. semitectum*. Its structure is like estrogens, so it is also known as estrogenic mycotoxin and may cause obvious estrogenic effects in humans and animals. ZEN is frequently found in corn, wheat, barley, sorghum, and rye. Its production is favored by high humidity and low temperature conditions (Alshannaq and Yu, 2017). Mycotoxins can be detected by precisely collecting samples and using a broad range of analytical methods, depending on matrix, sample size and number, time, available equipment by adopting techniques like ELISA, HPLC or LC-MS/MS (Zhang and Banerjee, 2020).

9.5 EMERGING FOODBORNE PARASITES

Bacterial, viral, and fungal infections are very common and gained much attention unlike parasitic infections. It is probable that 75% of emerging human pathogens are zoonotic (Woolhouse, 2002), and previously, it was thought that parasitic infections are only limited to a few geographic regions due to their host specificity, requirement of intermediate hosts and specific environmental factors. These limitations have been broken by international travelers and by transport of refrigerated food commodities (Orlandi et al., 2002).

9.5.1 WATERBORNE PARASITES TRANSMITTED BY CONTAMINATED FOOD

Parasites are mainly transmitted to the human body from contaminated water. The parasites spend most of their life stages in water and through this water they contaminate various foodstuffs such as fruits, vegetables, and shellfish. It is often difficult to associate an outbreak with a particular food item, and therefore, if the foodborne route is suspected, it becomes more difficult to identify how the food implicated became contaminated (Slifko et al., 2000). Among emerging water borne parasitic infections associated with food major are caused by *Cryptosporidium*, *Cyclospora cayetanensis*, *Fasciola*, *Fasciolopsis*, and *Giardia*.

9.5.2 MEAT BORNE PARASITE INFECTIONS

Among the foodborne parasites, most commonly reported include *Toxoplasma gondii*, *Trichinella* spp., *Sarcocystis* spp., and *Taenia* spp. Usually, humans become infected by eating undercooked meat that is mostly contaminated with cyst stage of parasites. There are lots of precautionary measures adopted to save people from helminths by inspecting the meat in the slaughterhouse and in laboratories while there are no specific tests to check toxoplasmosis and sarcocystosis. However, appropriate cooking of meat or other preservation techniques like smoking, curing, and drying are effective to get rid of these parasites (Dorny et al., 2009).

9.5.3 PARASITES TRANSMITTED BY SEAFOOD

Fish and amphibians can be contaminated with trematodes, cestodes, nematodes, and pentastomids that may cause zoonotic infections if consumed undercooked. However, these parasites may also spread through aquaculture, transportation, tourism, distribution systems of aquatic foods to bring local and international markets close and altered culinary practices (Keiser and Utzinger, 2005). Parasites have been detected in food by using various techniques like direct visualization by Zeihl–Eelsen stain or modified trichrome stain to allow visualization of protozoal oocysts, spores of microsporidium, and identification by microscopy (Doaa, 2012). Moreover, immunological assay of ELISA has been used for parasitic evaluation.

9.6 CONCLUSION

Food is a necessity of life and there is a close relationship between health and food consumption. Quality food to be consumed, particularly microbiological quality is a major factor affecting the consumers' health. Foodborne pathogenic microorganisms include various genera associated with food substances. Pathogenic microorganisms usually transfer from infected living organisms to food and water. Mycotoxins (aflatoxin, ochratoxin A, fumonisins, and ZEN) are cancerogenic metabolites in infected foods and responsible for much higher numbers of human deaths than are foodborne bacteria and viruses. Parasites mainly include protozoa and helminths. Several techniques including ELISA, HPLC, LCMS can be used for the purpose of detection.

KEYWORDS

- **foodborne bacteria**
- **food pathogens**
- **foodborne fungi**
- **mycotoxins**
- **foodborne parasites**
- **foodborne viruses**

REFERENCES

Adams, M. R., & Moss, M. O., (2008). *Food Microbiology*. Cambridge: RSC Publishing.

Alshannaq, A., & Yu, J. H., (2017). Occurrence, toxicity, and analysis of major mycotoxins in food. *International Journal of Environmental Research and Public Health, 14*(6), 632.

Andersson, A., Rönner, U., & Granum, P. E., (1995). What problems does the food industry have with the spore-forming pathogens *Bacillus cereus* and *Clostridium perfringens*? *International Journal of Food Microbiology, 28*, 145–155.

Appleton, H., (2000). Control of foodborne viruses. *British Medical Bulletin, 56*(1), 172–183.
Arnesen, L. P. S., Fagerlund, A., & Granum, P. E., (2008). From soil to gut: *Bacillus cereus* and its food poisoning toxins. *FEMS Microbiology Reviews, 32*, 579–606.
Bacon, R. T., & Sofos, J. N., (2003). Characteristics of biological hazards in foods. In: Schmidt, R. H., & Rodrick, G. E., (eds.), *Food Safety Handbook* (pp. 157–195). New Jersey: John Wiley and Sons.
Baert, L., Debevere, J., & Uyttendaele, M., (2009). The efficacy of preservation methods to inactivate foodborne viruses. *International Journal Food Microbiology, 131*(2, 3), 83–94.
Bintsis, T., (2017). Foodborne pathogens. *AIMS Microbiology, 3*(3), 529.
Bozkurt, H., Phan-Thien, K. Y., Van, O. F., Bell, T., & McConchie, R., (2021). Outbreaks, occurrence, and control of norovirus and hepatitis a virus contamination in berries: A review. *Critical Reviews in Food Science and Nutrition, 61*(1), 116–138.
Buchanan, R. L., Goris, L. G. M., Hayman, M. M., et al., (2017). A review of *Listeria monocytogenes*: An update on outbreaks, virulence, dose-response, ecology, and risk assessments. *Food Control, 75*, 1–13.
Carter, A. T., & Peck, M. W., (2015). Genomes, neurotoxins and biology of *Clostridium botulinum* group I and group II. *Research in Microbiology, 166*, 303–317.
Cliver, D. O., & Riemann, H. P., (2002). *Foodborne Diseases*. Gulf Professional Publishing.
Croxen, M. A., Law, R. J., Scholz, R., et al., (2013). Recent advances in understanding enteric pathogenic *Escherichia coli*. *Clinical Microbiology Reviews, 26*, 822–880.
Doaa, E. S. S., (2012). Detection of parasites in commonly consumed raw vegetables. *Alexandria Journal of Medicine, 48*(4), 345–352.
Dorny, P., Praet, N., Deckers, N., & Gabriël, S., (2009). Emerging foodborne parasites. *Veterinary Parasitology, 163*(3), 196–206.
Eygue, M., Richard-Forget, F., Cappelier, J. M., Pinson-Gadais, L., & Membré, J. M., (2020). Development of a risk-ranking framework to evaluate simultaneously biological and chemical hazards related to food safety: Application to emerging dietary practices in France. *Food Control, 115*, 107279.
FDA, (2012). *The Bad Bug Book: Foodborne Pathogenic Microorganisms and Natural Toxins*. https://www.fda.gov/files/food/published/Bad-Bug-Book-2nd-Edition-%28PDF%29.pdf.
Garcia, A., Fox, J. G., & Besser, T. E., (2010). Zoonotic enterohemorrhagic *Escherichia coli:* A one health perspective. *ILAR Journal, 51*, 221–232.
Girma, K., Tilahun, Z., & Haimanot, D., (2014). Review on milk safety with emphasis on its public health. *World Journal of Dairy and Food Sciences, 9*(2), 166–183.
Girones, R., Ferrus, M. A., Alonso, J. L., et al., (2010). Molecular detection of pathogens in water – the pros and cons of molecular techniques. *Water Research, 44*(15), 4325–4339.
Githui, W. A., Hawken, E. S., Juma, E. S., et al., (2000). Surveillance of drug resistant tuberculosis and molecular evaluation of transmission of resistant strains in refugee and non-refugee populations in North Eastern Kenya. *International Journal of Tuberculosis and Lung Disease, 4*, 947–955.
Godfroid, J., Scholz, H., Barbier, T., Nicolas, C., & Wattiau, P., (2011). Brucellosis at the animal/ecosystem/human interface at the beginning of the 21st century. *Preventive Veterinary Medicine, 102*, 118–131.
Grass, J. E., Gould, L. H., & Mahon, B. E., (2013). Epidemiology of foodborne disease outbreaks caused by *Clostridium perfringens*, United States, 1998–2010. *Foodborne Pathogens and Disease, 10*, 131–136.

Grohmann, G., & Lee, A., (2003). Viruses, food and environment. In: Hocking, A. D., (ed.), *Foodborne Microorganisms of Public Health Significance* (pp. 615–634). New South Wales: Australian Institute of Food Science and Technology, Inc.

Grove, S. F., Lee, A., Lewis, T., Stewart, C. M., Chen, H., & Hoover, D. G., (2006). Inactivation of foodborne viruses of significance by high pressure and other processes. *Journal of Food Protection, 69*(4), 957–968.

Hochel, I., Rüzicková, H., Krásny, L., et al., (2012). Occurrence of *Cronobacter* spp. in retail foods. *Journal of Applied Microbiology, 112*, 1257–1265.

Humphrey, T., O'Brien, S., & Madsen, M., (2007). Campylobacters as zoonotic pathogens: A food production perspective. *International Journal of Food Microbiology, 117*, 237–257.

ICMSF, (1996). *Micro-organisms in Foods 5, Characteristics of Microbial Pathogens*. New York: Kluwer Academic/Plenum Publishers.

Jaradat, Z. W., Mousa, W. A., Elbetieha, A., et al., (2014). *Cronobacter* spp. – opportunistic foodborne pathogens. A review of their virulence and environmental-adaptive traits. *Journal of Medical Microbiology, 63*, 1023–1037.

Jayarao, B. M., & Henning, D. R., (2001). Prevalence of foodborne pathogens in bulk tank milk. *Journal of Dairy Sciences, 84*, 2157.

Kandhai, M. C., Reij, M. W., Puyvelde, K., et al., (2004). A new protocol for the detection of *Enterobacter sakazakii* applied to environmental samples. *Journal of Food Protection, 67*, 1267–1270.

Karshima, N. S., (2013). The roles of veterinarians in the safety of foods of animal origin in Nigeria: A review. *Journal of Animal Production Advances, 3*, 57–68.

Keiser, J., & Utzinger, J., (2005). Emerging foodborne trematodiasis. *Emerging Infectious Diseases, 11*(1), 1507–1514.

Koopmans, M., Von, B. C. H., Vinjé, J., De Medici, D., & Monroe, S., (2002). Foodborne viruses. *FEMS Microbiology Reviews, 26*(2), 187–205.

Lee, R. M., Lessler, J., Lee, R. A., et al., (2013). Incubation periods of viral gastroenteritis: A systematic review. *BMC Infectious Diseases, 13*(1), 446.

Liaqat, A., Zahoor, T., Randhawa, M. A., & Shahid, M., (2019). Characterization and antimicrobial potential of bioactive components of sonicated extract from garlic (*Allium sativum*) against foodborne pathogens. *Journal of Food Processing Preservation, 43*(5), 13936.

Lien, K. W., Wang, X., Pan, M. H., & Ling, M. P., (2019). Assessing aflatoxin exposure risk from peanuts and peanut products imported to Taiwan. *Toxins, 11*(2), 80.

McCabe-Sellers, B. J., & Beattie, S. E., (2004). Food safety: Emerging trends in foodborne illness surveillance and prevention. *Journal of the American Dietetic Association, 104*(11), 1708–1717.

Moosavy, M. H., Hasanzadeh, M., Soleymani, J., & Mokhtarzadeh, A., (2019). Determination of aflatoxin M1 using an aptamer-based biosensor immobilized on the surface of dendritic fibrous nano-silica functionalized by amine groups. *Analytical Methods, 11*(30), 3910–3919.

Orlandi, P. A., Chu, D. M. T., Bier, J. W., & Jackson, G. J., (2002). Parasites and the food supply. *Food Technology, 56*(4), 72–79.

Pitt, J. L., (2000). Toxigenic fungi and mycotoxins. *British Medical Bulletin, 56*(1), 184–192.

Qi, D., Fei, T., Liu, H., Yao, H., Wu, D., & Liu, B., (2017). Development of multiple heart-cutting two-dimensional liquid chromatography coupled to quadrupole-orbitrap high resolution mass spectrometry for simultaneous determination of aflatoxin B1, B2, G1, G2, and ochratoxin A in snus, a smokeless tobacco product. *Journal of Agricultural and Food Chemistry*, *65*(45), 9923–9929.

Rajkowski, K. T., & Smith, J. L., (2001). Update: Food poisoning and other diseases induced by *Bacillus cereus*. In: Hui, Y. H., Pierson, M. D., & Gorham, J. R., (eds.), *Foodborne Disease Handbook* (pp. 61–76). New York: Marcel Dekker, Inc.

Robilotti, E., Deresinski, S., & Pinsky, B. A., (2015). Norovirus. *Clinical Microbiology Reviews*, *28*(1), 134–164.

Scallan, E., Hoekstra, R. M., Angulo, F. J., et al., (2011). Foodborne illness acquired in the United States – major pathogens. *Emerging Infectious Diseases*, *17*, 7–15.

Schaffner, N., Zumstein, J., & Parriaux, A., (2004). Factors influencing the bacteriological water quality in mountainous surface and groundwaters. *Acta Hydrochimica et Hydrobiologica*, *32*, 225–234.

Slifko, T. R., Smith, H. V., & Rose, J. B., (2000). Emerging parasite zoonoses associated with water and food. *International Journal for Parasitology*, *30*(1), 1379–1393.

Treagus, S., Wright, C., Baker-Austin, C., Longdon, B., & Lowther, J., (2021). The foodborne transmission of hepatitis E virus to humans. *Food and Environmental Virology*, 1–19.

Villar, L. M., De Paula, V. S., Diniz-Mendes, L., Lampe, E., & Gaspar, A. M. C., (2006). Evaluation of methods used to concentrate and detect hepatitis A virus in water samples. *Journal of Virological Methods*, *137*(2), 169–176.

Wangia, R. N., Githanga, D. P., Xue, K. S., Tang, L., Anzala, O. A., & Wang, J. S., (2019). Validation of urinary sphingolipid metabolites as biomarker of effect for fumonisins exposure in Kenyan children. *Biomarkers*, *24*(4), 379–388.

Woolhouse, M. E., (2002). Population biology of emerging and re-emerging pathogens. *Trends in Microbiology*, *10*(10), 3–7.

Wu, F., Groopman, J. D., & Pestka, J. J., (2014). Public health impacts of foodborne mycotoxins. *Annual Review of Food Science and Technology*, *5*, 351–372.

Wu, Y., Wen, J., Ma, Y., Ma, X., & Chen, Y., (2014). Epidemiology of foodborne disease outbreaks caused by *Vibrio parahaemolyticus*, China, 2003–2008. *Food Control*, *46*(1), 197–202.

Yugo, D. M., & Meng, X. J., (2013). Hepatitis E virus: Foodborne, waterborne and zoonotic transmission. *International Journal of Environmental Research and Public Health*, *10*(10), 4507–4533.

Zahoor, T., Liaqat, A., & Azhar, N., (2018). Safety and Quality aspects of animal origin foods. In: Khan, M. I., & Sameen, A., (eds.), *Animal Sourced Foods for Developing Economies: Preservation, Nutrition, and Safety* (pp. 207–224). Florida: CRC Press.

Zhang, K., & Banerjee, K., (2020). A review: Sample preparation and chromatographic technologies for detection of aflatoxins in foods. *Toxins*, *12*(9), 539.

Zhang, S., Zhou, S., Gong, Y. Y., Zhao, Y., & Wu, Y., (2020). Human dietary and internal exposure to zearalenone based on a 24-hour duplicate diet and following morning urine study. *Environment International*, *142*, 105852.

CHAPTER 10

Microbial Pathogens and Mycotoxins in Food and Control Measures

RAVEENDRAN RESHMA, RADHAKRISHNAN PREETHA, and DEVARAJAN THANGADURAI

ABSTRACT

Foodborne pathogens (FBP) produce many severe health issues, which is due to the consumption of contaminated food and water. Foodborne bacterial diseases are classified into foodborne infections and foodborne intoxications. The most common pathogenic bacteria involved in foodborne diseases (*Salmonella* sp., *Shigella* sp., *Listeria monocytogenes*, *Staphylococcus aureus*, *Clostridium botulinum*, *Yersinia enterocolitica*, *Vibrio* spp., *Escherichia coli*, and *Bacillus cereus*) are discussed in this chapter. The chapter also described viruses (Hepatitis A, Rotavirus, Norovirus (NoV), and Hepatitis E), parasites (*Giardia*, *Toxoplasma*, *Cryptosporidium*, *Ascaris*, and *Cyclospora*), and mycotoxins (Aflatoxin, Ochratoxin A, Patulin, Zearalenone (ZEN), Fumonisin, Sterigmatocystin, and Citrinin). In addition to that, the growth conditions and characteristics of each pathogenic organism, food involved in the disease outbreaks, most common clinical symptoms of foodborne illness, identification tools, and strict hygiene control measures to bring down the risk of transmission of foodborne diseases are also explained in detail.

10.1 INTRODUCTION

Foodborne diseases have significant health impacts on the public. Foodborne diseases can transmit via food contaminated with pathogenic microorganisms.

Food Microbial and Molecular Biology: From Fundamentals to Applications. Saher Islam, Devarajan Thangadurai, Jeyabalan Sangeetha, Zaira Zaman Chowdhury (Eds.)

However, several cases of foodborne diseases were not reported and tabulated properly since many people do not consult doctors or physicians during mild or moderate foodborne illnesses (Clelia, 2017). Cases were reported or documented only if it was affected severely or a large number of populations were affected at a time. The majority of reported cases and outbreaks were due to bacterial foodborne pathogens (FBP) followed by viral pathogens (WHO 2008). The most of outbreaks and cases allotted to bacterial diseases such as Staphylococcal intoxication, Salmonellosis, and *Clostridium perfringens* gastroenteritis (Newell et al., 2010). Foodborne diseases (FBDs) due to *Listeria* and *E. coli* are also reported frequently. As per Communicable Disease Center (CDC), USA, foodborne disease stated as a circumstance in which one or more people that sense an identical illness, especially intestinal illness after consuming a typical food, and etymological evaluation indicate that the diet as a primary source of disease (Banwart, 1998).

FBDs are caused by food contaminated with pathogens and its multiplication and establishment inside the human body after ingestion, or by consumption of food products in which pathogens already established and produced toxins (Bintsis, 2017). Based on the above criteria the foodborne bacterial diseases are classified into two: foodborne infections and foodborne intoxications. Some examples of foodborne infections and foodborne intoxications are given in Figure 10.1 (Frazier and Westhoff, 2008).

FOODBORNE BACTERIAL DISEASE

INTOXICATION

- **Staphylococcal intoxication**
- **Botulism**

INFECTIONS

- ***Clostridium perfringens***
- ***Salmonellosis***
- ***Enteropathogenic Escherichia***
- ***Bacillus cereus gastroenteritis***
- ***Yersiniosis***
- ***Vibrio cholera***
- ***Arizona infection***
- ***shigellosis***

FIGURE 10.1 Foodborne bacterial diseases.

Example for foodborne intoxication include botulism, which is caused by the toxin formed by *Clostridium botulinum* in the food component and Staphylococcal intoxication which caused by the presence of toxin produced by *Staphylococcus aureus*. Above mentioned intoxications are non-communicable in nature and the incubation periods may vary. The factors responsible for bacterial intoxication are inadequate cooking and improper handling. The toxins produced in the food before consumption or toxins produced after establishing pathogen in human intestine cause damage to the lining of cells in the GI tract, especially in the colon. This is followed by the symptoms of diarrhea and abdominal cramps, or even severe dysentery. The symptoms of foodborne diseases also include vomiting, nausea, diarrhea, weakness, and respiratory failure. Foodborne intoxication can classify into fungal intoxications, bacterial intoxications, plant toxicants, chemical toxicants, and poisonous animals (Frazier and Westhoff, 2008).

Foodborne infection or foodborne illness occurs due to establishment of bacteria in the human intestinal tract the human body due to the consumption of contaminated foods. It is a contagious disease and it can spread from one person to another and many bacterial, viruses, and parasitic diseases were reported under this category (Frazier and Westhoff, 2008). They can invade or multiply in the lining of the human intestinal tract. The incubation periods may vary from hours to days. Factors responsible for the bacterial infection are inadequate cooking, cross-contamination, and poor personal hygiene.

From the 1990s onwards, three foodborne bacterial organisms (*Salmonella* sp., *Campylobacter* sp., and *E. coli*) were reported frequently in food manufacturing units (Bacon and Sofos, 2003). Salmonella infection occurs due to the consumption of improperly cooked food such as egg, poultry, and vegetables (Bintsis, 2017). *E. coli* is a bacterium mostly found in the gut of homeothermic organisms (WHO, 2018). Another important foodborne bacterial pathogen, *Listeria monocytogenes*, was reported in improperly pasteurized milk and cheeses, ice cream, processed or raw poultry/vegetables, and seafoods (Frazier and Westhoff, 2008). *Cryptosporidium*, Calicivirus, *Cyclospora*, and Norovirus (NoV) are termed as the new foodborne pathogen (Tauxe, 1997). NoV and Hepatitis A virus (HAV) causes a large number of viral outbreaks worldwide (Koopmans and Duizer, 2004). In addition to viral/bacterial FBP, parasitic protozoa are also responsible for causing food and waterborne disease. *Cryptosporidium* sp., *Cyclospora cayetanensis*, and *Giardia intestinalis* are the primary source for diarrhea and gastrointestinal (GI) infections related to parasites (CDC, 2015). In addition to those toxins produced by fungi (mycotoxins) are a major concern in food safety. Mycotoxins are toxic metabolites produced by fungi infesting food products, especially in nuts and cereals, which are stored under at high moisture levels and elevated temperature (Jay et al., 2005). Aflatoxin is considered as the main carcinogenic mycotoxin. This chapter provides a detailed description about foodborne bacterial, viral, and parasitic diseases as well as mycotoxins.

10.2 DIFFERENT TYPES OF FOODBORNE BACTERIAL DISEASES

Bacteria are the primary source of foodborne disease and are having different nature and characteristics. Some bacteria are thermostable; they are *Staphylococcus aureus* and *Clostridium botulinum*. Some of them are thermostable but have the capability for spore formation; examples are *Bacillus aureus*, *Clostridium perfringens*, and *Bacillus subtilis*. Symptoms and severity depend on the type of bacteria that contaminated the foods (Table 10.1) (CDC, 2019). Good personal hygiene, proper cooking of foods and proper sanitization can control most bacterial infections.

10.2.1 FOODBORNE INTOXICATION

10.2.1.1 STAPHYLOCOCCUS AUREUS

Staphylococcus aureus occurs in the form of a clump of grass and short chains. They can cause clotting of blood plasma and form a thermostable nucleus, and in the case of oxygen requirement, it is non-obligatory in nature (Frazier and Westhoff, 2008). They do not produce a stale smell in many food substances or make them not attractive since they are fermentative and proteolytic (Frazier and Westhoff, 2008). The growth and toxin production of the pathogen depends upon storage temperature (4°C to 46°C) of food substances. Under aerobic conditions, if pH is low, it shows minimal growth. Under aerophilic and anaerobic conditions, the organism prefers the water activity of 0.86 and 0.90, respectively (Frazier and Westhoff, 2008). The toxin produced by *Staphylococcus aureus* is an enterotoxin because it causes injury to the GI tract (Frazier and Westhoff, 2008). Staphylococcal enterotoxins (SEs) are complex protein, which has the molecular weight ranges between 26,000 and 30,000 kD. Enterotoxins can lead to swelling of the inner lining of the GI tract. The toxin produced at the temperature ranges between 15.6°C and 46.1°C. Staphylococcal intoxication produced by *Staphylococcus*

TABLE 10.1 Foodborne Bacterial Diseases, Infection Period, and Symptoms (Kumar et al., 2019)

Microorganism	Disease	Infection Period	Food Involved	Symptoms
Salmonella species	Salmonellosis	6 to 7 hours	Contaminated water, raw milk, shrimp	Diarrhea, fever, headache, stomach pain
Shigella	Shigellosis	1 to 3 days	Salads, sandwiches	Abdominal pain, diarrhea, headache
Listeria monocytogenes	Listeriosis	3 to 70 days	skimmed milk, ice cream, and meat products	Headache, liver abscess, meningitis
Staphylococcus aureus	Staphylococcus infection	30 minutes to 8 hours	Fish, fish products, poultry, bakery foods, meat, meat products	Vomiting, weakness, abdominal cramps
Clostridium botulinum	Botulism	2 to 48 hours	Canned vegetables, fruits, fish	Nausea, vomiting, dizziness, headache
Yersinia enterocolitica	Yersiniosis	24 hours to 30 hours	Raw milk, ice cream, and inadequately cooked meats	Diarrhea, nausea, vomiting, and fever abdominal pain
Vibrio species	Cholera	2 to 48 hours	Fresh marine foods, fish, and fish products	Vomiting, mild fever, prostration, diarrhea, abdominal pain
Arizona hinshawi	Arizona infection	2 to 46 hours	Pastry, ice cream, poultry	Vomiting, headache, nausea, red throat
E. coli	*Escherichia coli* enteritis (EEC)	8 to 24 hours	Salads, apple juices, sausage	Headache, fever, abdominal cramps
Bacillus cereus	Emetic syndrome	1 to 6 hours	Noodles, cooked rice, pastry	Nausea and vomiting
	Diarrheal syndrome	8 to 16 hours	Milk and milk products, meat, sauce	Abdominal cramps, fever, bloating, nausea

aureus. The indications of Staphylococcal intoxications are nausea, diarrhea, abdominal cramps, vomiting, headache, weakness, muscle soreness (Kumar et al., 2019). Many foods responsible for causing *Staphylococcus* food poisoning are custard, bakery foods filled with cream, poultry, fish products, meat products, and so on. The most common outbreaks of Staphylococcal food poisoning are due to improper cooking of foods (Frazier and Westhoff, 2008). The outbreaks of food poisoning by *Staphylococcus* can be prevented by proper cooking and by avoiding food contamination. Food contamination is usually reduced by using conventional sanitation methods, practicing the use age of pasteurized milk instead of raw milk, and keeping workers or employees away from food when they have a Staphylococcal infection. The staphylococci growth can inhibit the sufficient refrigeration of foods and some cases by increased to more acidic pH. It can be inhibited by using bacteriostatic compounds such as serine or other bactericidal antibiotics (Frazier and Westhoff, 2008).

10.2.1.2 CLOSTRIDIUM BOTULINUM

It is a rod-shaped gram-negative spore-forming bacterium. Botulism is caused by the consumption of food, which contains highly toxic and soluble neurotoxin produced by *Clostridium botulinum*. *Clostridium botulinum* toxins are classified into six different types and are classified from A to G (Rood et al., 2018). Types A, B, E, F, and G cause human botulism, type C cause botulism in animals, and type D causes food poisoning in cattle. Type F strains are common in all types of food (Wen et al., 2004; Yibar et al., 2018). Based on cultural and physiological characters of *Clostridium botulinum* strains are usually distinguished into three groups:

- Group 1 which involves types A and B production
- Group 2 involves types E, B, and F production, and
- Group 3, which involves types C and D production, these are not proteolytic

The temperature required for the growth of group 1 is 30°C to 40°C; group 2 is 20°C to 37°C and for group 3 and 4, it is 30°C to 37°C. The group 1 organism which digests milk, group 2, which does not digest the milk, but they coagulate milk into soft curd, group 3 organism which does not change milk and group 4 which slowly digest the milk (Frazier and Westhoff, 2008).

This organism prefers a neutral pH. The toxin production and growth rate of *Clostridium botulinum* depend on the type of food and temperature at optimum pH. The production of toxins in the food will be inhibited at pH 4.5 or below. *Clostridium botulinum* cannot be growing at a water activity of 0.93 or less. The optimal temperature for growth and production of proteolytic strains is 35°C, and for a non-proteolytic strain is 26°C to 28°C. The toxin produced by microorganism gets absorbed in small intestine and cause numbness of the smooth muscles in the body. This toxin remains thermally stable and will get destroyed at a temperature above 90°C or by gamma radiation of 7 to 8 M rad (Frazier and Westhoff, 2008). The incubation period is about 12–48 hours (CDC, 1983). Food substances responsible for causing botulism are heat-processed canned foods of low acidity such as canned vegetables, fruits, fish, and fish products, meat, and meat products (Frazier and Westhoff, 2008). Nausea, vomiting, dizziness, headache are the symptoms of botulism. In fatal cases, death occurs 3 to 6 days after consumption of food with toxins. Administration of antitoxin is the method used for the treatment of botulism. Botulism can be prevented by using approved heat processes for canned foods or by avoiding the use of spoiled canned food (Frazier and Westhoff, 2008).

In the case of an adult, botulism symptoms may occur between 12 to 72 hours after the ingestion of toxin-containing foods. The symptoms include paralysis of muscles, respiratory failure, vomiting, nausea, headache, double vision, and death. Depending on the host resistance and other circumstances, the duration may vary from 1 to 10 days or more (Jay et al., 1995). This disease diagnosed by using specific antiserum, and exotoxin is responsible for all the symptoms. Infant botulism is considered as a particular disease. The infant botulism is distinct from foodborne botulism because in this case, toxins are produced due to the growth of the organism in the intestinal tract and *in vivo* release of the toxin. The symptoms of infant botulism include difficulty in milk feeding, weakness, and constipation. *Clostridium difficile* produces lethal toxins in the intestinal tract of infants (Gillian et al., 1983).

10.2.2 FOODBORNE INFECTIONS

10.2.2.1 CLOSTRIDIUM PERFRINGENS

In 1945, the United States reported the first case of *Clostridium perfringens* gastroenteritis and identified more commonly than previously. The organism is a spore-forming, rod-shaped, gram-positive, immobile, and anaerobic (Mc Clane et al., 2006, 2013; Li et al., 2007). The organism's optimal growth temperature is about 43°C to 47°C, and the maximum temperature at which it can grow is 55°C. The temperature for the controlled growth is 15°C, to 20°C pH below 5.0 or above 9.0 is not suitable for the growth of the organism. The spores of this organism can be detected in sewage, animal wastes, and soil (Mc Clane et al., 2006, 2013; Li et al., 2007). The proteinaceous foods that are partially boiled or roasted, salads, gravies, sauces are the foods involved in outbreaks of *Clostridium perfringens*. The spores of *C. perfringens* can be destroyed by using Gamma radiation. The enterotoxin released in the gut and extreme fluid gathering was observed inside the body's intestinal tract during the cells' sporulation. The enterotoxin is deactivated by heating at 60°C for about 10 minutes and also heat sensitive. The enterotoxin was detected with an indirect ELISA test (Mc Clane et al., 1984). To prevent illnesses due to *C. perfringens*, avoid the use of contaminated food, inhibit the growth of microbe by proper cooking, and avoid prolonged thawing of refrigerated food before consumption (Brown and Twedt, 1972). The outbreaks of *C. perfringens* food infection can also be controlled by following practices:

- The cooking will kill the vegetative cells and the spores of some strains
- Good personal hygiene
- Illness produced by *C. perfringens* can be avoided by preventing the germination and multiplication of spores
- The maximum growth temperature is about 50°C to 52°C, there is no growth and the vegetative cells get destroyed beyond 52°C; So, keeping food at this temperature can prevent the growth of the above pathogen

The symptoms usually appear within 8 to 15 hours, and the symptoms include diarrhea, fever, abdominal pain, and vomiting (Navarro et al., 2018).

10.2.2.2 VIBRIO SPECIES

Vibrio species is a gram-negative curve-shaped bacterium and exist in the family of Vibrionaceae (Ryan and Ray, 2004; Thompson et al., 2005; Faruque and Nair, 2008). They are not spore-forming bacteria. *Vibrio* species mostly found in the aquatic environment. *Vibrio cholera* is the most toxigenic species. It causes a large number of outbreaks worldwide. There are different kinds of *Vibrio* species; among these 12 species are reported as pathogenic to humans. Some species are responsible for causing foodborne infections in humans, they are:

- *Vibrio cholera*
- *Vibrio parahaemolyticus*
- *Vibrio vulnificus*
- *Vibrio alginolyticus*

10.2.2.2.1 *Vibrio cholera*

Vibrio cholera is a gram-negative, curve-shaped bacterium (Mandal et al., 2011). *Vibrio cholera* is responsible for causing cholera in humans (Frazier and Westhoff, 2008; Rodriguez and Kahwaji, 2020). Cholera is transmitted through contaminated food and water via fecal-oral transmission. The primary host for *Vibrio cholera* is the humans (Rodriguez and Kahwaji, 2020). The organism grows at a temperature of 10°C to 43°C at the pH range of 7.6. The incubation periods are 2 to 3 days (Azman et al., 2012). The studies or the survey reveals that they are equitably common in estuary water. So they are considered as the autochthonous estuarine bacteria in the Chesapeake Bay (Frazier and Westhoff, 2008). Mainly the seafood is associated with cholera. The other foods involved contaminated fruits and vegetables, meat, and grains (Azman et al., 2012). The symptoms include abdominal cramps, nausea, dehydration, diarrhea, and vomiting (Azman et al., 2012; CDC, 2018). Preventive measures are washing hands properly, drink boiled water or safe water, proper cooking of foods, and suitable hygienic needs (CDC, 2018).

10.2.2.2.2 *Vibrio parahaemolyticus*

Vibrio parahaemolyticus is a curve-rod shaped, gram-negative bacterium, facultatively anaerobic, and heat sensitive. It is commonly found in the ocean and coastal areas/water and has also been found in fresh marine foods, especially in fish products, shellfish, and other seafood. This organism is absorbed on the chitin particles and copepods (Kaneko and Colwell, 1973). At the same time, they are not found in the open oceanic region (Joseph et al., 1982), and they cannot tolerate the hydrostatic pressure of ocean depth (Sakazaki, 1983). The organism can grow in the presence of 1 to 8% NaCl, and it gets destroyed in freshwater (Bean et al., 1990). The organism cannot grow at a temperature of 4°C, and the growth occurs between 5°C to 9°C. The pH required for the growth was reported in the range of 7.2 to 7.3 at 3% NaCl. The organism is sensitive to the cold temperature, and the minimum growth was observed at 10°C to 13°C. The optimum growth temperature reported between 30°C and 35°C (Banwart, 1998). The pH range for the growth was reported in the range of 4.8 to 11 and the optimal pH range reported was 8.6 (Beuchat, 1973, 1975; Sakazaki, 1983). The optimum water activity documented for growth was 0.992. The organism has three types of antigens: O, H, and K (Beuchat et al., 1983). The D value or decimal reduction time varies from 0.8 to 65.1 minutes at 49°C and 47°C, respectively (Delmore and Chrisley, 1979). *V. parahaemolyticus* quickly replicate in oysters, usually on hot days (Lund, 2008). The period of incubation is 2 to 48 hours, and it often within 12 hours. Vomiting, mild fever, prostration, diarrhea, abdominal pain is the symptoms. *Vibrio parahaemolyticus* foodborne illness can be controlled by using heat treatment. So, in order to prevent the disease, thorough cooking of seafood, sterilization using radiation before the storage of seafood, and reduced use of chilled seafood are recommended (Frazier and Westhoff, 2008).

10.2.2.2.3 Vibrio vulnificus

Vibrio vulnificus is a curve-shaped gram-negative bacterium. This is mostly found in the marine environment (Oliver, 2005). These organisms are secluded from seawater and seafoods (Frazier and Westhoff, 2008). This organism growing in the pH range of 5 to 10 and at the temperature of 8°C to 43°C. The symptoms of *Vibrio vulnificus* are fever, vomiting, diarrhea, chills, and fatigue. After the consumption of seafood, the incubation period is about 12 to 72 hours. This organism is highly invasive and releases cytolysis and hemolysis (Frazier and Westhoff, 2008). This organism causes fatal infections (Hsuch et al., 2004; Bross et al., 2007) and wound infections (Yun and Kim, 2018). The preventive measures are proper food handling and personal hygiene (Nicholas, 2011).

10.2.2.2.4 Vibrio alginolyticus

Vibrio alginolyticus is a facultative anaerobic gram-negative bacterium typically found in marine flora (Li et al., 2009). *Vibrio alginolyticus* is the most common pathogenic among *Vibrio* species. This organism is mainly present in the blowfish and produce tetrodotoxin (Noguchi et al., 1987). Also reported as a typical inhabitant of seawater causes ear infections and tissue damages in humans (Ardic and Ozyurt, 2004; Chart, 2012).

10.2.2.3 LISTERIA MONOCYTOGENES

Listeria monocytogenes is responsible for causing human listeriosis, a potentially fatal foodborne infection (Allerberger and Wagner, 2010). *Listeria monocytogenes* are motile, non-spore-forming, and gram-positive. It has the capability of growing at 4°C. An increase in the rate of listeriosis infection has been reported in several European countries. This pathogen mainly affects those above the age of 65 and not associated with sexuality, culture, social, and economic aspects (Allerberger and Wagner, 2010). It can be commonly seen in nature and can be found in the soil and deteriorate vegetation. Though the sources mentioned above, it can get into animals, including sheep and chicken. Due to this infection, abortion, and breast inflammation occurs in the cattle (Frazier and Westhoff, 2008). The cycles for infection of Listeriosis are shown in Figure 10.2. *Listeria monocytogenes* segregated from sewage, milk, water, wastes of animals and humans. In 1929, listeriosis cases were reported for the first time in humans, especially in pregnant women, newborn babies, and the fetus. The disease takes place mostly in immunocompromised individuals, distinguished as meningitis, septicemia, and meningoencephalitis (Grayo et al., 2008). Like all other gram-positive organisms, this organism also can preserve in the soil (Frazier and Westhoff, 2008). *Listeria monocytogenes* can cultivate in the media, such as tryptose soy, tryptose broth, and brain heart infusion (Jones and D'Orazio, 2013). The essential nutrition required for growth includes riboflavin, biotin, thioctic acid, valine, isoleucine, and glutamine. The strains of the organism can grow in pH ranging from 4.0 to 9.7 and temperature from 1°C to 41°C. The optimum pH requirement is 4.4 to 4.2 for *Listeria*'s growth (Banwart, 1998). The strains of this organism exist in non-fat milk, dairy products, and other meat products.

The D values for *Listeria monocytogenes* at 61°C is 124s and at 65°C is 16.2s, respectively. There are many proven and suspected cases of foodborne listeriosis reported in different countries over the years. The two studies encourage that the elimination of organisms can be achieved in milk by using pasteurization. *L. monocytogenes* and *L. inocula* were isolated from some of the food products and from the environment (Jones and D'Orazio, 2013). Antibiotics such as ampicillin, rifampicin, coumermycin, and the combination of aminoglycoside antibiotics are used for the treatment of listeriosis. *L. monocytogenes* has psychotropic growth characteristics; it helps in enriching and can achieve a prolonged period incubation for various media with the incubation temperature of 4°C to 5°C (Steinberg et al., 1988).

L. monocytogenes produces two forms of illness; febrile gastroenteritis and invasive systemic disease (Malik et al., 2014). The former condition is commonly related to healthier individuals, and after consumption, the symptoms occur within one week (Liu, 2006). The symptoms include slight fever, abdominal cramps, and diarrhea. However, invasive systemic disease related to immunopathological challenged peoples such as pregnant women, newborn babies, older people, and fetuses (Cox, 1989).

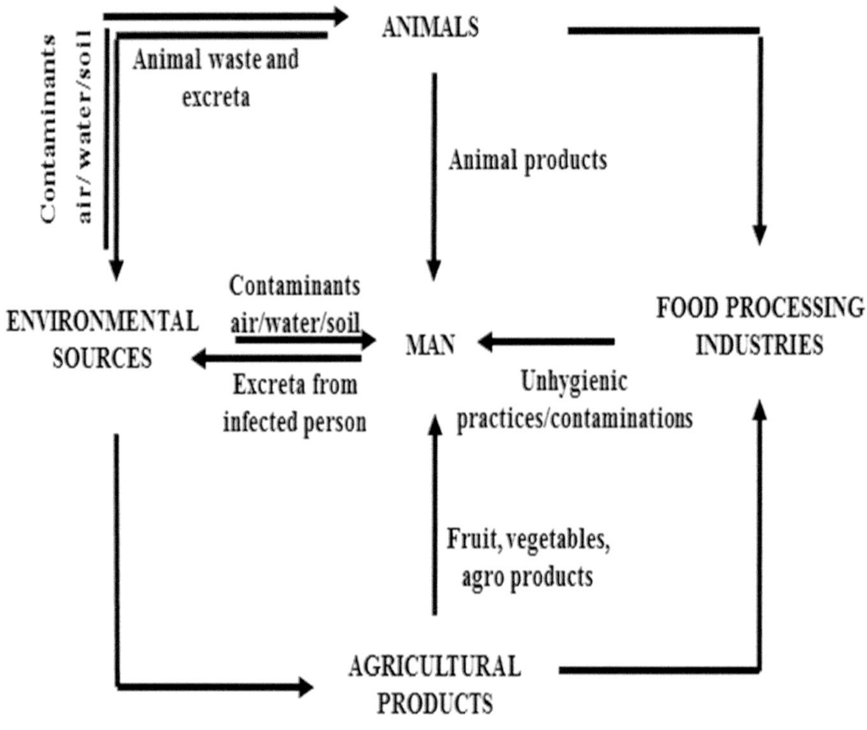

FIGURE 10.2 Cycles for infection of Listeriosis.

10.2.2.4 SALMONELLAE

Salmonellae coming under the family of *Enterobacteriaceae*. The ingestion of *Salmonellae* species causes salmonellosis (Sheikhzadeh et al., 2016). They are facultative anaerobic motile, rod-shaped gram-negative, and non-sporulating bacteria (Stephanie et al., 2019). *Salmonellae* infections are also called food poisoning, caused by a large number of serovars (Frazier and Westhoff,

2008). Over 2000 known serovars are reported for the organism, but only 50 commonly occurred and has O (somatic) and two phases of H (flagellar) antigen (Frazier and Westhoff, 2008). The isolates of *Salmonellae* can detect by using a serological procedure that is Kauffman-white scheme, and represent the organism by using the number, and letters of the different antigenic sites are O (Somatic), Vi (Capsular) and H (Flagellar) (CDC, 1982; Girma, 2015). The infecting bacteria multiplied in the food, and increasing the level of infection produce a large number of outbreaks in many countries. The ideal temperature for growth is 35°C to 37°C. *Salmonellae* cells can destroy at 63°C and susceptible to a pH of 4.5 or below 4.5. These organisms do not replicate at water activity in the range of 0.94. The D value ranges from 0.06 to 11.3 minutes at 60°C, which depending on the serotype and food. For a long time, the *Salmonellae* cells can survive under frozen and dried stages. The different *Salmonella* species that cause foodborne illnesses are *Salmonella enteritidis*, *Salmonella typhimurium*, *Salmonella* DT104, and *Salmonella heidelberg* (Frazier and Westhoff, 2008). The infection occurs through the transmission of bacteria through food which contain fecal matter. The foods responsible for causing the infection are raw meat, contaminated fruits, vegetables, chicken, fish, shrimp, cream-filled desserts, and toppings (Banwart, 1998). After the consumption of contaminated foods, symptoms occur within a period of 6 to 48 hours (CDC, 1983), and the infection lasts for 3 to 7 days. Stomach pain, dysentery, nausea, chills, pyrexia, and headache are the symptoms of *Salmonellae* infections. Children and adults having aged 60 or above are more susceptible to *Salmonellae* infections (Frazier and Westhoff, 2008). Depending on the type of disease-causing organism, *Salmonella* infection is generally divided into two kinds (Makendi et al., 2016; Worley et al., 2018):

- Typhoidal serotype causes systemic diseases such as typhoid fever. Typhoid fever in humans is caused by *Salmonella paratyphi* and *Salmonella typhi*. *Salmonella* infection commonly occurs through fecal contaminated foods. Typhoid fever is a consequential disease, and it can even lead to death. The symptoms of typhoid fever are, after the ingestion of infected food, the symptoms occur within 1 to 2 weeks and go on for 3 to 60 days. Symptoms are high fever, vomiting, muscle pain, abdominal cramps, and body weakness (Colomba et al., 2006).
- Non-Typhoidal serotype is the most accepted type of serotype and causes infection in humans and animals. Serotypes of *Salmonella* responsible for causing non-typhoidal infection, examples are *Salmonella enteritidis* and *Salmonella javiana* (Makendi et al., 2016; Worley et al., 2018).

The controlling measures for *Salmonella* infection are:

- Thoroughly cooking of all raw meats and meat products
- Make use of pasteurized milk
- Wash hands properly before and after deals with raw meats and other products of poultry
- Use clean surfaces and vessels for preparing foods
- Wash vessels, knives, and chopping board properly
- Destruction of organisms by heating

10.2.2.5 BACILLUS CEREUS

Bacillus cereus is a gram-positive spore-forming anaerobic bacterium. The organism commonly found in clay, dirt, and the foods that contaminated with dust and soil (McDowell et al., 2020). The spores of *B. cereus* are present in both plant products and animal products. The *B. cereus* can detect by using mannitol-egg yolk polymyxin agar plates. The selective medium is made by the addition of Polymyxin B, which inhibits gram-negative bacteria. *Bacillus cereus* is typically mannitol-negative and lecithinase-positive. No pH change occurs since it is mannitol negative; therefore the organism forms the pink color colonies and a dense precipitate appears in this medium around the colonies because of lecithin utilization (Banwart, 1998). The ideal temperature for growth is 30°C, and minimal temperature for germination is 10°C, and the maximum growth temperature is about 49°C. They grow best at a pH range of 4.9 to 9.3. The D value for the spores of *B. cereus* is 2 to 3.1 minutes at 100°C in skimmed milk, and for the phosphate buffer, the D value is 8 minutes at 100°C (Frazier and Westhoff, 2008). The incubation period is 30 minutes to 6 hours (McDowell et al., 2020). Two disorders associated with *B. cereus* are diarrheal syndrome and emetic syndrome (Bintsis, 2017). The diarrheal syndrome is due to toxin production at the time of growth of the organism (*B. cereus*) in the small intestine (Granum, 1994; Bintsis, 2017). The incubation period for the diarrheal disorder is 8 to 16 hours. The diarrheal syndrome symptoms are abdominal cramps, bloating, fever, vomiting, and nausea (Bintsis, 2017). In this case, the foods involved in the diarrheal disease outbreak were the sauce, meat, custards, and milk. The control measures for the diarrheal disorder are chilling the foods at low temperatures. The period of incubation for the emetic syndrome is 1 to 6 hours (Bintsis, 2017). The emetic syndrome occurs due to emetic toxin produced by *B. cereus* while growing in the food (McDowell et al., 2020). The symptoms of emetic syndrome are nausea and vomiting. The foods involved in the particular outbreak of emetic syndrome include noodles, pastry, cooked rice, and mashed potatoes ((Frazier and Westhoff, 2008; Bintsis, 2017).

10.2.2.6 ESCHERICHIA COLI ENTERITIS (EEC)

Escherichia coli is a rod-shaped gram-negative non-obligatory anaerobic bacterium. It is the most crucial organism in food microbiology. The desirable temperature for growth is 37°C, and it can survive at temperatures ranging from 10°C to 40°C. The optimal pH range for *E. coli* is about 7.0 to 7.5, minimal pH at which it can survive is 4, and maximal pH is 8.5 (Frazier and Westhoff, 2008). These are thermally susceptible organisms and can be killed by proper cooking under 63°C. *E. coli* is considered as an indicator organism to assure food safety since it has fecal origin and it is seen in the lower intestine of warm-blooded organisms. Most of the strains of *E. coli* are harmless and only pathogenic forms cause food poisoning. Investigations reveal the presence of pathogenic strain, *Escherichia coli* 0157:H7 in several peculiar foods such as poulets, apple juices, salads, uncooked vegetables, and so on. After the ingestion of infected foods, the symptoms occur within 3 to 9 days and continue to 4 to 10 days. Dysentery, stomach pain, emesis, pyrexia are the indications of *Escherichia coli* illness. The mechanism of this infection includes damage to the inner lining of the stomach and cause bleeding. The *EEC* toxin produces erythrocyte failure, thrombus in the renal which leads to renal diseases (Frazier and Westhoff, 2008). Preventive measures are (Kaur et al., 2010) as follows:

- Usage of pasteurized milk
- Proper washing of palm, vessels, and work station after deals with raw flesh
- The illness can prevent by proper cooking of red meats and other foods
- By maintaining personal cleanliness, prevention of fecal and oral contamination

In human, the consumption of food contaminated with EEC causing diseases is categorized into two:

- The first category of EEC strain produces an enterotoxin and causes enterotoxigenic illness in humans. These strains produce two heat-stable enterotoxins (S.T.) and a heat-labile (L.T.) toxin
- The second category that produces a cytolysin

Based on virulence genes, *EEC* is subdivided into different groups of pathotypes. They are Enteroinvasive *Escherichia coli (EIEC)*, Enterohemorrhagic *Escherichia coli* (*EHEC*), Enteroaggregative *Escherichia coli* (*EAEC*), Enteropathogenic *Escherichia coli* (*EPEC*) and Enterotoxigenic *Escherichia coli* (*ETEC*) (Miri et al., 2017).

10.2.2.6.1 Enterotoxigenic Escherichia coli (ETEC)

The bacteria get attached to the gut of the somatic cell using pilus and producing thermally unstable toxin (L.T.) or thermally stable toxin (S.T.) (Qadri et al., 2005; Croxen et al., 2013). Thermally stable toxin gets destroyed at 60°C, and there is no interference in biological functions (Robertson et al., 1983). When toxin combines with cellular protein, the absorbency of the membrane increase and cause abnormality in the concentration of electrolytes in the body, leading to dysentery. In several countries, the serotype 027:H7 recorded as a vital cause of this disease (CDC, 2020). *ETEC* in food can be detected by using DNA hybridization (Moseley et al., 1980; Patamaroj et al., 1983).

10.2.2.6.2 Enteropathogenic Escherichia coli (EPEC)

In worldwide, the places with poor hygienic conditions *EPEC* strains cause dysentery in infants (Browne, 1987). Several serotypes (Oll:H12; 055:H6) are involved in outbreaks in different countries. The strains of this pathogen attach to the epithelial cells, but they do not produce any toxins. These results in severe injury and demolish the villus and following in malassimilation and dysentery. The incubation period is 9 to 12 hours, and the illness lasts for 12 days (Lee et al., 2012).

10.2.2.6.3 Enteroinvasive Escherichia coli (EIEC)

EIEC is highly invasive, but they do not produce harmful toxins (Sagar, 2018). The strains of *EIEC* cause dysentery. The bacterial strains can expand and get attached to the somatic cells in the gut, intruded, and the cell-to-cell movement spread the infection to the abdomen. The incubation period occurs within 10 to 18 days. The illness duration lasts for 4 to 7 days. The indications are fever, vomiting, muscle pains, chills, and dysentery (Doyle, 2003).

10.2.2.6.4 Enterohemorrhagic *Escherichia coli* (EHEC)

Hemorrhagic colitis and hemolytic uremic syndrome (HUS) are the infections caused by *EHEC* strains (Nguyen and Sperandi, 2012). Animals, mostly cattle, are the primary carriers for this infection. Especially in sensitive individuals, the consumption of infected food in fewer quantities can produce the disease (Lee et al., 2012). The *EHEC* produces Vero toxins, intimin, and enterohemolysin. Gastroenteritis occurs due to the ingestion of *EHEC* through food and water. The protein intimin facilitates intimate adhesion resulting in lesion formation. Thus destroy the absorptive villi, resulting in malabsorption. Bacteria then produce toxins (Stx) that damage colon epithelial cells resulting in hemorrhagic colitis. Toxins (Stx1 or Stx2) are assimilated into the circulatory system and destruct the capillary vessels of the gut, cerebrum, and cause renal damage (Jay, 1998).

10.2.2.6.5 Enteroaggregative *Escherichia coli* (EAEC)

The pathogen enters the body through contaminated food and water, improper cooking of food (Jensen et al., 2014). The bacteria enter the intestine through various contaminated food and attached to the small intestine, and produces cytotoxin. The illness duration lasts for 3 to 14 days. The indications are hemorrhagic colitis and dysentery in children, also cause urinary infections (Jensen et al., 2014). The period of incubation is 8 to 48 hours (CDC, 2020).

10.2.2.7 YERSINIA ENTEROCOLITICA

Yersinia enterocolitica is a rod-shaped bacterium that belongs to the family of *Yersiniaceae* and has designated as an emerging foodborne illness in humans (CDC, 2020). It has isolated from the waste of many animals and their intestinal tracts. The main reservoir for the infection is pigs. *Yersinia pseudotuberculosis* is usually seen in the animal's food (Frazier and Westhoff, 2008). The strains can grow at a temperature ranging from 0°C to 44°C and the most optimum temperature for growth is 25°C–29°C. In the case of milk and uncooked meat, the growth occurs at 1°C. The cells can grow under 5% NaCl and a pH of 4.5 (Bhunia, 2008). *Yersinia enterocolitica* cells are sensitive to pasteurization. Gastroenteritis was reported in humans by virulent strains of *Y. enterocolitica*. Symptoms are diarrhea, nausea, vomiting, fever and abdominal pain (Banwart, 1998). Indication generally occurs within 24–30 hours after the ingestion of infected foods. The infection mainly occurs in immunocompetent hosts, and the symptoms for this category include endocarditis, pneumonia, septicemia, and meningitis (Aziz and Yelamanchili, 2021). Foods involved in this infection are milk and milk products, and inadequate cooking of flesh (Banwart, 1998). *Y. enterocolitica* cell is thermally susceptible, and hence proper pasteurization or using heat treatment makes food free from infection. The illness caused by *Y. enterocolitica* can be prevented by proper sanitation and by avoiding the consumption of unpasteurized milk and the low temperature cooked meat (Frazier and Westhoff, 2008).

10.2.2.8 SHIGELLA

Shigella is the organism that causes the disease shigellosis; it also is known as bacillary dysentery. The four different kinds of *Shigella* species, *Shigella dysenteriae*, *Shigella hoydii*, *Shigella flexneri*, and *Shigella sonnet* are reported to cause disease. An enterotoxin (Shiga toxin) is produced by *S. dysenteriae* (Brown et al., 2017). The favorable temperature for the growth of the strain is 37°C. The source of direct infection is poor personal cleanliness, infected food, as well as inadequately cooked food. Foods involved in the *Shigella* outbreaks are sandwiches and salads because during the preparation there is hand contact, and moreover, the usage of raw vegetables also increases the issue (Frazier and Westhoff, 2008). After the consumption of infected food and water, a bacterium goes into the gut and small intestine and reaches the colon. Then the bacteria invade epithelial cells of the colon. Bacteria produce large numbers of virulence proteins such as IpaA, IpaB, IpaC, and IpaD, and lesA (Jay, 1995). After the consumption of contaminated food, the symptoms occur within 1–3 days. Symptoms precede for six days; in some instances, symptoms remain for three weeks. Indications are stomach pain, dysentery, sputum, and pustule, migraine, pyrexia, and sometimes dysentery along with blood. The controlling measures include avoiding handling food by contaminated hand or the person who is suspected of the disease and by washing vegetables with chlorinated water (Frazier and Westhoff, 2008).

10.3 DIFFERENT TYPES OF FOODBORNE VIRUSES

The wide varieties of viruses transmitted through foods. All these viruses belong to specific families, and various types of infections correlated with these viruses are neural diseases, paralysis, and fever. According to the CDC, USA, around 45 million people get sick, and many numbers of people hospitalized, and many of them die by infecting different viruses (Bosch et al., 2016). The most commonly reported foodborne diseases are Gastroenteritis and Hepatitis. The main types of reported foodborne viruses are Rotavirus, NoV, Hepatitis A, and E (Table 10.2).

TABLE 10.2 Viral Diseases Transmitted Through Food (Kumar et al., 2019)

Viruses	Foods Involved	Incubation Period	Symptoms
Rotavirus	Raw meat and water	1 to 3 days	Diarrhea
Norovirus	Fresh oysters, cold foods, salads, and drinking water	24 hours to 48 hours	Vomiting, fever, diarrhea
Hepatitis A	Fruits, vegetables, milk	15 to 50 days	Nausea, vomiting, fatigue, abdominal pain, jaundice
Hepatitis E	Pig liver, sausages, boar, and deer meats	2 to 10 weeks	Neurological disorder

10.3.1 ROTAVIRUS

Rotavirus exists in the family of *Reoviridae*. In underdeveloped countries, rotaviruses are responsible for diarrheal diseases among infants, and many deaths were also reported. The incubation period varies from 24 to 72 hours (Carvalho-Costa et al., 2006). In young children, a significant risk factor for rotavirus diseases is the usage of milk powder prepared using low-quality water and salads (CDC, 2015). A small dosage of the viruses is highly infectious in humans and animals (Neethirajan et al., 2017). In several countries, the outbreak of foodborne infections is high (Greening and Cannon, 2016). Therefore, highly sensitive technology was used to detect these viruses in order to ensure food safety (Neethirajan et al., 2017). The rotavirus infection test methods include immune electron microscopy, latex agglutination methods, and ELISA (Jay, 1995).

10.3.2 NOROVIRUS (NoV)

NoV belongs to the Caliciviridae family. NoV can survive in the environment and are highly infectious, even if it is in a small dose. As per the CDC report, 50% of outbreaks in foodborne infections reported are due to NoV in the USA (Neethirajan et al., 2017). In USA, this virus causes around 21 million illnesses annually (Neethirajan et al., 2017). NoV can resist heat up to 60°C. Identification of NoV is essential for public health and safety. Raw oysters, cold foods, salads, and drinking water are the food source for the infection (CDC, 2003). After ingestion of the virus, the symptoms begin within 24 hours to 48 hours, and last for 1 to 3 days. The symptoms are fever, nausea, dehydration, abdominal cramps. The infection can be prevented by frequently washing hands, adequate washing of fruits and vegetables, and proper washing of clothes (CDC, 2019).

10.3.3 HEPATITIS A VIRUS (HAV)

Hepatitis A virus (HAV) belongs to *the Picornaviridae* family, and it causes liver disease. It is a single-stranded RNA virus (Sánchez, 2013). Its genomic categorization can be classified into three genetic constitutions (Neethirajan et al., 2017). In worldwide, millions of outbreaks occur due to HAV (Adefisoye et al., 2016). The indication of the infection occurs within 15 days to 50 days. The infection symptoms are liver disease, darker urine, icterus, stomach ache, and body weakness. The virus circulates from one person to another through the excreta and the ingestion of contaminated food and water (Neethirajan et al., 2017). Even at low doses, the HAV is highly contagious (Sánchez, 2013).

10.3.4 HEPATITIS E VIRUS (HEV)

Hepatitis E virus (HEV) belongs to the family of *Hepeviridae*, and it is a single-stranded RNA virus (Chen and Yuan, 2014). HEV produces a wide range of extra-hepatic indications, including acute thyroiditis, glomerulonephritis, neurological disorders, encephalitis, and neuralgic amyotrophy (Kamar and Abravanel, 2015). In the different locations of the world, the HEV outbreaks were reported several times. The mortality rate was higher in pregnant women than in teenagers and adults (Chen and Yuan, 2014). HEV transmits through fecal contamination through the food and water (Di Bartolo et al., 2015). Hepatitis E occurs after the ingestion of raw or inadequately cooked fresh meats such as sausages, meat, and liver of boar and deer (Van der Poel, 2014).

10.4 DIFFERENT TYPES OF FOODBORNE PARASITES

The parasitic protozoa are responsible for causing food and waterborne disease. *Cryptosporidium*, *Cyclospora*, *Giardia*, and *Toxoplasma* are organisms with more significant concern in the diet (Orlandi et al., 2002). *Toxoplasma* is the leading parasitic protozoa in the food sources. Protozoan parasites which do not replicate in the food sources and it can survive in the wet food and the humid environmental condition. The three main classifications of parasites that cause disease in humans are protozoa, helminths, and ectoparasites. The general control measures are good hygiene practices; minimize the spreading of cysts by proper washing of fruits and vegetables and frequently cleaning the utensils (Doyle, 2003).

10.4.1 GIARDIA

The species of *Giardia* infect animals as well as humans and cause giardiasis. The infection transmits through fecal contamination and oral transmission. The dosage for infection of *Giardia* is between 10 and 100 cysts (Rentdorff and Holt, 1954). Worldwide this is the most frequently isolated parasite. They are infective cysts and it can survive up to two weeks in the cold and humid environment. This cyst will destroy under a heating regime of 71.7°C for 15 seconds, and the cysts were destroyed at –18°C for 1 hour. The incubation period varies from 1 to 2 weeks. The symptoms include diarrhea, bloating, and flatulence. The disease period lasts for more than five days (Dawson, 2005).

10.4.2 TOXOPLASMA

Toxoplasma gondii is the most common human infection in developing countries (Flegr, 2007). In human's *Toxoplasma gondii* causes various clinical illness (McLeod et al., 2020). Toxoplasmosis is the parasitic infection caused by *Toxoplasma gondii* (CDC, 2015). *T. gondii* oocysts can be destroyed at a temperature greater than 66°C. This parasite can be mostly found in the feces of cats and half-cooked meat. Toxoplasma infection occurs in humans through the consumption of contaminated food containing oocysts (Montoya and Liesenfeld, 2004). The incubation period in humans is 5 to 23 days (Hussain et al., 2017). The disease, toxoplasmosis can be detected by desolation and microscopic examination of organisms after staining (Newell et al., 2010). In humans, the symptoms of toxoplasmosis are fever, body pain, headache, and fatigue, in some people, eye infections may occur (CDC, 2015). Many studies reveal that this infection produces personality changes in infected humans (Flegr, 2013), neurological disorders and mental disorders in humans (De Barros et al., 2017).

10.4.3 CYCLOSPORA

Cyclospora mostly found in animals. *Cyclospora cayetanensis* is the only species found in anthropoids. The *Cyclospora* species are responsible for causing Cyclosporiasis. *Cyclospora* primarily termed as an anthropoid pathogen (Ashford, 1979). After the consumption of oocysts contained food, the incubation period is from 2 to 11 days. The oocysts can survive in water at 4°C and 37°C for 4 days and 7 days, respectively. *Cyclospora* can be able to resist many disinfectants. Diarrhea, weight loss, abdominal cramps, vomiting, nausea, and fever indicate *Cyclospora*. Cyclosporiasis can be treated by using trimethoprim-sulfamethoxazole (Dawson, 2005).

10.4.4 CRYPTOSPORIDIUM

Cryptosporidium is the source of causing diarrhea in humans and animals. In 1976, the first case of human Cryptosporidiosis was reported. The incubation period is about 2 to 10 days. *Cryptosporidium parvum* is responsible for causing human illness. The typical symptom of *C. parvum* infection is diarrhea, abdominal cramp, and vomiting. A complete recovery is possible for the person with a properly functioning immune system (Dawson, 2005). The oocysts can survive in the saltwater, fresh, and brackish water for various months at 15°C to 30°C (Doyle, 2003).

10.4.5 ASCARIS

Ascaris lumbricoides is the most widespread human intestinal roundworm (Harhay et al., 2010; Burton et al., 2013). *Ascaris lumbricoides* is the species affects mainly on humans and cause ascariasis. Humans are the only host for this infection (Doyle, 2003). Ascariasis is the main cause for malabsorption, growth retardation and damaged immune regulation (Hagel, 2003; Papier, 2014). Infection occurs in humans through consumption of contaminated food and water, unhygienic practices, improper washing and cooking of foods. This causes major problems in liver and lungs (Doyle, 2003). The symptoms of ascariasis include vomiting, wheezing, nausea, fever, and abdominal problems (Bhunia, 2008).

10.5 DIFFERENT TYPES OF MYCOTOXINS IN FOODS

Mycotoxins are the primary metabolites of fungi and are highly toxic and cause harmful effects to human beings and animals. Mycotoxicosis developed through the consumption of food contaminated with mycotoxins (Bennet and Klich, 2003).

10.5.1 AFLATOXIN

The strains of *Aspergillus flavus* and *Aspergillus parasiticus* are responsible for producing aflatoxin. The two types of aflatoxin are reported B1 and G1. Aflatoxins (AFs) are toxic to different animals, including humans and other species (Istambouli, 2016). The main two elements that influence the production of Aflatoxin is moisture content and temperature condition. The foods mainly reported for the presence of AFs are corn, maize, fruits, milk, rice, soybean, and sweet potatoes (Lorbeer et al., 2000; Magnoli et al., 2006; Waller et al., 2007; Rooney-Lathram et al., 2008). *Flavobacterium aurantiacum* has been reported to remove aflatoxin metabolite B1 (AFB_1) from various foods. Aflatoxin causes the disfunctioning of immune system, liver damages, and abnormality in fetus development in humans (Amaike and Nancy, 2011; Kensler et al., 2011; Roze et al., 2013).

10.5.2 OCHRATOXIN A

The strains of *Penicillium viridicatum*, *Penicillium cyclopium*, *Penicillium variabile*, and *Penicillium commune* produces Ochratoxin A. This toxin is well recognized and has the highest toxicity. Ochratoxin A produced at the water activity of 0.95 and the temperature of 30°C. This toxin is thermostable. The food from which the isolation of Ochratoxin A was reported includes wheat, dough, bread, and milk products (Mishra et al., 2015). Ochratoxin A causes liver damages and abnormality in fetus development (Chu, 2003).

10.5.3 PATULIN

Patulin is produced by *Penicillium patulum*, *Penicillium expansum*, *Penicillium claviforme*, few *Aspergillus* species, *Byssochlamys nivea*, and *Byssochlamys fulva* (Frazier and Westhoff, 2008). The food commonly involved in the isolation of Patulin is apple juice and apple cider. The patulin level reported in the apple crush up to 440 μg/L. The temperature for the growth and the production of patulin is 0°C to 24°C (Mandappa et al., 2017). In humans, Patulin causes GI infections, liver damages, kidney damages, and vomiting (Mostrom, 2016).

10.5.4 ZEARALENONE (ZEN)

Mainly five types of readily occurring ZEN were reported and which was produced by *Fusarium tricinctum* and *Fusarium graminearum*. Foods associated with this toxin are wheat, cereals, crops, grains, benne. The toxin appears as a scintillate bluish-green color under high wave actinic radiation, and the toxin appears as greenish under low wave actinic radiation (Jay, 2008). In humans, ZEN causes hormonal imbalances and also affect the reproductive system (Rogowska et al., 2019).

10.5.5 STERIGMATOCYSTIN

Aspergillus nidulans, *Aspergillus rugulosus*, and *Aspergillus versicolor* are given rise to Sterigmatocystin toxin (Jay, 2008). This toxin is similar to Aflatoxin in the case of structure as well as physiological nature (Frazier and Westhoff, 2008). This toxin produces liver cancer in creatures. The toxin appears as a bright reddish-brown color under actinic radiation. The toxin discovered in caffeine seed, cereals, and grains (Jay, 2008). According to IARC classification Sterigmatocystin is considered as a possible human carcinogen (Viegas et al., 2019)

10.5.6 FUMONISIN

Fumonisin is a mycotoxin produced by the fungus *Fusarium verticillioides*, *Fusarium proliferatum* and *Fusarium* species (Rheeder et al., 2002; Smith,

2018). The foods involved in the fumonisin infection are mostly grains and other grain products (Cendoya et al., 2018). IARC characterize FB1 as a group 2B possible carcinogen to humans. *F. proliferatum* and *F. verticillioides* are most commonly found in maize (Marasas, 2001). Temperature and water potential are the main factors influence on the growth and the production of fumonisin (Magan, 2007). The maximum amount of fumonisin produced at 25°C to 15°C at water activity of 0.995 (Cendoya et al., 2017). Fumonisin causes liver cancer and esophageal cancer in humans (Kamle et al., 2019).

10.5.7 CITRININ

Citrinin is formed by *Penicillium viridicatum* and *Penicillium citrinum*. The toxin present in white rice, grains, cereals, stale bun, and bran (Abramson et al., 2001). The toxin appears as citrus yellowish under the high wave actinic radiation. Citrinin formed from seeds of caffeine and cocoa (Jay, 2008). The toxin accumulates in the kidney and causes kidney failures in humans and animals also produce liver damages and GI tract infections (Doughari, 2015).

10.6 CONCLUSION

In the food safety sector, microbial contamination and their after-effects to the public are the crucial issues to be resolved. This chapter summarizes many diseases that occur due to the consumption of food contaminated with bacteria, viruses, parasites, and mycotoxins. Moreover, it describes their mode of action in the host and its control measures. The routes of transmission of disease are through food as well as water. Most of the diseases are transmitted from humans to humans and animals to humans. The methods used to prevent foodborne infection are also discussed in this chapter. The preventive techniques include the practice of personal hygiene; avoid the use of improperly cooked meat/contaminated food and water, usage of pasteurized milk, quick refrigeration or freezing of perishable foods, and so on. There is a well-built relation between public health and food safety. Hence, scientific knowledge about foodborne disease and its control measures will maintain an excellent public health system.

KEYWORDS

- **clinical symptoms**
- **foodborne illness**
- **foodborne infections**
- **foodborne intoxications**
- **foodborne pathogens**
- **mycotoxins**
- **pathogenic bacteria**

REFERENCES

Abramson, D., Usleber, E., & Märtlbauer, E., (2001). Immunochemical method for citrinin. *Methods Mol. Biol.*, *157*, 195–204.

Adefisoye, M. A., Nwodo, U. U., Green, E., & Okoh, A. I., (2016). Quantitative PCR detection and characterization of human adenovirus, rotavirus and hepatitis A virus in discharged effluents of two wastewater treatment facilities in the Eastern Cape, South Africa. *Food and Environmental Virology*, *8*(4), 262–274.

Allerberger, F., & Wagner, M., (2010). Listeriosis: A resurgent foodborne infection. *Journal of Clinical and Microbiology and Infection*, *16*, 16–23.

Amaike, S., & Nancy, P. K., (2011). *Aspergillus flavus*. *Annu. Rev. Phytopathol.*, *49*, 107–133.

Ardic, N., & Ozyurt, M., (2004). *Vibrio alginolyticus*'a bağli bir otit olgusu [Case report: Otitis due to *Vibrio alginolyticus*]. *Mikrobiyol. Bul.*, *38*(1, 2), 145–148.

Ashford, R. W., (1979). The occurrence of an undescribed coccidian in man in Papua New Guinea. *Annals of Tropical Medicine and Parasitology*, *73*, 497–500.

Audurier, A., & Martin, C., (1989). Phage typing of *Listeria monocytogenes*. *Int. J. Food Microbiol.*, *8*, 251–257.

Aziz, M., & Yelamanchili, V. S., (2021). *Yersinia enterocolitica*. In: *StatPearls*. StatPearls Publishing.

Azman, A. S., Rudolph, K. E., Cummings, D. A., & Lessler, J., (2013). The incubation period of cholera: A systematic review. *J. Infect.*, *66*(5), 432–438.

Bacon, R. T., & Safos, J. N., (2003). Characteristics of biological hazards in foods. *Food Safety Handbook* (pp. 157–195). New York: John Wiley and Sons.

Banwart, G. J., (1998). *Handbook of Basic Food Microbiology: Foodborne Agents Causing Illness* (p. 774). Springer International publishing.

Bean, N. H., Griffin, P. M., Goulding, J. S., & Ivey, C. B., (1990). Foodborne disease outbreaks, 5-year summary, 1983–1987. *J. Food Protect*, *53*, 804–817.

Bennett, J. W., & Klich, M., (2003). Mycotoxins. *Clinical Microbiology Reviews*, *16*(3), 497–516.

Beuchat, L. R., (1973). Interacting effects on pH, temperature, and salt concentration on growing and survival of *Vibrio parahaemolyticus*. *Appl. Microbiol.*, *25*, 844–846.

Beuchat, L. R., (1975). Environmental factors affecting the survival and growth of *Vibrio parahaemolyticus*. A review. *J. Milk Food Technol.*, *38*, 476–480.

Bhunia, A. K., (2008). *Foodborne Microbial Pathogens: Mechanisms and Pathogenesis* (p. 276). New York: Springer-Verlag Publishing.

Bintsis, T., (2017). Foodborne pathogens. *AIMS Microbiology*, *3*(3), 529–563.

Bosch, A., Pintó, R. M., & Guix, S., (2016). Foodborne viruses. *Current Opinion in Food Science*, *8*, 110–119.

Bross, M. H., Soch, K., Morales, R., & Mitchell, R. B., (2007). *Vibrio vulnificus* infection: Diagnosis and treatment. *Am. Fam. Physician*, *76*(4), 539–544.

Brown, D. F., & Twedt, R. M., (1972). Assessment of the sanitary effectiveness of holding temperatures on beef cooked at low temperature. *Appl. Microbiol.*, *24*, 599–603.

Brown, J. D., Willcox, S. J., Franklin, N., et al., (2017). *Shigella* species epidemiology and antimicrobial susceptibility: The implications of emerging azithromycin resistance for guiding treatment, guidelines and breakpoints. *Journal of Antimicrobial Chemotherapy*, *72*(11), 3181–3186.

Browne, R. R. M., (1987). Traditional Enteropathogenic *Escherichia coli* of infantile diarrhea. *Rev. Infect. Dis.*, *9*(1), 28–53.

Burton, B., Clint, C., & Oeltmann, T., (2013). *Human Parasitology*, *Intestinal Nematodes* (p. 291). Amsterdam: Academic Press.

Carvalho-Costa, F. A., Assis, R. M., Fialho, A. M., Bóia, M. N., Alves, D. P., Martins, C. M., & Leite, J. P., (2006). Detection and molecular characterization of group A rotavirus from hospitalized children in Rio de Janeiro, Brazil, 2004. *Mem. Inst. Oswaldo Cruz*, *101*(3), 291–294.

Cendoya, E., Monge, M. D. P., Chiacchiera, S. M., Farnochi, M. C., & Ramirez. M. L., (2018). Influence of water activity and temperature on growth and fumonisin production by *Fusarium proliferatum* strains on irradiated wheat grains. *Int. J. Food Microbiol.*, *266*, 158–166.

Centers for Disease Control and Prevention (CDC), (1982). *Salmonella Surveillance*, *Annual Summary (CDC)*, 1980. Atlanta (G.A.).

Centers for Disease Control and Prevention (CDC), (1983). Foodborne disease surveillance, annual summary 1981. Atlanta (G.A.) *Morb. Mortal. Wkly. Rep.*, *39*(SS01), 15–23.

Centers for Disease Control and Prevention (CDC), (2003). Outbreaks of Gastroenteritis associated with noroviruses on cruise ships – United States – 2002. *JAM*, *289*(2), 167–169.

Centers for Disease Control and Prevention (CDC), (2004). *Summary of Human Vibrio Isolates Reported to CDC*. Atlanta (G.A.).

Centers for Disease Control and Prevention (CDC), (2013). *About Parasites*. http://www.cdc.gov/parasites/about.html (accessed on 11 February 2022).

Centers for Disease Control and Prevention (CDC), (2015). *Parasites-Food*. http://www.cdc.gov/parasites/about.html (accessed on 11 February 2022).

Centers for Disease Control and Prevention (CDC), (2018). *Cholera – Vibrio cholerae infection*. http://www.cdc.gov/cholera /index.html (accessed on 18 February 2022).

Centers for Disease Control and Prevention (CDC), (2019). *Norovirus*. http://www.cdc.gov/norovirus/index.html (accessed on 18 February 2022).

Centers for Disease Control and Prevention (CDC). (2007). Preliminary food net data on the incidence of infection with pathogens transmitted commonly through food-10 states. *Morb. Mortal Wkly. Rep.*, *57*(14), 366–370.

Centers for Disease Control and Prevention Yellow Book, (2020). *Health Information for International Travel* (p. 78). New York: Oxford University Press.

Chart, H., (2012). *Vibrio*, *Mobiluncus*, *Gardnerella* and *Spirillum*: Cholera; vaginosis; rat-bite fever. *Medical Microbiology*, 314–323.

Chen, Q., Yuan, L., Wan, J., Chen, Y., & Du, C., (2014). Colorimetric detection of hepatitis E virus based on reverse transcription loop-mediated isothermal amplification (RT-LAMP) assay. *J. Virol. Methods*, *197*, 29–33.

Chu, F. S., (2003). Mycotoxins toxicology. *Handbook of Encyclopedia of Food Sciences and Nutrition* (pp. 4096–4108). Academic Press.

Clelia, A., (2017). Food spoilage and food safety: Is there a link? The microbiological quality of food. *Foodborne Spoilers* (pp. 283–300). Woodhead Publishing.

Colomba, C., Saporito, L., Infurnari, L., Tumminia, S., & Titone, L., (2006). Typhoid fever as a cause of opportunistic infection: Case report. *BMC Infect. Dis.*, *6*(1), 38.

Cox, L. J., (1989). A perspective on listeriosis. *Food Technology*, *43*(12), 52–59.

Croxen, M. A., Law, R. J., Scholz, R., Keeney, K. M., Wlodarska, M., & Finlay, B. B., (2013). Recent advances in understanding enteric pathogenic *Escherichia coli*. *Clin. Microbiol.*, *26*, 822–880.

Dawson, D., (2005). Foodborne protozoan parasites. *Int. J. Food Microbiol.*, *103*(2), 207–227.

De Barros, J. L. V. M., Barbosa, I. G., Salem, H., Rocha, N. P., Kummer, A., Okusaga, O. O., Soares, J. C., & Teixeira, A. L., (2017). Is there any association between *Toxoplasma gondii* infection and bipolar disorder? A systematic review and meta-analysis. *J. Affect Disord.*, *209*, 59–65.

Delmore, R. P. Jr., & Crisley, F. D., (1979). Thermal resistance of *Vibrio parahaemolyticus* in clam homogenate. *J. Food Prot.*, *42*, 31–134.

Di Bartolo, I., Angeloni, G., Ponterio, E., Ostanello, F., & Ruggeri, F. M., (2015). Detection of Hepatitis E virus in pork liver sausages. *Int. J. Food Microbiol.*, *193*, 29–33.

Doughari, J. H., (2015). The occurrence, properties and significance of citrinin mycotoxin. *Journal of Plant Pathology and Microbiology*, *6*, 321.

Doyle, E. M., (2003). *FRI Briefings: Foodborne Parasites: A Review of the Scientific Literature Review*. Food Research Institute, University of Wisconsin-Madison. https://fri.wisc.edu/files/Briefs_File/parasites.pdf (accessed on 18 February 2022).

Faruque, S. M., & Nair, G. B., (2008). *Vibrio cholerae: Genomics and Molecular Biology*. Caister Academic Press. https://doi.org/10.21775/9781910190777.

Flegr, J., (2007). Effects of toxoplasma on human behavior. *Schizophr. Bull.*, *33*(3), 757–760.

Flegr, J., (2013). Influence of latent toxoplasma infection on human personality, physiology and morphology: Pros and cons of the toxoplasma-human model in studying the manipulation hypothesis. *The Journal of Experimental Biology*, *216*, 127–133.

Frazier, W. C., & Westhoff, D. C., (2008). *Food Microbiology* (p. 560). New Delhi: Tata McGraw Hill Education Publishing.

Frazier, W. C., & Westhoff, D. C., (2008). *Handbook of Food Science Series: Food and Nutrition Science* (Vol. 4). McGraw Hill Publication.

Gillian, P. H., Brown, L., & Berman, R. E., (1983). Differentiation of *Clostridium difficile* toxin from *Clostridium botulinum* toxin by the mouse lethality test. *Appl. Environ. Microbiol.*, *45*, 347–349.

Girma, G., (2015). Prevalence, antibiogram and growth potential of *Salmonella* and *Shigella* in Ethiopia: Implications for public health: A review. *Research Journal of Microbiology*, *10*(7), 288–307.

Granum, P. E., & Lund, T., (1997). *Bacillus cereus* and its food poisoning toxins. *FEMS Microbiol. Lett.*, *157*(2), 223–228.

Grayo, S., Lott-Desroches, M. C., & Dussurget, O., (2008). Rapid eradication of *Listeria monocytogenes* by moxifloxacin in a murine model of central nervous system listeriosis. *Antimicrobial Agents and Chemotherapy*, *52*(9), 3210–3215.

Greening, G. E., & Cannon, J. L., (2016). Human and animal viruses in food (including taxonomy of enteric viruses). *Viruses in Foods*, *2016*, 5–57.

Greening, G. E., (2006). Enteric viruses – a new hazard in our food? *Food Aust.*, *58*, 589–592.

Hagel, I., Lynch, N. R., Puccio, F., et al., (2003). Defective regulation of the protective IgE response against intestinal helminth *Ascaris lumbricoides* in malnourished children. *J. Trop Pediatr.*, *49*, 136–142.

Harhay, M. O., Horton, J., & Olliaro, P. L., (2010). Epidemiology and control of human gastrointestinal parasites in children. *Expert Review of Anti-Infective Therapy*, *8*(2), 219–234.

Hsueh, P. R., Lin, C. Y., Tang, H. J., et al., (2004). *Vibrio vulnificus* in Taiwan. *Emerging Infectious Diseases*, *10*(8), 1363–1368.

Hussain, M. A., Stitt, V., Szabo, E. A., & Nelan, B., (2017). *Toxoplasma gondii* in the food supply. *Pathogens*, *6*(2), 21.

Istamboulie, G., Paniel, N., Zara, L., Reguillo, G. L., Barthelmebs, L., & Noguer, T., (2016). Development of an impedimetric aptasensor for the determination of aflatoxin M1 in milk. *Talanta*, *146*, 464–469.

Jay, J. M., Loessner, M. J., & Golden, D. A., (1995). *Handbook of Modern Food Microbiology: Food Science* (pp. 97–107). New York: Springer Publishing.

Jay, J. M., Loessner, M. J., & Golden, D. A., (2008). *Handbook of Modern Food Microbiology: Food Science* (p. 790). New York: Springer Publishing.

Jensen, B. H., Poulsen, A., Rasmussen, S. H. R., et al., (2017). Genetic virulence profile of enteroaggregative *Escherichia coli* strains isolated from Danish Children with either acute or persistent diarrhea. *Front Cell Infect. Microbiol.*, *7*, 230.

Jones, G. S., & D'Orazio, S. E., (2013). *Listeria monocytogenes*: Cultivation and laboratory maintenance. *Curr. Protoc. Microbiol.*, *31*, 9B.2.1–9B.2.7.

Joseph, S. W., Colwell, R. R., & Kaper, J. B., (1982). *Vibrio parahaemolyticus* and related halophilic *vibrios*. *Crit. Rev. Microbiol.*, *10*, 77–124.

Kamar, N., Abravanel, F., Lhomme, S., Rostaing, L., & Izopet, J., (2015). Hepatitis E virus: Chronic infection, extra-hepatic manifestations, and treatment. *Clin. Res. Hepatol. Gastroenterol.*, *39*, 20–27.

Kamle, M., Mahato, D. K., Devi, S., Lee, K. E., Kang, S. G., & Kumar, P., (2019). Fumonisins: Impact on agriculture, food and human health and their management strategies. *Toxins (Basel)*, *11*(6), 328.

Kaneko, T., & Colwell, R. R., (1973). Ecology of *Vibrio parahaemolyticus* in Chesapeake bay. *J. Bacteriol.*, *113*, 24–32.

Kaur, P., Chakraborti, A., & Asea, A., (2010). Enteroaggregative *Escherichia coli*: An emerging enteric foodborne pathogen. *Interdisciplinary Perspectives on Infectious Diseases*, *10*, 1–11.

Kensler, T. W., Roebuck, B. D., Wogan, G. N., & Groopman, J. D., (2011). Aflatoxin: A 50-year odyssey of mechanistic and translational toxicology. *Toxicol. Sci.*, *120*, 28–48.

Koopmans, M., & Duizer, E., (2004). Foodborne viruses: An emerging problem. *International Journal of Food Microbiology*, *90*, 23–41.

Kumar, A., Matinée, M., Dhiman, A., Kumar, A., & Sharma, T. K., (2019). Aptamer technology for the detection of foodborne pathogens and toxins. *Journal of Advanced Biosensors for Health Care Applications*, *2*, 45–69.

Lee, H. M., Kwon, J., Choi, J. S., et al., (2013). Rapid detection of norovirus from fresh lettuce using immunomagnetic separation and a quantum dots assay. *J. Food Prot.*, *76*(4), 707–711.

Li, J., Sayeed, S., & McClane, B. A., (2007). Prevalence of enterotoxigenic *Clostridium perfringens* isolates in Pittsburgh (Pennsylvania) area soils and home kitchens. *Appl. Environ. Microbiol.*, *73*, 7218–7224.

Li, X. C., Xiang, Z. Y., Xu, X. M., Yan, W. H., & Ma, J. M., (2009). Endophthalmitis caused by *Vibrio alginolyticus*. *J. Clin. Microbiol.*, *47*(10), 3379–3381.

Liu, D., (2006). Identification, subtyping and virulence determination of *Listeria monocytogenes*, an important foodborne pathogen. *Journal of Medical Microbiology*, *55*, 645–659.

Lorbeer, J. W., Ransom, V. E., & Tuffley, J. J., (2000). *Nature and Source of Inoculums of Aspergillus niger Causing the Aspergillus Black Mold Disease of Onions in New York* (pp. 1–6). Geneva, NY: New York State Integrated Pest Management Grants Program.

Lovell-Read, S. R., & De Carvalho, L. P. S., (2019). *Salmonella*, meet mycobacteria. *J. Exp. Med.*, *216*(4), 721–722.

Lund, B. M., (2008). Properties of microorganisms that cause foodborne disease. In: *The Microbiological Safety of Food in Healthcare Settings*. Blackwell Publishing.

Magan, N., (2007). Fungi in extreme environments. *Mycota*, *4*, 85–103.

Magnoli, C., Astoreca, A., Ponsone, L., Fernandez-Juri, M. G., Chiacchiera, S., & Dalcero, A., (2006). Ochratoxin A and the occurrence of ochratoxin A-producing black *aspergilli* in stored peanut seeds from Cordoba, Argentina. *J. Sci. Food Agric.*, *86*, 2369–2373.

Makendi, C., Page, A. J., Wren, B. W., et al., (2016). A phylogenetic and phenotypic analysis of *Salmonella enterica* serovar Weltevreden, an emerging agent of diarrheal disease in tropical regions. *PLoS Negl. Trop. Dis.*, *10*(2), e0004446.

Malik, A., Erginkaya, Z., Ahmad, S., & Erten, H., (2014). *Food Processing: Strategies for Quality Assessment*. Springer: New York.

Mandal, S., Mandal, M. D., & Pal, N. K., (2011). Cholera: A great global concern. *Asian Pacific Journal of Tropical Medicine*, *4*(7), 573–580.

Mandappa, I. M., Basavaraj, K., & Manonmani, H. K., (2018). Analysis of mycotoxins in fruit juices. In: *Fruit Juices: Extraction, Composition, Quality and Analysis* (pp. 763–777). Elsevier.

Marasas, W. F., (2001). Discovery and occurrence of the fumonisins: A historical perspective. *J. Environ Health Perspect*, *109*(S2), 239–243.

Mayr, C., Strohe, G., & Contzen, M., (2009). Detection of rotavirus in food associated with a gastroenteritis outbreak in a mother and child sanatorium. *Int. J. Food Microbiol.*, *135*(2) 179–182.

McClane, B. A., & Strouse, R. J., (1984). Rapid detection of *Clostridium perfringens* type A enterotoxin by enzyme-linked immunosorbent assay. *J. Clin. Microbiol.*, *19*, 112–115.

McClane, B. A., Robertson, S. L., & Li, J., (2013). *Clostridium perfringens*. In: Doyle, M. P., & Buchanan, R. L., (eds.), *Food Microbiology* (pp. 465–489). Washington D.C.

McClane, B. A., Uzal, F. A., Fernandez, M. M. E., Lyerly, D., & Wilkins, T., (2006). The enterotoxic *Clostridia*. In: Dworkin, M., Falkow, S., Rosenberg, E., Schleifer, K. H., & Stackebrandt, E., (eds.), *The Prokaryotes* (pp. 698–752). New York: Springer International Publishing.

McDowell, R. H., Sands, E. M., & Friedman, H., (2020). *Bacillus cereus*. Treasure Island (FL): StatPearls Publishing.

McLeod, R., Cohen, W., Dovgin, S., Finkelstein, L., & Boyer, K. M., (2020). Human *Toxoplasma* infection, the model apicomplexan – perspectives and methods. *Toxoplasma gondii*, 117–227.

McLeod, R., Kieffer, F., et al., (2009). Why prevent, diagnose and treat congenital toxoplasmosis? *Memorias do Instituto Oswaldo Cruz*, *4*(2), 320–344.

Miri, S. T., Dashti, A., Mostaan, S., Kazemi, F., & Bouzari, S., (2017). Identification of different *Escherichia coli* pathotypes in north and north-west provinces of Iran. *Iran J. Microbiol.*, *9*(1), 33–37.

Mishra, R. K., Hayat, A., Catanante, G., Ocana, C., & Marty, J. L., (2015). A label-free aptasensor for ochratoxin A detection in cocoa beans: An application to chocolate industries. *Analytica Chimica Acta*, *889*, 106–112.

Montoya, J. G., & Liesenfeld, O., (2004). Toxoplasmosis. *Lancet*, *363*(9425), 1965–1976.

Moseley, S. L., Huq, I., Alim, A. R. M. A., Samadpour-Motalebi, M., & Falkow, S., (1980). Detection of enterotoxigenic *Escherichia coli* by DNA colony hybridization. *J. Infect. Dis*, *142*, 892–898.

Mostrom, M., (2016). *Mycotoxins: Toxicology*. In: *Encyclopedia of Food and Health* (pp. 43–48).

Navarro, M. A., McClane, B. A., & Uzal, F. A., (2018). Mechanisms of action and cell death associated with *Clostridium perfringens* toxins. *Toxins (Basel)*, *10*(5), 212. doi: 10.3390/toxins10050212.

Neethirajan, S., Ahmed, S. R., Chand, R., Buozis, J., & Nagy, E., (2017). Recent advances in biosensos development for foodborne virus detection. *Nanotheranostics*, *1*(3), 272–295.

Newell, D. G., Koopmans, M., Verhoef, L., et al., (2010). Foodborne diseases – the challenges of 20 years ago still persist while new ones continue to emerge. *International Journal of Food Microbiology*, *139*(1), S3–S15.

Nicholas, D., (2011). *V. vulnificus* oysters: Pearls and perils. *Clinical Infectious Diseases*, *52*, 788–792.

Noguchi, T., Hwang, D. F., Arakawa, O., et al., *(1987)*. *Vibrio alginolyticus*, a tetrodotoxin-producing bacterium, in the intestines of the fish Fugu vermicularis vermicularis. *Marine Biology*, *94*(4), 625–630.

Nyugen, Y., & Sperandi, V., (2012). Enterohemorrhagic *E. coli* (EHEC) pathogenesis. *Front Cell Infect. Mirobiol.*, *2*, 90.

Ojeda, R. J. A., & Kahwaji, C. I., (2020). *Vibrio cholerae*. In: *StatPearls*. Stat Pearls Publishing.

Oliver, J. D., (2005). Wound infections caused by *Vibrio vulnificus* and other marine bacteria. *J. Epidemiol. Infect.*, *133*(3), 383–391.

Orlandi, P. A., Chu, D. M. T., Bier, J. W., & Jackson, G. J., (2002). Parasites and the food supply. *Food Technology*, *56*(4), 72–81.

Papier, K., Williams, G. M., Luceres-Catubig, R., et al., (2014). Childhood malnutrition and parasitic helminth interactions. *Clin. Infect. Dis.*, *59*(2), 234–243.

Patamaroj, U., Seriwatana, J., & Echeverria, P., (1983). Identification of enterotoxigenic *Escherichia coli* isolated from swine with diarrhea in Thailand by colony hybridization, using three enterotoxin gene probes. *Journal of Clinical Microbiology*, *18*(6), 1429–1431.

Qadri, F., Svennerholm, A. M., Faruque, A. S., & Sack, R. B., (2005). Enterotoxigenic *Escherichia coli* in developing countries: Epidemiology, microbiology, clinical features, treatment, and prevention. *Clin. Microbiol. Rev.*, *18*, 465–483.

Rendtorff, R. C., & Holt, C. J., (1954). The experimental transmission of human intestinal protozoan parasites: 1V. Attempts to transmit *Entamoeba coli* and *Giardia lamblia* cysts by water. *Am. J. Hyg.*, *60*, 327–338.

Rheeder, J. P., Marasas, W. F. O., & Vismer, H. F., (2002). Production of fumonisin analogs by *Fusarium* species. *Appl. Environ. Microbiol.*, *68*(5), 2101–2105.

Robertson, D. C., Dreyfus, L. A., & Frantz, J. C., (1983). Chemical and immunological properties of *Escherichia coli* heat-stable enterotoxin. *Prog. Food Nutr. Sci.*, *7*, 147–156.

Rogowska, A., Pomastowski, P., Sagandykova, G., & Buszewski, B., (2019). Zearalenone and its metabolites: Effect on human health, metabolism and neutralization methods. *Toxicon*, *162*, 46–56.

Rood, J. I., Adams, V., Lacey, J., et al., (2018). Expansion of the *Clostridium perfringens* toxin-based typing scheme. *Anaerobe*, *53*, 5–10.

Rooney-Latham, S., Janousek, C. N., Eskalen, A., & Gubler, W. D., (2008). First report of *Aspergillus carbonarius* causing sour rot of table grapes (*Vitis vinifera*) in California. *Plant Dis.*, *92*, 651.

Roze, L. V., Hong, S. Y., & Linz, J. E., (2013). Aflatoxin biosynthesis: Current frontiers. *Annu. Rev. Food Sci. Technol.*, *4*, 293–311.

Ryan, K. J., & Ray, C. G., (2004). *Handbook of Sherris Medical Microbiology: An Introduction to Infectious Diseases* (p. 997). New York: Medical Publishing Division, McGraw Hill.

Sagar, S., Kumar, S., Jaiswal, A., Kumar, A., & Koli, D. K., (2018). Mechanism of biofilm formation by *E. coli*. *Chem. Sci. Rev. Lett.*, *7*(28), 1040–1052.

Sakazaki, R., & Bryan, F. L., (1979). *Vibrio* infections. *Foodborne Infections and Intoxications* (pp. pp. 173–209).

Sánchez, G., (2013). *Handbook of Hepatitis A Virus in Food: Detection and Inactivation Methods* (p. 47). New York: Springer.

Sheikhzadeh, E., Chamsaz, M., Turner, A. P. F., Jager, E. W. H., & Beni, V., (2016). Label-free impedimetric biosensor for *Salmonella typhimurium* detection based on poly [pyrrole-co-3-carboxyl-pyrrole] copolymer supported aptamer. *Biosens Bioelectron.*, *15*(80), 194–200.

Smith, G. W., (2018). *Fumonisins*. In: *Handbook of Veterinary Toxicology, Basic and Clinical Principles*. https://doi.org/1016/B978-0-12-370467-2.X5095-0.

Steinbruegge, E. G., Maxcy, R. B., & Liewen, M. B., (1988). Fate of *Listeria monocytogenes* on ready to serve lettuce. *J. Expt. Biol. Med.*, *27*, 123–131.

Tauxe, R. V., (1997). Emerging foodborne diseases: An evolving public health challenge. *Emerg. Infect. Dis.*, *3*(4), 425–434.

Thompson, F. L., Gevers, D., & Thompson, C. C., (2005). Phylogeny and molecular identification of *Vibrios* on the basis of multilocus sequence analysis. *Applied and Environmental Microbiology*, *71*(9), 5107–5115.

Van, D. P. W. H., (2014). Food and environmental routes of hepatitis E virus transmission. *Curr. Opin. Virol.*, *4*, 91–96.

Venkitanarayanan, K. S., & Doyle, M. P., (2003). *Escherichia coli:* Occurrence. In: *Handbook of Encyclopedia of Food Sciences and Nutrition*, 2149–2152.

Viegas, C., Almeida, B., Monteiro, A., et al., (2019). Settled dust assessment in clinical environment: Useful for the evaluation of a wider bioburden spectrum. *International Journal of Environmental Health Research*, *31*, 160–178.

Waller, J. M., Bigger, M., & Hillocks, R. A., (2007). *Handbook of Pest Management in Organic Farming, Coffee Pests, Diseases and Their Management*. Cambridge: CABI Publishing.

Wen, Q., & McClane, B. A., (2004). Detection of enterotoxigenic *Clostridium perfringens* types A isolates in American retail foods. *Appl. Environ. Microbiol.*, *70*(5), 2685–2691.

Wheeler, C., Vogt, T. M., Armstrong, G. L., et al., (2005). An outbreak of hepatitis A associated with green onions. *N Engl. Jour. Med.*, *353*(9), 890–897.

World Health Organization (WHO), (2008). *Foodborne Disease Outbreaks: Guidelines for Investigation and Control.* https://apps.who.int/iris/handle/10665/43771 (accessed on 18 February 2022).

World Health Organization (WHO), (2018). *Escherichia coli.* https://www.who.int/news-room/fact-sheets/detail/e-coli (accessed on 11 February 2022).

Worley, J., Meng, J., Allard, M. W., Brown, E. W., & Timme, R. E., (2018). *Salmonella enterica.* Phylogeny-based on whole-genome sequencing reveals two new clades and novel patterns of horizontally acquired genetic elements. *mBio, 9*(6), e02303–2318.

Yibar, A., Cetin, E., Ata, Z., Erkose, E., & Tayar, M., (2018). *Clostridium perfringens* contamination in retail meat and meat-based products in Bursa, Turkey. *Foodborne Pathog. Dis., 15*(4), 239–245.

Yun, N. R., & Kim, D. M., (2018). *Vibrio vulnificus* infection: A persistent threat to public health. *Korean J. Intern. Med., 33*(6), 1070–1078.

CHAPTER 11

Food Microbial Hazards, Safety, and Quality Control: A Strategic Approach

AYSHA SAMEEN, AMNA SAHAR, FARWA TARIQ, USMAN MIR KHAN, TAYYABA TARIQ, and BUSHRA ISHFAQ

ABSTRACT

Food is any material or substance eaten or drunk to provide energy and nutrients for the body's growth, development, and maintenance. It is usually of an animal or plant origin consisting of nutrients like fats, proteins, carbohydrates, vitamins, and minerals. Food can be considered safe if it is free from all hazardous substances that can affect consumers' health. In pre-historic times, the food safety chain was very short; it compromised hunter-gatherers and their families. When societies grew complex and larger, this chain became more diffused due to an expanded international trade network and long shipping distances. This chain can be a reason for introducing unfamiliar food safety hazards during food transportation. This situation places a high burden of responsibility on traders, government bodies, and international organizations. This chapter covers the hazards, their types, foodborne diseases, and strategies to ensure food safety and quality (FSQ). Developing countries face challenges due to raised standards of FSQ management worldwide. Different food quality and safety assurance programs are discussed as well like quality management systems (QMS), HACCP certification, ISO 9000 family, good manufacturing practices (GMP)/good hygiene practices (GHP), total quality management (TQM), good working practices (GWP), good lab practices (GLP), etc. Moreover, the role of some novel processing technologies is also focused in this

Food Microbial and Molecular Biology: From Fundamentals to Applications. Saher Islam, Devarajan Thangadurai, Jeyabalan Sangeetha, Zaira Zaman Chowdhury (Eds.)

regard. The effectiveness of a quality and safety assurance system is based on the relationship of external, internal, and structural factors. Food quality assurance (QA) and certification has an important role in building the trust of the customer. The industry needs to select a suitable management system for evaluating and measuring the performance of the company.

11.1 INTRODUCTION

Food is any material or substance eaten or drunk to provide energy and nutrients for the body's growth, development, and maintenance. It is usually of an animal or plant origin consisting of nutrients like fats, proteins, carbohydrates, vitamins, and minerals (Abdulmumeen et al., 2012). Food is the basic precondition for survival but food quality and safety issues embrace us all time (Jeukendrup and Gleeson, 2018). According to the Food and Agriculture Organization of the United Nations (FAO), food quality can be defined as "a complex characteristic of food that determines its value or acceptability to consumers." Food quality has different elements such as food's nutritional value, organoleptic, and functional properties, standardized food, food safety, declared net quantity, appearance like shape, size, color, flavor, texture aroma, shelf-life, viscosity, adulteration, and packaging, etc. (Nielsen, 2010; McLoughlin and Miura, 2017). Food Suitability is a term that is defined by Codex Alimentarius as the assurance that food is fit for human consumption. Not only organizations and government bodies have devised standards for food quality, but customers have some expectations as well, and they vary from customer to customer (Raspor, 2010; Demirci et al., 2016).

Basic quality control and quality assurance (QA) programs are used by food industries like other industries for achieving food quality, e.g., ISO 9000 Quality Management System Standard, HACCP system, GMPs, etc. (Nguz, 2007; Hubbard, 2012; Demirci et al., 2016).

There is a need to improve food safety regulations and systems to ensure the safety of consumers by taking multidimensional approaches at all levels of the food chain, from farm to fork (Unnevehr and Huirne, 2002; Gorris, 2005; Jeukendrup and Gleeson, 2018). Despite all these implementations, foodborne diseases (FBDs) are still a major concern worldwide, carrying a considerable morbidity, disability, and mortality burden. Foodborne diseases cause mostly gastrointestinal problems, but they can cause immunological, gynecological, neurological issues and can even lead to multi-organ failure and cancers (WHO, 2008).

11.2 FOOD HAZARDS

Food hazard can be characterized by any physical, chemical or biological agent that has the potential to exert harmful effects on human health (Peter et al., 2013).

11.2.1 PHYSICAL HAZARDS

Physical hazards are any poisonous or deleterious foreign objects incorporated into food that have clinical evidence of injury or traumatic injury including perforation of tissues present in the oral cavity and gastrointestinal tract if ingested. This includes hard or sharp foreign objects and choking hazards. Contaminants like dirt, hair, insects, wood splinters, glass, and metal are included in the category of physical hazards (Das et al., 2019). Potential sources of these hazards include:

- Foreign objects that are present in raw materials
- Objects are broken from containers, machinery, and equipment present at the processing plant
- Objects linked with maintenance operations such as a piece of glass from a bulb

Analytical considerations for the detection of physical hazards in food can be the visual examination of raw material and other objects involved in processing, separation of food components from heavier foreign objects by sedimentation, filtration, and other such processes (Hui et al., 2002). Food quality and hygiene are regulated by food laws ensuring food should do not cause any harm to the consumer (Silano and Silano, 2017; Hurst et al., 2018). The state is responsible for providing a legislative framework that defines certain conditions involved in the processing and provision of food (Alston and Quinn, 2017; Kaltenborn et al., 2017).

11.2.2 CHEMICAL HAZARDS

These hazards include agrochemicals, heavy metals, sanitizing chemicals, and some naturally occurring chemicals. Agrochemicals are used in the production of crops like fertilizers, growth regulators, miticides, fungicides, and insecticides. They can also be added to food during post-harvest handling and processing at industries. Heavy metals can be incorporated into the water from geogenic sources, waste material of industries, organic fertilizers, domestic effluents and pharmaceutical sources. Chemicals for sanitization and cleaning can be hazardous, if not used correctly (Pretty, 2012). Naturally occurring hazards include allergens like weeds, alkaloids, phytohaemagglutinin, mycotoxins like aflatoxins (AFs) and mushroom toxins. Mycotoxins are secondary metabolites from fungi contaminating fruits and vegetables like spices, dried fruits, soybeans, cereals, corn, tree nuts, and peanuts. Mycotoxins are mainly produced by fungal species of *Stachybotrys*, *Aspergillus*, *Penicillium* and *Fusarium* (Hurst et al., 2018).

11.2.3 BIOLOGICAL HAZARDS (BIOHAZ)

Biological hazards (BIOHAZ) are the microbes like viruses, bacteria, and parasites responsible for foodborne diseases. Moreover, some toxins produced by fungi are biological hazard (Merna and Al-Thani, 2011; EFSA Panel on BIOHAZ, 2013). Some microorganisms associated with foods are as given below:

1. **Bacteria**: *Staphylococcus aureus*, *Bacillus cereus*, *Clostridium* sp., *Listeria monocytogenes*, *Escherichia coli*, *Shigella*, and *Salmonella*.
2. **Viruses**: Coronaviruses, Adenoviruses, Parvoviruses, Coxsackie viruses, Enteroviruses, Astroviruses, Rotaviruses, Norwalk-like virus and Hepatitis A.
3. **Parasites**: *Isospora*, *Sarcocystis*, *Toxoplasma*, *Entamoeba*, *Giardia*, and *Cryptosporidium* (Sun et al., 2019).

Microorganisms can be shifted to food due to incidental contamination from dust, soil and surroundings, poor production, inappropriate handling practices, use of contaminated water, unsanitary handling of food, sewage fluids, infected animals and humans (Bultman et al., 2013). Freshly harvested food have few to millions of bacteria/gram and from none to hundreds of mold spores (Yang et al., 2017). When grains, flour, and other food commodities are stored in a high moisture environment, it can lead to the growth of microorganisms leading to alteration of product properties (Okoye and Oni, 2017).

11.3 FOODBORNE DISEASES

Foodborne illness is any illness resulting due to consumption of food or water contaminated with pathogenic bacteria, parasites, and viruses, mycotoxins, poisonous mushrooms, and some species of beans. Individuals of all life stages are at the risk of such diseases, especially children and older adults (WHO, 2007). Major FBDs is brucellosis, cholera, traveler's disease, leptospirosis, salmonellosis, shigellosis, tuberculosis, botulism, etc. Poor hand hygiene practices are one of the major causes of the spread of the disease (Rolando, 2011). Food preparation without washing hands after coughing, sneezing, touching open sores, after using the toilet can lead to foodborne disease if consumed. Toxins can be produced by the bacteria present in food before its consumption (Payne-Palacio and Theis, 2004).

11.3.1 EFFECT OF FOODBORNE DISEASES

Children who are malnourished and have weak immune systems are more at risk of foodborne diseases and diarrhea leading to increased child mortality (WHO, 2007). Foodborne diseases can lead to malnutrition; this is a vicious cycle of mortality and disease (Prüss-Ustün et al., 2011). Foodborne diseases are one of the major reasons for the increased mortality rate in children under 5 years. This rate is higher in developing countries (72/1,000 live births) than developed ones (6/1,000 live births) (UNICEF, 2010). Unsafe food consumption leads to more than 200 diseases, from diarrhea to cancers. Foodborne diseases affect almost 600 million people worldwide and 420,000 die from these illnesses every year. Hepatitis A that spreads mostly through uncooked or raw seafood can lead to severe liver damage. Mycotoxins can be present in staple foods like cereals or corn and long-term exposure to them can lead to compromised immunity and normal development or even certain cancers. Persistent organic pollutants (POPs) such as polychlorinated biphenyls (PCBs) and dioxins can lead to disturbances in the reproductive and immune system, hormonal imbalance and cancers. Exposure to heavy metals like mercury, cadmium, and lead causes kidney and neurological damage (WHO, 2019).

11.4 STRATEGIES TO ENSURE GLOBAL FOOD QUALITY AND SAFETY

11.4.1 FOOD PREPARATION PRACTICES AT HOME

Unhygienic practices during domestic food preparations are also a key factor in promoting FBDs. If a person deals with food with contaminated hands then he/she is surely putting all his/her family at the risk of infection. The use of unsafe water for cleaning, food processing and cooking increases the chances of exposure to hazards (Pruss-Ustun et al., 2011). Consume cold foods cold and hot foods hot. Be familiar with the signs of unsafe food to immediately identify it and discard it. These signs include bad smell, unusual taste, mold, and rancidity (King and Burgess, 1993). The following steps can ensure FSQ at the domestic level:

- Fresh food should be consumed or before the expiry date (if mentioned)
- Do not buy foods with any signs of spoilage or deterioration
- Store foods by keeping in view the right temperature
- Consume foods as early as possible after preparing
- Use safe and clean water, it can be boiled before use if chances of contamination there
- Wash hands with water and soap before preparing foods, consuming them, interacting with animals and visiting the toilet
- Food must be covered
- Food should be cooked thoroughly to minimize the chances of infection if certain hazard present
- Wash raw food and utensils before food preparation
- Avoid sneezing, coughing into food
- Keep trash bins always closed
- Keep animals away from the kitchen and dining area (King and Burgess, 1993 Wilna and Abdulkadir, 2016)

11.4.2 RESTAURANTS AND STREET VENDORS

As the world population is increasing like a snowball, new technologies are introduced in every aspect of life to make it easy and comfortable. In this advanced era, people are mostly fond of eating junk foods; they are not committed to cook foods in their houses and try to eat in streets from restaurants, hotels, motels, and particularly from street vendors (Troedsson, 2009). This trend is mostly present among teenagers, young, and baby boomers whose taste buds' desires rich, spicy, crunchy, junk, food especially meat and meat-based products. Most people are talking rubbish as a snack providing nothing but empty calories and low-quality food leading to serious health complications (Kaushik et al., 2011). Street-vended foods (SVFs) that encompass a big range of ready to eat beverages and fruits prepared and sold in streets are growing global concern. Their final preparation is done when the customer arrives, orders then purchases and takes away (Winarno and Allain, 1991). Fruits and vegetables sold outside the markets and streets come under this umbrella. Although these foods are affordable and accessible but they can contain hazards as food safety requirements are often neglected (WHO, 2010). This can lead to unprecedented incidences of foodborne illnesses making food safety a global concern and developmental issue leading to financial consequences (Troedsson, 2009). It is very tough to keep a check on SVFs and regulate them in populated areas (Winarno and Allain, 1991; WHO, 2010).

Every restaurant and hotel has at least more than five types of meat cuisines. When meat base products demand is huge, workers in hotels and restaurants do not pay enough attention to meat hygiene, personal hygiene, which later becomes the cause of numerous FBDs depending upon their causal agents (virus and bacteria) (Bas et al., 2006). Lack of safety skills and hygiene knowledge and lack of cooking and handling skills such as temperature, time, personal hygiene, reheating, and thawing of frozen products, etc., can lead to serious problems (Osaili et al., 2013). Food business operators shall ensure hygiene maintenance, training of food handlers, appropriate handling of food and equipment, environmental hygiene and personal hygiene care during all stages of production, processing, and distribution of food. Clean and uncontaminated water must be used and proper waste management should be ensured (Azanza et al., 2000). Local authorities and food inspectors must visit SVFs and restaurants regularly to improve food quality (Winarno and Allain, 1991). A comprehensive approach is required to implement it, with full attention and strict follow up on a regular basis for quality maintenances (Gomes-Neves et al., 2011).

11.4.3 FAO STRATEGIES TO PROMOTE GLOBAL FOOD SAFETY

FAO focuses on adequate food control programs worldwide for ensuring that all individuals on this planet are food secure in terms of quantity, quality, affordability, and safety. This is the key factor for promoting public health. In this regard, Codex Alimentarius Commission (CAC) implemented the Codex Alimentarius General Principles of Food Hygiene in 1995 for providing a strong foundation for food safety and hygiene and highlighting the main control points at every step in the food chain (Codex Alimentarius, 2012). It further suggested that Hazard Analysis Critical Control Point (HACCP) system must be implemented for improving food suitability and safety for human consumption. Critical control points (CCP) and food hazards are identified by HACCP. CCP are the points on which measures should be taken to eliminate or prevent hazards or to decrease them to an acceptable limit. So by applying HACCP throughout the food chain, effective use of resources, food safety and time management of food safety concerns can be obtained. The teamwork approach is the key element necessary for the systems to be successfully implemented. The HACCP method is compatible with all the other QMS, such as the International Organization for Standardization's ISO 9000 series (FAO, 1997).

11.4.4 WHO STRATEGIES TO PROMOTE GLOBAL FOOD SAFETY

The world has become a global village and the opportunities for international trade are more than ever before. Food and food products trade between continents and countries puts the world at a challenge to maintain food quality and safety across the borders. A single or few countries cannot maintain this safe chain, because this is not a national concern and there is a need to address it through closer linkages at the international level among food safety authorities (WHO/FAO, 2011). In 2000, WHO was urged by the 53rd World Health Assembly for implementing a strategy to improve food safety practices and reduces the burden of foodborne diseases worldwide. In 2007, 50 countries adopted the Beijing Declaration on Food Safety, a prestigious international forum. In 2006, an international meeting was held by the collaboration of multiple international partners and the WHO Department of Food Safety and Zoonoses to launch an initiative for providing a strategic framework for estimation of the worldwide burden of foodborne illnesses (Pruss-Ustun et al., 2011). Moreover, it was identified that there is a need of exchanging facts and information on food safety issues and making access possible in case of food safety emergencies. For this purpose, International Food Safety Authorities Network (INFOSAN) was launched by WHO with the collaboration of FAO for facilitating the collaboration of food safety authorities at the national as well as the international level (WHO/FAO, 2011). WHO has the mandate to strengthen systems of food safety, good manufacturing practices (GMPs) promotion and educated consumers and retailers for proper handling of food. One of the most critical interventions for preventing foodborne diseases is training food handlers and educating consumers. In 2006, a food hygiene message was developed by WHO to ensure the safety of customer and industry. These five ways of safer food are given below:

- Keep clean (this covers the personal hygiene as well as an area for food preparation)

- Raw and cooked food should be separated
- Cook thoroughly (this can reduce the risk of infection from food hazards)
- Store food at a suitable temperature, and
- Use safe raw materials and water

11.4.5 CHALLENGES FOR DEVELOPING COUNTRIES

Developing countries have more burdens of foodborne diseases due to poor food safety conditions. Their food systems are not well developed and implementation is poorly monitored. They face various challenges while ensuring FSQ such as rapid population growth, urbanization, poor hygiene and sanitary practices, pre, and post-harvest food losses due to lack of resources, insufficient scientific expertise and adequate facilities, lack of coordination between government departments dealing with food and environment hygiene issues (FAO undated; Henson, 2003; WHO, 2007). The quality and safety of food are negatively impacted due to these factors. Suitable food safety control strategies should be the core focus of food legislation worldwide. Government should take on a regulatory and advisory role in this matter and provide supporting infrastructure to industries and food handlers at any stage from farm to fork. This responsibility of maintaining food safety is shared between the government, industry, and consumers. Government is responsible for food legislation, research, and acquiring information, guidance to industry, consumer education, and provision of health care services. Industry and trade stakeholders must ensure good handling of food during processing and distributing, training of staff especially food handlers and appropriate labeling. Consumers must try to educate themselves about the products they are eating, safe domestic food handling practices and community participation (Henson, 2003; Wilna and Abdulkadir, 2016).

Every country has devised legislation regarding food trade and import. If developing countries want to participate in the international market, then they must fulfill the import necessities of the recipient country like the Food and Drug Administration (FDA) regulates the import of food in the USA. European Union (EU) countries have their trade requirements. Moreover, World Trade Organization (WTO) agreement emphasizes on recognition of international food standards (IFS) in international markets. Phytosanitary, sanitary, certificate of origin is an essential part of import documents. The purpose of these regulations is to ensure harmony in world food markets and providing equal chances of participation in international trade to developing countries (Henson, 2003; FAO, 2020). In 2004, WHO, and FAO developed the Codex Trust Fund for assisting developing countries to improve their food quality and safety in terms of facility and training support systems (WHO, 2004). The rules of WTO also states that developed countries must help developing countries for achieving food safety status at the international level. This will increase the trade between developing and developed countries, the better economic condition of developing counties and less outbreak of FBDs (WHO, 2007; Wilna and Abdulkadir, 2016).

11.5 AN ANALYTICAL MODEL OF FOOD QUALITY

Food will be considered of good quality if it satisfies the needs and expectations of consumers. Analytical model of food quality comprises of consumer requirements that include (Peri, 2006):

1. **Safety Requirements**: Normally mean the absence of harmful substances in food. In case of risk factors, consumer health can be affected and industry can be punished by law.
2. **Commodity Requirements**: Account for the genuineness and authenticity of the product. Voluntary regulations, law, and customary practices set these requirements. The product should be according to claims, labels, and regulating bodies.
3. **Nutritional Requirements**: These are of key importance because food is taken to fulfill the nutritional needs of the body. Functional foods have additional health benefits beyond energy provision. Strong reactions are there in case of any fraud in this regard leading to product rejection.
4. **Sensory Requirements**: These are the vital mean of interaction between consumers and products. Sensations are transformed to perceptions by the brain at a space closely connected to other brain functions like memory, values, emotions, culture, etc. These complex interactions are responsible for our memory, ideas, emotions, and sensory reactions associated with food.
5. **Production Context Requirements**: These are related to tradition, origin, and culture of food/area where it is produced or use of organic agriculture. These requirements have a strong emotive and psychological effect on consumers. Their appeal to consumers depends on "how," "where" and "when" of a food product rather than on "what."
6. **Ethical Requirements**: These are related to the system of values conditioning behaviors of a consumer. They include organic agriculture, the defense of biodiversity, defense of the environment, the

well-being of animals, etc. However, unfortunately, there are fewer ways of verifying for a consumer that ethical requirements are met or not. There is no way to assess, either biological agriculture was respected or not during the production of an apple. There are more chances of deceit and fraud. That is why guarantee requirements are becoming more popular.

7. **Guarantee Requirements**: These are intended to develop the trust of consumers. Traceability and certification of producers are the instruments that offer consumer guarantee. Nothing can guarantee the industry more than consumers' personal trust in the people supplying them with food. Consumers' perception of reliability and professional competence of producer is the basis of their credibility.
8. **Product/Packaging System Requirements**: These are essential for product recognition, advertisement, and use. They include esthetic requirements as well, relating to its presentation and information provided on the label. The products that are easy to handle are preferred by customers and their desire for ease is the most fertile ground for advertising and marketing experts.
9. **Product/Market System Requirements**: These include product's availability at the right place, right time, in desired amount and affordable price.

11.6 FOOD STANDARDS

Food standards have also been established along with the food laws and regulations to ensure the quality and safety of food. Food standards are not legally required until or unless they are part of the food regulations. One of the best examples of food standards is Codex Standards. For the implementation of the joint Food Standards Program of World Health Organization (WHO)/Food and Agricultural Organization (FAO), CAC has been given the authority (Havinga, 2006). Codex Alimentarius can be defined as a set of standards for the safety, quality, and suitability of food. They consist of codes of good hygiene practice such as general food hygiene principles, standards for the usage of animal drugs in foods, standards for the determination of maximum residue levels (MRL) when a pesticide is used on food products, and standards for the regulation of food additives. Food standards promote international trade of food and are approved by countries all over the world. The main purpose of developing food standards are the safety of consumers (Newslow, 2013).

11.7 QUALITY MANAGEMENT SYSTEMS (QMS)

Quality management systems (QMS) are essential in all departments of the food-related industry to make sure the safety and quality of food (Orris and Whitehead, 2000). Major QMS in the food sector are IFS, safe quality food (SQF), global food safety initiative (GFSI), International Organization for Standardization – ISO 22000:2005 and British Retail Consortium (BRC) (Baert et al., 2005). To compete with the largest food manufacturers in the international market, large numbers of business holders in the food sector are adopting QA systems (Beatty, 2006; Karipidis et al., 2008). Management-related aspects in food industries are focused on by TQM and ISO. On the other hand, HACCP, and GMP cover technical aspects (Hoogland et al., 1998). Guidelines are covered by GMPs whereas total quality management (TQM) makes use of rewards and assessments. The only QMS comprised of a complete 14 stage plan is HACCP (van der Speigel et al., 2003).

11.7.1 GLOBAL FOOD SAFETY INITIATIVE (GFSI)

The GFSI is collaborated by CIES also known as Food Business Forum, which is an organization of the world's biggest retailers and it was initiated in 2000. The primary mission of GFSI is the continuous advancement of the systems required for the management of food safety to restore the faith of consumers in the safest food delivery. It is primarily aimed at consumer protection and to gain and sustain consumers' trust. Other major objectives of this initiative are to provide a distinguished global platform for the world's largest retailers to connect and exchange information, statistics, and approved food hygiene practices. It is also aimed at increasing the cost efficiency of the supply chain of food products by commonly agreed upon GFSI standards. The fundamental principle of GFSI is the fact that insurance of food safety should not be a competitor in nature because the internationalization of the market may result in a single problem affecting the entire food supply chain (CIES, 2008). In 2007, a guidance document of GFSI was published which consists of globally accepted criteria for the comparison and evaluation of food safety schemes. Four major standards of GFSI are IFS, BRC, SQF, and HACCP. The criteria for all of these food safety standards are defined by experts in food safety. It is primarily aimed at the safe production of food products consequently reducing cost and audits in the supply chain (CIES, 2008).

11.7.2 IFS AND BRC STANDARDS

BRC standard and IFS is based on the HACCP system, GHP/GMP principles and ISO 9001 (Drabas and Wojciechowski, 2006). These standards include the requirements as given below:

- The requirement to do threat analysis of allergies associated hazards
- Work efficiency monitoring
- Cover facial hair with masks during working
- Carry out products' application tests
- Documentation of stocks of raw material and final products
- Use of FIFO principle "first-in, first-out"
- Establishment of a system allowing the industry to get information about GMO
- Elaboration of all procedures used in case of unusual circumstances
- Identification and elaboration of the area of the industry where hazardous material like glass occur
- Applying of metal detectors
- Devising a procedure for qualification, verification, and approval of suppliers
- Ban smoking in the entire premises of industry

11.7.3 SAFE QUALITY FOOD (SQF)

A system was developed in 1995 by the Australian government and farmers' associations 1995 for controlling whole and SQF system agro-food chain. It later remained as SQF 2000. The base of this system is requirements established by ISO 9000 and Codex Alimentarius. Since farmers' associations were involved in the development of these systems, therefore, they can be applied in primary production a well. SQF is managed by Food Marketing Institute (FMI) in Washington since 2003. Due to variation in processes, size, products, and circumstances using a single standard are impractical. Therefore, several standards of SQF have been made (Safe Quality Food Institute, 2008):

- SQF 1000 for smaller businesses and primary production
- SQF 2000 for big industries
- SQF 3000 for restaurants and retail

11.7.4 HACCP CERTIFICATION

Dutch food-processing industry launched a standard of certification called hazard analysis and critical control point system (HACCP) in 1996 with the name of "Criteria for the assessment of an operation HACCP-system." It was renamed "Requirements for a HACCP-based food safety system" after six years (SCV, 2008). HACCP is a proactive, systematic, and preventative approach to food safety that relies on identifying and controlling hazards in the food chain from harvesting to consumption. It can be assured by controlling the source, design of the product, process control, good hygiene maintenance during production, processing of food, distribution, storing, trading, preparation, and eating. This system consists of 7 steps and 12 principals that are given below (1–12 are principles and 1–7 are steps) (Burgess et al., 2019; Das et al., 2019):

- Assembling HACCP team
- Product description (describe food and its distribution)
- Identification of intended use and target customers
- Construct a flow diagram
- Verification of flow diagram
- Conduction of hazard analysis
- Determination of the critical control points (CCPs)
- Establishment of critical limit(s)
- Establishment of a system to monitoring the control of the CCP
- Identify where CCPs are lacking by monitoring
- Establishment of procedures for verifying the efficiency of the HACCP system
- Documentation of all the procedures and records

11.7.5 STANDARDS FROM THE ISO 9000 FAMILY

ISO 9000 is a family of standards of quality management that is implemented in various organizational systems. Their application was allowed in a diverse range of enterprises regardless of their operating branches. Standards comprised by the ISO 9000 family are following:

- ISO 9000, this standard supports the basis as well as terminology of QMS (ISO 9000:2000)
- ISO 9001, this standard is specifically implemented in enterprises as well as it proposes requirements concerned with the quality management system (ISO 9001:2000)

- ISO 9004, it is associated with the specification of the guidelines to improve the already implemented system of the company or organization (ISO 9004:2000)
- ISO 19011, it provides the recommendations regarding audit (ISO 19011:2002)

Although standards mentioned are optional still customers, contractors, as well as recipients, urge their supplier to adopt systems fulfilling with standards of ISO 9000 because they can then be assured that their suppliers work in a specific, distinguished manner by keeping in view the following eight principles:

- Customer-centric
- Leadership (leaders establish the unity of the aim and operation of the organization)
- Engrossment of the workers
- Process approach
- System approach to management
- Continuous improvement
- Decision making based on facts
- Mutually beneficial cooperation with suppliers

Apart from these, it streamlines the qualification procedure for customers as well as suppliers along with a reduction of cost associated with these operations. Management of organization and resources, product realization and measurement processes, analysis, and improvements are the main areas that are being covered by a quality management system based on standards of ISO 9000 (Knaflewska and Pospiech, 2007).

11.7.5.1 ISO 22000:2005

This standard was published in September 2005. The goal of it is to merge the codes of the quality management system being used in the food industries (Fabisz-Kijowska and Kijowski, 2006). However, it is optional as it surpasses the agenda of GHP/GMP and HACCP. Its range encompasses (ISO 22000:2005):

- The prerequisite program (PRP), i.e., the above-mentioned GHP/GMP principles and GAP (good agricultural practice), GVP (good veterinarian practice), GPP (good production practice), GDP (good distribution practice), GTP (good trading practice)

- The HACCP system
- The identification system (traceability system)
- The quality management system ISO 9001:2000

ISO 22000:2005 incorporates both the HACCP and quality management system (ISO 9001:2000). There are also cross-references between ISO 22000, ISO 9004 and terms and definitions from ISO 9000. The implementation of the most effective system of food quality and safety that was designed into the current framework of the management may be beneficial for both organizations as well as interested parties. Additionally, its implementation into the enterprise may be independent of the other existing management system (Fabisz-Kijowska and Kijowski, 2006).

11.7.6 GOOD MANUFACTURING PRACTICES (GMPs)/GOOD HYGIENE PRACTICES (GHP)

Good Hygiene Practice (GHP) denotes all food processing procedures must be carried out in hygienic conditions to guarantee food safety. GMP refers to the procedures to be undertaken and conditions to be fulfilled to ensure that food production, packaging, raw materials and other related materials having contact with food are conducted properly to ensure safe food production at the end of the process (Turlejska, 2003). Manufacturing practices include well-defined and controlled protocols. It includes validated processes that show their compliance with specifications as well as consistency. These practices are well controlled and are carefully evaluated for any change made to them. Necessary validations are performed for all those changes, having an impact on food quality. Procedures and all related guidelines are written in a language that is clear-cut and easy to comprehend. Operators are well trained to perform as well as document all the procedures (good documentation practices (GDocP)) (Abdellah et al., 2015).

During manufacturing, records are well maintained either manually or by using different instruments. These records show that all necessary steps were taken during a procedure and all instructions were followed and food was of required quantity and quality. Any deviances in the procedure are also documented (De Oliveira et al., 2016). A complete history of a batch and manufacture records (including distribution) are maintained in comprehensible form. Systems have been developed that can trace any manufactured product any time from sale or supply. The distribution of the food reduces the risk factor related to the quality. If any complaints are registered for a marketed food, these are well investigated and the cause for this quality defect is identified to make sure this negligence is not being repeated in future (Freimuller et al., 2015; He et al., 2015). The GMPs and any of their PRPs should be carefully monitored to make sure if they are adequate for food safety requirements or not. Auditing and inspection activities should be effective and authenticated. The area covered by GMP requirements (Turlejska, 2003) comprises of following mentioned below:

- The working site, environment, and infrastructure
- Equipment and machines
- Disinfecting and washing procedures
- Water supplies
- Waste management
- Pest control and protection
- Training of staff
- Personnel training and hygiene
- Documentation and record keeping

11.7.7 SIX SIGMA

In the middle of the 1980s, the concept of six sigma emerged from the work of Motorola. The method utilizes the application of statistical tools in business progressions, with the motive to eliminate risks. A key goal of Six Sigma is quality improvement, enhancing the pace of production and making the product more economical. Six Sigma focuses on quality improvement by waste removal; waste is identified and removed, which is of no value to the customer. The technique helps the company to carry out operations more efficiently at lower costs, improve quality, enhance customer satisfaction and upsurge profitability (Werkema, 2004; Tjahjono and Ball, 2010; Paiva, 2013). The Greek letter "Sigma (σ)" is used by the statisticians to quantify the change in a process. It is the measure of a company's performance. 3 to 4 defects are aimed to be found per million manufactured products by the organizations that use the Six-Sigma method.

In this method, a system is implemented that makes measurements and analyzes the process, any deviations from the normal behavior are analyzed, and steps are taken to avoid any abnormal behaviors as much as possible. A cycle known as DMAIC (Define, Measure, Analyze, Improve, Control) is utilized as a guide for professionals, to apply on the most innovative projects and highly ambitious targets set by the company (Figure 11.1). The cycle can be described as follows:

1. **Define**: Definition of the problem and the situation that needs improvement, keeping in view the company's targets and strategic ambitions.
2. **Measure**: Application of established and consistent methods for information and data handling and processing.
3. **Analyze**: Analysis of the obtained information from the measurement tools to identify the loopholes and gaps in the current performance of the system and developing a method to reduce the flaw to achieve the desired goal. The analysis must be done with the aid of statistical methods and an approach.
4. **Increment**: Install processes, utilize the project management tools and planning techniques for newly established processes.
5. **Control**: Implement effective controls to improve the cycle and create a continuously improving cycle (Werkema, 2004; Kwak and Anbar, 2006; Tjahjono and Ball, 2010; Paiva, 2013).

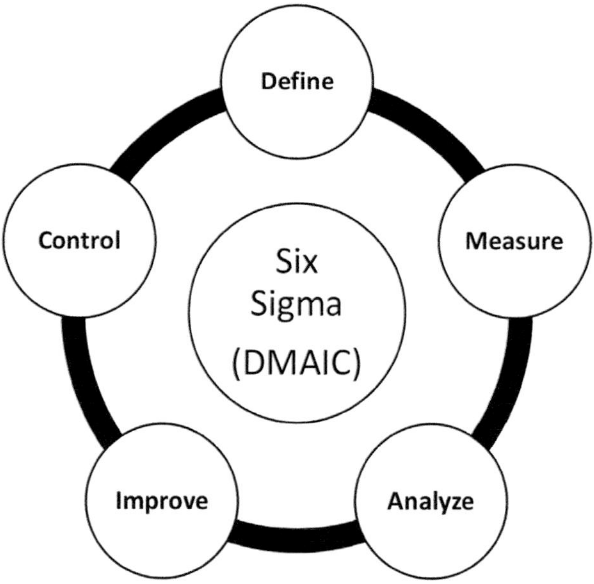

FIGURE 11.1 DMAIC cycle (Six-Sigma approach).

11.7.8 ALLIED QUALITY ASSURANCE PUBLICATION (AQAP)

There should be cooperation between user, ordered, and supplier in the field of QA of the product. There should be cooperation in the quality systems are described in ISO 9000 series. Adjustments must be made in quality issues to reduce the risk level associated. The representative must supervise the QA system of the deliverer, especially in the areas where risks are more. Assessment and analysis of risk that is associated with the deliverer and product to ensure that the necessities of the contract will be fulfilled (Wierzbicki, 2003).

11.7.9 TOTAL QUALITY MANAGEMENT (TQM)

Term TQM was first introduced in North America during the 1980s. This term refers to the management approach to improve quality that was utilized in Japan for acquiring long-term success. TQM possesses both quality concepts and management principles including empowerment of people, strategic planning, leadership, customer focus process management and improvement (Dark et al., 2015; Schilling and Neubauer, 2017; Tricker, 2019).

These concepts and principles first evolved during the 20th Century with the considerable involvement of many experts related to the field of quality management. During the 1980s and 1990s, many food businesses in North America approved the TQM approach and proposed the framework for using it in their system of Quality Management, for acquiring maximum advantage in the global market (Child, 2015; Demirci et al., 2016; Wallace et al., 2018; Tricker, 2019). Key characteristics of TQM are given below:

- Quality must meet the requirement of customers
- There occur inter-departmental quality problems
- Quality control is finding, monitoring, and eliminating reasons for quality problems
- Quality assurance relies on management systems effective audit and reviews and Prevention
- There should be managed quality
- Focuses must be on prevention not cure

11.7.9.1 IMPLEMENTING QMS

Before implementation of QMS, the two important requirements will be ensured:

1. **Customers' Requirements**: An organization's ability to provide desired product and service to meet their expectations and needs.
2. **Organization's Requirements**: Optimum cost with effective utilization of available resources like materials, technology, working force and information.

A quality management system has four distinct components: quality control, quality planning, QA and improvement. The major constraints in implementing advanced FSQ management programs are lack of human resources and finance due to high cost, less personnel, lack of knowledge, experience, and time restrictions. The analysis of key factors influencing the application of these integrated or advanced systems can be helpful to understand the association between certain elements like internal (communication, staff involvement, leadership), external (relationships with authorities, suppliers, and customers), structural (ownership structure and size) and also the potential with which management practices are used. After resolving all these elemental concerns an agri-industry has the preparedness to follow the standards of the advanced quality system. Those organizations that follow TQM are fully aware of customer's needs and try to provide consistent quality products by delivering when needed and priced for value. Such organizations develop cost-effective processes by using process management techniques and are capable of fulfilling customer's expectations. With the implementation of ISO 9004, TQM can also be initiated and this will be a further changing step in an organization for improvement and quality. In Figure 11.2, the principles of TQM are presented (Rotaru et al., 2005).

The accountability for implementing and regulating food laws is allocated to the agencies regulating the government. These enforcement actions are divided into two classes:

- First, they comprise of inspection and auditing of establishments that are involved in processing, handling, and food storage to make sure that necessary sanitary and well-controlled conditions are applied; audits are utilized by a few regulatory agencies enforcing HACCP-based regulations (Marriott et al., 2018), and
- Second, include food analysis and inspection for detrimental substances to make sure conformity to acknowledged limits along with tolerances.

Despite efforts made by government agencies for enforcing food laws as well as regulations, the misbranded foods that are harmful for human use sometimes penetrate food distribution or supply chain (Kotsanopoulos and Arvanitoyannis, 2017). Whenever a harmful or misbranded food is noticed, a detrimental agent or substance to be found in that food, it gives confirmation of probability of harmful agent or substance to found there in food, or a definite foodborne illness occurs. Therefore, agencies regulating government and food companies take mandatory action for protecting customers against these violations. These circumstances often result in food being recollected from the marketplace. Additionally, if there is confirmation of food adulteration or misbranded food substances due to negligence by some food company (Hubbard, 2012; Kotsanopoulos and Arvanitoyannis, 2017).

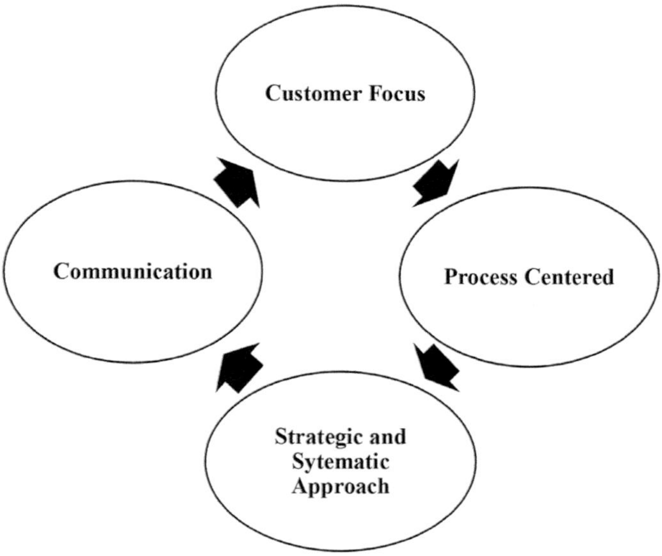

FIGURE 11.2 Principles of TQM.

11.7.10 GOOD WORKING PRACTICES (GWP)

Working practices may be regulated by compliance between employers and workers' representatives, or the practices, which are considered without any formal agreement. The processes that are used to produce goods must act in accordance with precise procedures of manufacturing and marketing authorizations, to obtain the indispensable standard of the products (Bidwell et al., 2013). Protocols must be followed and if required must be recorded while taking care of products and handling the raw materials including processes like quarantine, sampling, cleaning storage, dispensing, labeling, packaging, processing, and distribution. Deviation from the authorized protocol must be avoided to the fullest extent. Approved methods must be followed to carry out any discreet process in case of divergence from normality (Karmacharya, 2014). The batch number should be used to label and identify the equipment and materials; it is an encouraging practice to mention the materials and capacity of the equipment. Mention the stage of the production process where it seems reasonable and applicable. In many cases recording of previously processed product is also useful (Sonnemann et al., 2018). Only authorized and relevant personnel should be allowed inside the production facility and intra process controls are applied and carried out inside the production area (Press, 2003). Contamination of both the raw material at the start of a process and the finished product must be avoided. Strict measures must be put in place to avoid cross contaminations on both technical and organizational level, like keeping a check on gaseous, vaporized, and liquid effluents, which can cause contamination of the product. Standard operating procedures (SOPs) must include periodical checks of the production units to check for the sources of contamination and production areas should be regularly scrutinized from time to time for environmental sustainability (Patel and Chotai, 2013). Before starting packing operations ensure that documents used, work place, packaging lines are clean. No previous product or unnecessary documents, products or materials are lying around that are not required for the operation at hand (Cohen, 2014). Display the number of batches and the name of the product at each packing line or station. Seals should be applied to avoid any contamination as quickly as possible. In case of delay in labeling operation, proper care must be taken and proper methods should be applied to avoid any risk of mislabeling and mix up (Moerman and Wouters, 2016).

11.7.11 GOOD LAB PRACTICES (GLP)

Any quality system related to organizational conditions and process under which environmental safety and non-clinical health studies and research are planned, practically done, examined, documented, archived, and finally reported must be according to recommendations. Its maintenance is done for producing quality or standard test data also providing a tool for (Feigenbaum et al., 2002; Melethil, 2006). A chemical must be bought in minimum quantity, even if it seems cheap. Store chemicals within correctly labeled and well-labeled containers having suitable and identified hazard warning (either utilizing circled letter written with help of permanent marker or any of the labels with special hazard code having yellow color), also store chemicals according to the instructions by the supplier. All the chemicals containing containers should be properly sealed when transporting and store hazardous chemicals like volatile compounds within the fire-proof cabinet. On the other hand, if a chemical is non-inflammable or inflammable in little amounts, store in the fume cupboard. Flammable solvents having flashpoints less than 32°C (e.g., ethanol, acetone, propanols, diethyl ether, petroleum ether, isoamyl alcohol, butanols) their storage must be done in bottles with not greater than 500 ml unless within the fire-proof cabinet (Sikora, 2015; De Oliveira et al., 2016; Manning, 2018). Following GLP and discarding wastes properly can reduce the incidence of incorporation of these chemicals to water and food materials.

11.7.12 GOOD DOCUMENTATION PRACTICE (GDocP)

Good documentation practice (GDocP) is normally abbreviated as GDP, but it is recommended to abbreviate it as GDocP so that it might not be confused with "good distribution practice (GDP)," which is also being abbreviated as GDP. These are the practices that are used in the food, medical, pharmaceutical industries to create and maintain necessary documents. Some of these standards are codified by different proficient authorities while others are not, and these are known as cGMP (with emphasis on the "c," or "current"). Some proficient authorities' issue or adopt recommendations, which may include non-codified GDP/GDocP prospects. These guidelines and cGMP prospects in addition to the legal requirements, are inspected by authorities but not the Law, and if departures are seen, then necessary comments are made. In the last few years, it has been seen that these GDocP is also expanding to the cosmetic industries and ingredient manufacturers (Trafialek and Kolo Zyn-Krajewska, 2011; He et al., 2015).

11.7.13 GOOD DISTRIBUTION PRACTICES (GDP)

The distributed food products must be authorized according to the defined set of rules. Storage conditions are well checked and maintained throughout transportation. Chances of any contamination are minimized and adequate turnover of stored food is ensured. To minimize the quality risk of the product, safe, and secure areas are chosen for storage. It is ensured that a good quality product is delivered to the right addressee within the time limit provided. An effective tracing system should be used to identify any faulty product (Blanchfield, 2005; Trafialek et al., 2015).

11.8 PROCESSING TECHNOLOGIES FOR IMPROVING FOOD QUALITY AND SAFETY

Consumer's demands for healthy, fresh, and nutritious food have triggered the food industries to adopt new technologies for high product quality. These technologies can help in enhancing functional and nutritional characteristics, reducing the processing cost and reduction of the foodborne pathogen. Some of these technologies are discussed below (Considine et al., 2008).

11.8.1 HIGH TEMPERATURE SHORT TIME (HTST) PASTEURIZATION AND ULTRA-HIGH TEMPERATURE (UHT)

Traditional pasteurization was used for milk and milk products for 30 minutes in the large tank. Now the modified form of pasteurization is used called high temperature short time (HTST) pasteurization. In this type of pasteurization, metal plates are used for attaining the temperature of 161°F for 15 seconds. This increases the shelf life but the product still needs refrigeration. Ultra-high temperature (UHT) is an aseptic method done in aseptic containers above 135°C (275°F) for 2–3 seconds; packaging is also done by following food safety standards. No refrigeration is required after this type of treatment (IDFA, 2020).

11.8.2 HIGH PRESSURE PROCESSING (HPP)

This processing can inactivate enzymes and microorganisms and modify structures exerting no effect on nutritional and sensory attributes like texture, the color of products. This helps in controlling spoilage of food, ensuring food safety and increasing the shelf life of the product (Cheftel, 1995; Hogan et al., 2005; Patterson, 2005). Color changes in meat/poultry are there due to change in heme displacement/release, changes in myoglobin and ferrous atom oxidation. HPP can help in coping up with these color change issues (Cheftel and Culioli, 1997).

11.8.3 ULTRASONICATION

It is a technique that can pasteurize and preserve food products by inactivation of enzymes and micro-organism at low temperature improving food quality.

The baking industry is dependent on control of dough properties to achieve consistency and food quality. Ultrasonic measurements like attenuation and velocity at 100 kHz help in assessing flour quality (Alava et al., 2007). They can be used to determine the rheological properties of doughs made from different type of flour. It can help in discriminating different types of flours (Gomez et al., 2008; Chandrapala et al., 2012).

11.8.4 VACUUM FRYING

Traditional frying of potatoes and other foods can lead to acrylamide due to the high content of reducing sugars and amino acid asparagine. Acrylamide formation can be reduced by using the vacuum frying method. This method does not compromise the taste and nutritional properties of foods (Granda et al., 2004).

11.8.5 NANOTECHNOLOGY

Nano-particles can be used in the food industry to improve nutritionally and flow properties, color, flavor, increase shelf life, color, and extending shelf life. Healthy foods with reduced fat, sugar, and salt content can be produced by using this technology to prevent diseases. Nutraceuticals can be incorporated in carriers including phytosterols, beta-carotene, and lycopene are used in foods for preventing the accumulation of cholesterol (Mozafari et al., 2006). Food additive and food constituents enclosed in like Octenyl succinic anhydride-polylysine (Yu et al., 2009; Sekhon, 2010). The anti-cancer activity of curcumin was increased by hydrophobically modified starch encapsulation (Yu and Huang, 2010). TiO_2 nanoparticles can be used against foodborne pathogens like *L. monocytogenes*, *Vibrio parahaemolyticus* and *Salmonella choleraesuis* (Kim et al., 2003).

11.8.6 BIOMIMETICS

Food taste and perception are based on the five senses we are gifted by nature. Biomimetics as the 'abstraction of good design from nature' is to tap and analyze nature's potential of innovative solutions. They are human sense inspired technologies to given the decision about the quality of the final product (Ghasemi-Varnamkhasti et al., 2010).

11.8.7 MICROWAVE PROCESSING

Significant reduction in cooking consumption and cooking time can be achieved by using microwaves. Microwave technologies include sterilizing, heating, and drying leading to better food quality and safety. It can retain higher levels of bioactive compounds, the activity of antioxidant and vegetable color than conventional methods. It can inactivate enzymes and bioactive active components (Guo et al., 2017).

11.8.8 RADIO FREQUENCY ELECTRIC FIELD (RFEF) PROCESSING

Commercial heating methods generally focus on convention and conduction methods transferring heat from outer surfaces of products to inner ones leading to more time consumption overheating and dehydration. On the other hand, interaction between foods and electromagnetic energy for controlling bacteria, appearance, improving sensory and nutritional attributes (Di Rosa et al., 2019).

11.8.9 OHMIC HEATING

Ohmic heating also called electrical resistance heating, Joule heating, direct electrical resistance heating is a procedure of heating food products by passing current dissipating energy directly to food. Potential applications for ohmic heating include heating, pasteurization, sterilization, fermentation, dehydration, evaporation, and blanching. It is a faster heating method along with the sustainability of nutritional value and color properties. However, great start-up cost, lack of awareness and lack of its application in foods having oil and fat limit its use (Kaur and Singh, 2016).

11.8.10 PULSED ELECTRIC FIELD

This commercialized nonthermal intervention technology was introduced in 2005. This method has excellent microbial inactivation for liquid products. It can kill molds, yeasts, and vegetative bacteria. Pumpable and fluid can be pasteurized using PEF. It assures the retention of product freshness and quality compared to non-thermal procedures (Barbosa-Canovas and Swanson, 2019).

11.8.11 ULTRAVIOLET (UV) LIGHT IRRADIATION

Radiations from the ultraviolet (UV) region of the electromagnetic spectrum are used in UV processing. These radiations can reduce the growth of the microbial population by damaging their DNA. UV irradiation serves as a universal ligand to active cell surface receptors and to induce the assembly of signaling complexes at the plasma membrane. These complexes activate cytosolic protein kinase cascades that relay signals to the cell nucleus, thereby regulating the activity of a variety of transcription factors. Key cellular responses are up-regulation of matrix-degrading metalloproteinase and down-regulation of extracellular matrix biosynthesis. The resultant net deficit of structural integrity of skin connective tissue is one of the major factors that contribute to the phenotype of photoaging (Xu and Fisher, 2005).

11.9 RELATIONSHIPS BETWEEN THE QUALITY SYSTEMS IN FOOD INDUSTRY

Specific approaches are highly required in agri-food production industries to satisfy consumers' expected quality parameters (van der Speigel et al., 2003). It is very essential to integrate above mentioned various quality systems to improve the overall performance of the industry. HACCP is considered as an important part of quality systems because it not only assures the safety of food products but also ensure the better implementation of whole quality system (Efstratiadis and Arvanitoyannis, 2000). There is a clear distinction between assurance and management. The word assurance describes the product itself and includes various safety assurance systems (GHP, GMP, and HACCP) and also the QA control points. Thus, the term assurance not only refers to the safety but also the quality of the end product (Sikora and Strada, 2003).

The introduction and also the maintenance of other quality parameters (nutritional, convenience, and sensory values) of food in QA programs are not needed by law but desired by the consumers. The other term management indicates the entire organization concerns regarding product quality and includes QMS (ISO-9000, ISO-14000, etc.), and in addition, also contains TQM. Various implemented management and QA systems are ISO 9000, ISO 22000, ISO 14000 and ISO 18000. Within TQM, ISO 9001 can play a significant role by supporting procedures and systems, although it is a smaller part of TQM actions. By increasing the understanding relationship between safety and quality systems, the overall performance of QMS can be improved significantly. Combination or integration of such systems by food manufactures can assure the safety and quality aspects of food products. For example, when ISO 9001 in combination with HACCP principles are implemented in the food industry, many management and technological issues regarding quality and safety attributes of food products are reduced. Hence, ISO 9001 can be beneficial for implementing HACCP (Rotaru et al., 2005). Moreover, food manufacturers should apply HACCP principles as required by legislation while other systems can be applied voluntarily. In Table 11.1, integrated model regarding essential requirements for food organization is indicated. The assurance of safer food production is a mandatory requirement for the agri-production industry and can be achieved by following an organizational and systematic structure, controlling processes, activities, and resources mentioned by the quality and hygienic systems like ISO 9001 HACCP and 14000 series (Early, 1995). Food industries that do not adopt ISO 9001 protocols, the standards of ISO 22000:2005 are effective and can be offered as alternative food safety management system (FSMS).

11.10 FOOD QUALITIES AND CONSUMER PERCEPTION

The perceptions of consumers vary with time (Five-Shaw and Rowe, 1996). Commonly interlinked risks occur frequently and are less harmful. Moreover, these risk factors are known to be in control of individuals and are often underestimated (Alvensleben, 2002). Now a day's people are more curious about their health, specifically in terms of nutrition. It is assumed that this drastic change in consumer perception is mainly because of safety measures taken by manufacturers and awareness of consumers. Food manufacturers in the industry are working ambitiously to ensure food safety in products. Reliability and trust are the main factors on which effective communication regarding food risk and safety between people and manufacturers is based. There are various dimensions of trust such as honesty, competence, and concern with public welfare (Frewer and Miles, 2001).

The best way to develop trust is to actively provide information to consumers regarding food safety in peaceful times, and proper information should be provided on labels. No misinformation should be there during labeling and advertisement. The priority of manufacturers must be safety-sensitive consumers for communication. Moreover, they have to share all of their food quality and food safety initiatives actively with them by considering this as their ethical duty and responsibility (Rohr et al., 2005).

TABLE 11.1 Essentials of Food Quality and Safety Assurance in the Industry

Features	Requirements
Quality management	• The commitment of top management • Sharing of responsibilities, plan, material, and human resources • Effective quality system assurance
Production areas and equipment	• Guarantee the presence of all equipment and facilities, installation, and maintenance of equipment, sanitization, cleaning
Staff	• Development of a proper organizational framework • Description of executive positions • Training of staff • Development of attitude and behavioral competencies • Supervision of personal health and hygiene
Documentation	• Develop, approval, updating, distribution, maintenance of document
Regulations	• Follow contractual requirements • Apply mandatory requirements
Quality control	• Apply GLP and good sampling techniques • Inspect the analytical methods • Validate the process • Maintenance, checking, calibration of monitoring and measuring devices
Processing	• Process validation • Avoid cross-contamination • Procure quality raw materials • Fulfil label requirements • Ensure good quality of end product • Identification and tracking of batches
Consumer complaints	• Handle the complaints from customers • Documentation of recalls and withdrawals • Analyze the decisions
Self-inspection	• Conduct internal audit • Check corrective and compliance
Supplier relationships	• Identify key suppliers • Communicate open and clear • Share future plans and information • Start joint improvement and development activities

Source: Adapted from: Rotaru et al. (2015).

11.11 CONCLUSIONS

Ensuring food quality and food safety is a necessity today, especially for developing countries to take part in international trade. Both negative and positive experiences from all countries can be used for the betterment of systems globally. All types of hazards can lead to deleterious effects on the body; strategies should be devised to avoid them. FBDs exert a significant effect on the health, morbidity, mortality, and economic development of a country. Food safety must be ensured to avoid the risk of any hazards in food.

KEYWORDS

- **foodborne disease**
- **food hazards**
- **food microbiology**
- **food quality and safety assurance programs**
- **food quality challenges**
- **food quality improvement**
- **food standards**
- **global food quality**
- **global food safety**
- **hazardous substances**
- **novel technologies**
- **quality assurance**
- **quality management systems**

REFERENCES

Abdellah, A., Noordin, M. I., & Ismail, W. A. W., (2015). Importance and globalization status of good manufacturing practice (GMP) requirements for pharmaceutical excipients. *Saudi Pharmaceutical Journal*, *23*(1), 9–13.

Abdulmumeen, H. A., Risikat, A. N., & Sururah, A. R., (2012). Food: Its preservatives, additives and applications. *International Journal of Chemical and Biochemical Sciences*, *1*, 36–47.

Alava, J. M., Sahi, S. S., Garcia-Alvarez, J., Tur, A. O., Chavez, J. A., & Garcıa, M. J., (2007). Use of ultrasound for the determination of flour quality. *Ultrasonics*, *46*, 270e276.

Alston, P., & Quinn, G., (2017). The nature and scope of states parties' obligations under the international covenant on economic, social and cultural rights. In: *Economic, Social and Cultural Rights* (pp. 3–76). Routledge.

Alvensleben, R. V., (2002). Verbraucherbild—Verbraucherverantwortung—Verbrauchererziehung. Ziele und Fakten. *ZLR-Zeitschrift fur das Gesamte Lebensmittelrecht*, *2*, 139–150.

Amenta, V., Aschberger, K., Arena, M., Bouwmeester, H., Moniz, F. B., Brandhoff, P., Gottardo, S., et al., (2015). Regulatory aspects of nanotechnology in the agri/feed/food sector in EU and non-EU countries. *Regulatory Toxicology and Pharmacology*, *73*(1), 463–476.

Aung, M. M., & Chang, Y. S., (2014). Traceability in a food supply chain: Safety and quality perspectives. *Food Control*, *39*,172–184.

Azanza, M. P., Gatchalian, C. F., & Ortega, M. P., (2000). Food safety knowledge and practices of street food vendors in a Philippines university campus. *International Journal of Food Science and Nutrition*, *51*, 235–246.

Baert, K., Devlieghere, F., Jacxsens, L., & Debevere, J., (2005). Quality management systems. In: *The Food Industry in Safety in the Agri-Food Chain* (pp. 877–879). Wageningen: Academic Publishers.

Barbosa-Canovas, G. V., & Swanson, B., (2019). Nonthermal inactivation of endoproteases by pulsed electric field technology. In: *Pulsed Electric Fields in Food Processing: Fundamental Aspects and Applications* (p. 135).

Baş, M., Ersun, A. S., & Kıvanç, G., (2006). The evaluation of food hygiene knowledge, attitudes, and practices of food handlers' in food businesses in Turkey. *Food Control*, *17*(4), 317–322.

Beatty, J. R., (2006). The quality journey: Historical and workforce perspectives and the assessment of commitment to quality. *International Journal of Productivity and Reliability Management*, *1*, 139–167.

Bidwell, M., Briscoe, F., Fernandez-Mateo, I., & Sterling, A., (2013). The employment relationship and inequality: How and why changes in employment practices are reshaping rewards in organizations. *Academy of Management Annals*, *7*(1), 61–121.

Blanchfield, J. R., (2005). Good manufacturing practice (GMP) in the food industry. In: *Handbook of Hygiene Control in the Food Industry* (pp. 324–347). Woodhead Publishing.

Bonnell, A. D., (2012). *Quality Assurance in Seafood Processing: A Practical Guide* (pp. 23–54). Springer Science and Business Media.

Bultman, M. W., Fisher, F. S., & Pappagianis, D., (2013). The ecology of soil-borne human pathogens. In: *Essentials of Medical Geology* (pp. 477–504). Springer, Dordrecht.

Burgess, C. M., Arroyo, C., Bolton, D. J., Danaher, M., O'Connor, L., O'Mahony, P. J., & Tlustos, C., (2019). Food safety: A public health issue of growing importance. In: *Introduction to Human Nutrition* (Vol. 1, p. 388).

Chandrapala, J., Oliver, C., Kentish, S., & Ashokkumar, M., (2012). Ultrasonics in food processing – Food quality assurance and food safety. *Trends in Food Science and Technology*, *26*(2), 88–98.

Cheftel, J. C., (1995). High pressure, microbial inactivation and food preservation. *Food Science and Technology International*, *1*, 75–90.

Child, J., (2015). *Organization: Contemporary Principles and Practice*. USA: John Wiley and Sons.

CIES, (2008). *The Food Business Forum*. Available at: http://www.sqfi.com; http://www.ciesnet.com (accessed on 11 February 2022).

Codex Alimentarius, (2012). *About Codex*. Available at: http://www.codexalimentarius.org/about-codex/en/ (accessed on 11 February 2022).

Cohen, M. X., (2014). *Analyzing Neural Time Series Data: Theory and Practice*. USA: MIT Press.
Considine, K. M., Kelly, A. L., Fitzgerald, G. F., Hill, C., & Sleator, R. D., (2008). High-pressure processing–effects on microbial food safety and food quality. *FEMS Microbiology Letters*, *281*(1), 1–9.
Dark, G., McLean, D., & Weatherhead, S., (2015). *Kitchen Operations*. Australia: Pearson Higher Education.
Das, A. K., Nanda, P. K., Das, A., & Biswas, S., (2019). Hazards and safety issues of meat and meat products. In: *Food Safety and Human Health* (pp. 145–168). Cambridge: Academic Press.
De Oliveira, C. A. F., Da Cruz, A. G., Tavolaro, P., & Corassin, C. H., (2016). Food safety: Good manufacturing practices (GMP), sanitation standard operating procedures (SSOP), hazard analysis and critical control point (HACCP). In: *Antimicrobial Food Packaging* (pp. 129–139). Cambridge: Academic Press.
Demirci, M. N., Soon, J. M., & Wallace, C. A., (2016). Positioning food safety in halal assurance. *Food Control*, *70*, 257–270.
Di Rosa, A. R., Bressan, F., Leone, F., Falqui, L., & Chiofalo, V., (2019). Radiofrequency heating on food of animal origin: A review. *European Food Research and Technology*, 1–11.
Drabas, E., & Wojciechowski, A., (2006). IFS and BRC – International standards of food products quality assurance. *Kalejd. Miesn.*, *2*, 86–94.
EFSA Panel on Biological Hazards (BIOHAZ), (2013). Scientific opinion on the maintenance of the list of QPS biological agents intentionally added to food and feed. *EFSA Journal*, *11*(11), 3449.
Fabisz-Kijowska, A., & Kijowski, J., (2006). Food safety requirements by new international standard. *Mieso Wedl.*, *5*, 8–12.
FAO, Food and Agricultural Organization, (1997). *Hazard Analysis and Critical Control Point (HACCP) System and Guidelines for Its Application*. Available at: http://www.fao.org/docrep/005/y1579e/y1579e03.htm (accessed on 11 February 2022).
FAO, Food and Agricultural Organization. (2020). *The Importance of Food Quality and Safety for Developing Countries*. Available at: http://www.fao.org/docrep/meeting/x1845e.htm (accessed on 11 February 2022).
Five-Shaw, C., & Rowe, G., (1996). Public perceptions of everyday food hazards: A psychometric study. *Risk Analysis*, *16*(4), 487–500.
Frewer, L. J., & Miles, S., (2001). Risk perception, communication and trust. How might consumer confidence in the food supply be maintained?. In: Frewer, L. J., Risvik, E., & Schifferstein, H., (eds.), *Food, People and Society: A European Perspective of Consumers Food Choices* (pp. 401–413). Springer-Verlag: Berlin.
Ghasemi-Varnamkhasti, M., Mohtasebi, S. S., & Siadat, M., (2010). Biomimetic-based odor and taste sensing systems to food quality and safety characterization: An overview on basic principles and recent achievements. *Journal of Food Engineering*, *100*(3), 377–387.
Gomes-Neves, E., Cardoso, C. S., Araújo, A. C., & Da Costa, J. M. C., (2011). Meat handlers training in Portugal: A survey on knowledge and practice. *Food Control*, *22*(3, 4), 501–507.
Gomez, M., Oliete, B., Garcia-Alvarez, J., Ronda, F., & Salazar, J., (2008). Characterization of cake batters by ultrasound measurements. *Journal of Food Engineering*, *89*, 408–413.
Gorris, L. G. M., (2005). Food safety objective: An integral part of food chain management. *Food Control*, *16*, 801–809.
Granda, C., Moreira, R. G., & Tichy, S. E., (2004). Reduction of acrylamide formation in potato chips by low-temperature vacuum frying. *Journal of Food Science*, *69*(8), E405–E411.

Guo, Q., Sun, D. W., Cheng, J. H., & Han, Z., (2017). Microwave processing techniques and their recent applications in the food industry. *Trends in Food Science and Technology*, *67*, 236–247.

Habermas, J., (2015). *The Theory of Communicative Action: Lifeworld and Systems, A Critique of Functionalist Reason* (Vol. 2). USA: John Wiley and Sons.

Havinga, T., (2006). Private regulation of food safety by supermarkets. *Law and Policy*, *28*(4), 515–533.

He, T. T., Ung, C. O. L., Hu, H., & Wang, Y. T., (2015). Good manufacturing practice (GMP) regulation of herbal medicine in comparative research: China GMP, cGMP, WHO-GMP, PIC/S and EU-GMP. *European Journal of Integrative Medicine*, *7*(1), 55–66.

Henson, S., (2003). *The Economics of Food Safety in Developing Countries. ESA Working Paper No. 03–19*. Agricultural Development Economics Division, FAO. Available at: ftp://ftp.fao.org/docrep/fao/007/ae052e/ae052e00.pdf (accessed on 11 February 2022).

Hogan, E., Kelly, A. L., & Sun, D. W., (2005). High pressure processing of foods: An overview. *Emerging Technologies for Food Processing* (pp. 3–31). New York: Academic Press.

Hoogland, J. P., Jellema, A., & Jorgen, M. T. G., (1998). Quality assurance systems. In: Jongen, W. M. F., & Meulenberg, M. T. G., (eds.), *Innovative of Food Productions Systems: Product Quality and Consumer Acceptance*. Wageningen.

Hubbard, M. R., (2012). *Statistical Quality Control for the Food Industry*. Springer Science and Business Media.

Hurst, W. J., Finley, J. W., & deMan, J. M., (2018). Additives and contaminants. In: *Principles of Food Chemistry* (pp. 527–565). Switzerland: Springer.

IDFA, International Dairy Foundation. (2020). *Pasteurization*. Available at: https://www.idfa.org/pasteurization (accessed on 11 February 2022).

ISO 19011, (2002). *Guidelines for Quality and/or Environmental Management Systems Auditing*. Available at: https://www.academia.edu/36204649/Guidelines_for_quality_and_or_environmental_management_systems_auditing (accessed on 11 February 2022).

ISO 9000, (2000). *Quality Management Systems – Fundamentals and Vocabulary*. Available at: https://www.iso.org/files/live/sites/isoorg/files/archive/pdf/en/watermarksample.pdf (accessed on 11 February 2022).

ISO 9004, (2000). *Quality Management Systems – Guidelines for Performance Improvements*. Available at: https://www.iso.org/standard/28692.html (accessed on 11 February 2022).

Jeukendrup, A., & Gleeson, M., (2018). *Sport Nutrition* (pp. 18–30). Human Kinetics.

Johnson, R., (2014). *Food Fraud and Economically Motivated Adulteration of Food and Food Ingredients*. Congressional Research Service. Available at: https://fas.org/sgp/crs/misc/R43358.pdf (accessed on 11 February 2022).

Kaltenborn, M., Abdulai, A. G., Roelen, K., & Hague, S., (2017). *The Influence of Policy and Legal Frameworks on the Development of National Social Protection Systems*. IDS Working Paper.

Karipidis, P., Athanassiadis, K., Aggelopoulos, S., & Giompliakis, E., (2008). Factors affecting the adoption of quality assurance systems in small food enterprises. *Food Controls*, *20*(2), 93–98.

Karmacharya, J. B., (2014). Good manufacturing practices (GMP) for medicinal products. In: Basnet, P., (ed.), *Promising Pharmaceuticals* (pp. 101–148). Croatia: Intech.

Kaur, N., & Singh, A. K., (2016). Ohmic heating: Concept and applications – a review. *Critical Reviews in Food Science and Nutrition*, *56*(14), 2338–2351.

Kaushik, J. S., Narang, M., & Parakh, A., (2011). Fast food consumption in children. *Indian Paediatrics*, *48*, 97–101.

Kim, B., Kim, D., Cho, D., & Cho, S., (2003). Bactericidal effect of TiO_2 photocatalyst on selected foodborne pathogenic bacteria. *Chemosphere*, *52*, 277–281.

King, F. S., & Burgess, A., (1993). *Nutrition for Developing Countries* (2nd edn.), Oxford, UK: Oxford University Press.

Knaflewska, J., & Pospiech, E., (2007). Quality assurance systems in the food industry and health security of food. *Acta Scientiarum Polonorum Technologia Alimentaria*, *6*(2), 75–84.

Kotsanopoulos, K. V., & Arvanitoyannis, I. S., (2017). The role of auditing, food safety, and food quality standards in the food industry: A review. *Comprehensive Reviews in Food Science and Food Safety*, *16*(5), 760–775.

Kwak, Y. H., & Anbari, F. T., (2006). Benefits, obstacles, and future of six sigma approach. *Technovation*, *26*(5, 6), 708–715.

Marriott, N. G., Schilling, M. W., & Gravani, R. B., (2018). *Principles of Food Sanitation* (pp. 29–114). Springer.

McLoughlin, C., & Miura, T., (2017). *True Kaizen: Management's Role in Improving Work Climate and Culture*. Boca Raton: CRC Press.

Merna, T., & Al-Thani, F. F., (2011). *Corporate Risk Management*. USA: John Wiley and Sons.

Moerman, F., & Wouters, P. C., (2016). Emerging trends and methods in food factory design. In: *Innovation and Future Trends in Food Manufacturing and Supply Chain Technologies* (pp. 41–79). United Kingdom: Woodhead Publishing.

Mozafari, M. R., Flanagan, J., & Matia-Merino, L., (2006). Recent trends in the lipid-based nanoencapsulation of antioxidants and their role in foods. *The Journal of the Science of Food and Agriculture*, *86*, 2038–2045.

Nanda, V., (2016). *Quality Management System Handbook for Product Development Companies*. Boca Raton: CRC Press.

Nestle, M., (2013). *Food Politics: How the Food Industry Influences Nutrition and Health*. California: University of California Press.

Newslow, D., (2013). *Food Safety Management Programs: Applications, Best Practices and Compliance*. Boca Raton: CRC Press.

Nguz, K., (2007). Assessing food safety system in sub-Saharan countries: An overview of key issues. *Food Control*, *18*, 131–134.

Nielsen, S. S., (2010). *Food Analysis*. New York: Springer.

Northcutt, J. K., & Parisi, M. A., (2013). Major food laws and regulations. In: *Guide to US Food Laws and Regulations* (pp. 73–96). Chichester: Wiley Blackwell.

Okoye, J. I., & Oni, K., (2017). Promotion of indigenous food preservation and processing knowledge and the challenge of food security in Africa. *Journal of Food Security*, *5*(3), 75–87.

Orris, G. D., & Whitehead, A. J., (2000). Hazard analysis and critical control point (HACCP) as a part of an overall quality assurance system in international food trade. *Food Control*, *11*, 345–351.

Osaili, T. M., Jamous, D. O. A., Obeidat, B. A., Bawadi, H. A., Tayyem, R. F., & Subih, H. S., (2013). Food safety knowledge among food workers in restaurants in Jordan. *Food Control*, *31*(1), 145–150.

Paiva, C. L., (2013). Quality management: Important aspects for the food industry. *Food Industry* 191–218.

Patel, K. T., & Chotai, N. P., (2013). GMP requirements for "buildings and facilities" for API-comparison of schedule M, India and ICH guideline and approach for compliance to different regulatory expectations. *Pharma Science Monitor*, *4*(1).

Patterson, M., (1999). High-pressure treatment of foods. In: Robinson, R. K., Batt, C. A., & Patel, P. D., (eds.), *The Encyclopaedia of Food Microbiology* (pp. 1059–1065). New York: Academic Press.

Peri, C., (2006). The universe of food quality. *Food Quality and Preference*, *17*(1, 2), 3–8.

Peter, R., Mateja, A., & Mojca, J., (2013). Food chain safety management systems: The impact of good practices. In: *Advances in Food Process Engineering Research and Applications* (pp. 607–625). Boston, MA: Springer.

Press, D., (2003). *Guidelines for Failure Mode and Effects Analysis (FMEA) for Automotive, Aerospace, and General Manufacturing Industries*. Boca Raton: CRC Press.

Pretty, J., (2012). *The Pesticide Detox: Towards a More Sustainable Agriculture*. USA: Earthscan.

Prüss-Ustün, A., Vickers, C., Haefliger, P., & Bertollini, R., (2011). Knowns and unknowns on burden of disease due to chemicals: A systematic review. *Environ Health*, *10*(9), 1–15.

Raspor, M. A. M. J. P., (2010). Inconsistent terminology in food safety field: A permanent risk factor?. *Journal of Food and Nutrition Research*, *49*(4), 186–194.

Röhr, A., Lüddecke, K., Drusch, S., Müller, M. J., & Alvensleben, R. V., (2005). Food quality and safety – consumer perception and public health concern. *Food Control*, *16*(8), 649–655.

Safe Quality Food Institute, (2008). Available at: www.sqfi.com (accessed on 11 February 2022).

Schilling, E. G., & Neubauer, D. V., (2017). *Acceptance Sampling in Quality Control*. Boca Raton: CRC Press.

SCV, The Foundation for the Certification of Food Safety Systems, (2008). HACCP requirements. Available at: http://www.foodsafetymanagement.info (accessed on 11 February 2022).

Silano, M., & Silano, V., (2017). Food and feed chemical contaminants in the European Union: Regulatory, scientific, and technical issues concerning chemical contaminants occurrence, risk assessment, and risk management in the European Union. *Critical Reviews in Food Science and Nutrition*, *57*(10), 2162–2217.

Sonnemann, G., Tsang, M., & Schuhmacher, M., (2018). *Integrated Lifecycle and Risk Assessment for Industrial Processes and Products*. CRC Press.

Sun, Y., Wu, S., & Gong, G., (2019). Trends of research on polycyclic aromatic hydrocarbons in food: A 20-year perspective from 1997 to 2017. *Trends in Food Science and Technology*, *83*, 86–98.

Tjahjono, B., & Ball, P., (2010). Six sigma: A literature review. *International Journal of Lean Six Sigma*, *1*(3), 216–233.

Trafialek, J., & Kolozyn-Krajewska, D., (2011). Implementation of safety assurance system in food production in Poland. *Polish Journal of Food and Nutrition Sciences*, *61*(2), 115–124.

Tricker, R., (2019). *Quality Management Systems: A Practical Guide to Standards Implementation*. United Kingdom: Routledge.

Troedsson, H., (2009). *Forging Shared International Standards for Food Safety*. WHO speech. Available at: http://www.wpro.who.int/china/media_centre/speeches/speech_20090621.htm (accessed on 11 February 2022).

Turlejska, H., (2003). *Zasady GHP/GMP Oraz Systemu HACCP Jako Narzedzia Zapewnienia Bezpieczenstwa Zdrowotnego Ywnosci [GHP/GMP Principles and HACCP System as a Tolls for Food Safety Assurance]*. Poradnik Dla przedsiebiorcy.

UNICEF (United Nations Children's Fund). (2010). *Progress for Children: Achieving the MDGs with Equity*. New York, UNICEF.

Unnevehr, L. J., & Huirne, R. B. M., (2002). New approaches to food safety economics: Overview and new research directions. *Summary of Papers Presented at a Frontis Workshop on NEW Approaches to Food Safety Economics*. Wageningen, The Netherlands.

Van, D. S. M., Luning, P. A., Ziggers, G. W., & Jongen, W. M. F., (2003). Towards a conceptual model to measure effectiveness of food quality systems. *Trends in Food Science and Technology*, *14*, 424–431.

Wallace, C. A., Sperber, W. H., & Mortimore, S. E., (2018). *Food Safety for the 21st Century: Managing HACCP and Food Safety Throughout the Global Supply Chain*. USA: John Wiley and Sons.

Werkema, C., (2004). *Criando a Cultura Seis Sigma*. Nova Lima: Werkema Editora.

Wheatley, V. M., & Spink, J., (2013). Defining the public health threat of dietary supplement fraud. *Comprehensive Reviews in Food Science and Food Safety*, *12*(6), 599–613.

WHO, World Health Organization. (2019). *Food Safety*. Available at: https://www.who.int/news-room/fact-sheets/detail/food-safety (accessed on 11 February 2022).

WHO, (2004). Food safety in developing countries: Building capacity. *Weekly Epidemiology Record*, *79*(18), 173–180.

WHO, (2007). *Countries Urged to Be More Vigilant About Food Safety*. Available at: https://www.who.int/news/item/19-07-2007-countries-urged-to-be-more-vigilant-about-food-safety (accessed on 18 February 2022).

WHO, (2008). *WHO Initiative to Estimate the Global Burden of Foodborne Diseases*. A Summary Document. Geneva. Available at: https://apps.who.int/iris/bitstream/handle/10665/159844/9789241507950_eng.pdf (accessed on 11 February 2022).

WHO, (2010). *Basic Steps to Improve Safety of Street-Vended Food*. Available at: https://www.who.int/foodsafety/fs_management/No_03_StreetFood_Jun10_en.pdf (accessed on 18 February 2022).

WHO/FAO, (2011). *The International Food Safety Authorities Network (INFOSAN) Progress Report 2004–2010*. Available at: http://whqlibdoc.who.int/publications/2011/9789241501286_eng.pdf (accessed on 11 February 2022).

Wierzbicki, S., (2003). AQAP w Sokołów S.A. [AQAP at Sokołów S.A.] *Bezp. Hig.ywn.*, *6*(1), 3.

Wilna, H. O., & Abdulkadir, A. E., (2016). Food quality and food safety. In: Temple, N. J., & Stey, N., (eds.), *Community Nutrition for Developing Countries*. Canada: Athabasca University Press and UNISA Press.

Winarno, F. G., & Allain, A., (1991). Street foods in developing countries: Lessons from Asia. *Journal of Food, Nutrition and Agriculture*, *1*(1), 11–18.

Xu, Y., & Fisher, G. J., (2005). Ultraviolet (UV) light irradiation induced signal transduction in skin photoaging. *Journal of Dermatological Science*, *1*(2), S1–S8.

Yang, W., Li, D., & Mugambi, A., (2017). Spoilage microorganisms in cereal products. *Food Spoilage Microorganisms: Ecology and Control*. Boca Raton: CRC Press.

Yu, H., Huang, Y., & Huang, Q., (2009). Synthesis and characterization of novel antimicrobial emulsifiers from polylysine. The *Journal of Agricultural and Food Chemistry*, *58*, 1290–1295.

PART III

Sequencing Techniques and Bioinformatic Approaches in Food Safety, Quality, and Productivity

CHAPTER 12

Molecular Techniques in Food Safety and Quality Assurance

LAKSHMANAN NANDHINI, MURALIDHARAN PRIYADHARSHEE, RADHAKRISHNAN PREETHA, and DEVARAJAN THANGADURAI

ABSTRACT

Food safety is still a concern throughout the world despite the various efforts taken by the people involved in the food chain. Microbial hazards contribute to most food safety concerns since the microbes can enter at any stage of the food chain, cause foodborne illness and produce toxins that may have serious consequences on humans. Other than microbes, chemicals used during the processing stage, heavy metal and pesticide residue contamination play a significant part in the food safety concerns. Food safety and management system are established based on hazard analysis critical control points (HACCP), good manufacturing practices (GMPs), good hygiene practices (GHP) to manage and maintain food safety throughout the food chain. A risk analysis framework used to predict the source of the problem can prevent the issues that might arise during the process. Genetically modified (GM) foods and nanotechnology in the food industry have also become a growing concern in the food safety aspect due to insufficient toxicological studies. Proteomics, genomics, and bioinformatics can identify food quality traits, adulterants, allergens, pathogens, and xenobiotics present in the food product. This chapter discusses food safety/quality, especially microbial food safety, and techniques involved in food safety in detail.

Food Microbial and Molecular Biology: From Fundamentals to Applications. Saher Islam, Devarajan Thangadurai, Jeyabalan Sangeetha, Zaira Zaman Chowdhury (Eds.)

12.1 INTRODUCTION

Food safety and quality (FSQ) have become a priority in the field of research over the past few years (Weng and Neethirajan, 2017). Understanding the perception of consumers is a vital part that constitutes food quality. The knowledge of food contaminants and their effects on an individual can perform a significant part in food safety. Food quality and safety are properties of the products that cannot be verified by ordinary people and thus classified as credence attributes. Consumers decide upon the quality and safety of the product either by essential characteristics (appearance of food commodity) or by external attributes (brand name). Therefore, further clarification is required for consumers to confirm the safety and quality of food commodities. Growing consumers' expectations and the urge to meet the standards of regulatory bodies leads to the effectuation of quality assurance (QA) and safety management systems (Grunert, 2005; Sikora and Strada, 2005). This chapter deals with various attributes of FSQ, such as traceability, hazards, omics technology, and regulation of food standards.

12.2 FOOD QUALITY

Food quality determines the acceptability of the product to the consumers (FAO/WHO, 2003). It is essential for building a strong relationship based on trust between producers and consumers (Scalco et al., 2020). The quality criteria vary from one food industry to another; however, the basic parameters are the same. It is checked and controlled by the QA people. Quality management system, including quality control systems, plays a vital role in maintaining product quality. ISO 9000 provides information and process regarding these systems. Raw material inspection, product labeling, product traceability all come under food quality. Food quality is not only limited to sensory attributes, consumer perception, or overall liking (Moskowitz, 1995). Nowadays, consumers' choice of product changes from low quality to high-quality products and are willing to pay high for the same. Therefore, food quality has become an essential criterion for the industries to make market opportunities.

12.3 FOOD SAFETY

With the rising requirement for commercialized food, the safety of food has become a concern for food industries, administrative departments, and society. The term 'Food Safety' and 'Food Security' are closely linked and often misinterpreted. Food security ensures that all the people from different background can always access adequate, harmless, and healthy foodstuffs to avoid malnutrition (Jia and Jukes, 2013). The food safety stands for assurance that the consumption of the product will not lead to any ill effect on the individuals. The government should conduct a campaign on food safety to create greater awareness about the same (Jia and Jukes, 2013). Based on technical and commercial research, regulations for food safety were scrutinized regularly (Henson and Caswell, 1999). Education on food safety should start from school, which will significantly impact social awareness (Hoffmann et al., 2019). Food safety became a primary concern for the food industry because of more stringent regulations and increase awareness of the consumer.

12.4 TRACEABILITY

Traceability is an approach used to assure the quality of the food production and products. Ability to trace back the product from the final system ascertains both FSQ (Karippacheril et al., 2017). Traceability helps in establishing the legitimacy of the foodstuffs, and verify the demands given by the manufactures about the products are real. The product label should contain the lot number or batch number, which helps in trace backing the product until the raw material (Girardon et al., 2019). Traceability is a useful tool to authorize the reliability of food and food products. Barcode scanning is the commercial method for automated entry of inventory activities. The cost prohibits the widespread usage of RFID (Radio Frequency Identification) for the same purpose (Van Rijswijk and Frewer, 2008; Ramaswamy et al., 2019).

12.4.1 RELATIONSHIP BETWEEN TRACEABILITY, QUALITY, AND SAFETY

Traceability ensures food safety and brand protection. The food product recall system explains the need for food traceability in case of an outbreak of an emergency. If the claim is genuine, products are trace backed, checked, and recalled. Thus, all the food handlers should keep an account of sources from where they get the food and customers who consume the food products (one step forward, one step backward traceability), thus ensuring the safety and quality of the food product (King et al., 2017).

12.5 FOOD ADULTERATION AND AUTHENTICATION

Adulterants are substances imparted to a food product during processing that makes them harmful, or inferior in quality. Whereas food contaminants are the substances that are present in the food and not necessarily added to the products (Attrey, 2017). It may be due to microbes enter in food during food processing or toxic chemicals/ heavy metals enter food by mistake or even metal pieces from the machine. Food fraud is the study of adulteration done for economic benefits (Banerjee et al., 2017). The process of analyzing the product for its legal claim, whether it is contaminated or not, is called food authentication.

12.6 FOOD SAFETY – HAZARDS

Hazard is a factor that maybe present in the food or food products and could damage or cause injury to the person who consumes it. The three broad classifications of risks are, namely: biological, chemical, and physical hazards (Wester, 2018) (Figure 12.1).

FIGURE 12.1 Types of hazards.

12.6.1 PHYSICAL HAZARDS

The hazards that usually occur due to the potential availability of foreign materials in food and food products are called physical hazards (Raj et al., 2011; Potter and Hotchkiss, 2012). These hazards include foreign substances such as metal fragments, wooden splinters, and stones that are not generally found in food but introduced during primary processing, harvesting, and may cause health risks to consumers. These hazards cause varying degrees of injury and illness, and deaths may occur rarely. The physical contaminants usually involved are metal pieces, thorns, woods, glass, mesh, wire, rust, dirt, insects, plastics, natural materials and so on (Potter and Hotchkiss, 2012). Foreign material items fall into these classes:

- Those that are pointed and may lead to damage
- Those that are rigid and may induce tooth damage
- Those that can cause suffocation or congestion
- Those that transport microbiological hazards into food products

GMP and accurate study or analysis of risks decrease hazards in foods that have undergone processing. The latter approach is generally called "hazard analysis and critical control point" or HACCP (NACMCF, 1992).

12.6.1.1 METHODS TO CONTROL HAZARDS

- Monitoring of CCP (critical control points)
- Use of mesh or screens to remove insects and flies
- Following Good Hygiene Practices (hairnet, gloves)
- Use of metal detectors and magnets (Raj et al., 2011)

12.6.2 CHEMICAL HAZARDS

Adulteration of food products with chemicals can occur during any level of manufacturing, starting from the storage of raw ingredients to utilization of finished goods (Bouzembrak and Marvin, 2019). Consumption of these chemical hazards (toxic compounds) by animals or humans can cause adverse effects. The dosage of exposure and efficiency of the chemical are the factors that determine lethality. The result can be acute (short term) such as allergic responses or chronic (long term) due to bioaccumulation of compounds which have ill effects on humans (Bouzembrak and Marvin, 2019). The climatic changes, agricultural conditions, and economic factors have a more significant impact on chemical hazards like pesticides and mycotoxins (Bouzembrak and Marvin, 2019).

Various chemical agents that are potentially hazardous to health include insecticides, pesticides, fungicides, lubricants, paints, food additives (coloring agents, preservatives, antioxidants, emulsifiers) and environmental pollutants (Schmidt and Rodrick, 2003). The chemical contaminants are broadly classified into three classes and are:

- Those that are of natural origin, including plants, microorganisms, and animals
- Foods may become contaminated with toxicants during the production, processing, or storage levels; these toxicants include lead, PCBs (Polychlorinated biphenyl), dioxins, furans, PAHs (Polycyclic aromatic hydrocarbon), and pyrolysis products, and
- Substances intentionally added to foodstuffs that carry out the desired action. These include food additives that act as a preservative or antioxidant (Singh et al., 2019)

12.6.3 BIOLOGICAL HAZARDS (BIOHAZ)

Biological hazards (BIOHAZ) are majorly caused due to microbial contamination by various organisms such as bacteria, molds, insects, and fragments. The entry and the growth of the pathogenic microorganisms in the food products are the root causes of BIOHAZ that can cause serious illnesses. The source of natural hazards includes lack of personal hygiene, environmental microflora, and contaminated raw materials/water (Taulo et al., 2008). *Escherichia coli*, *Staphylococcus aureus*, and *Clostridium botulinum* are among the dominant bacteria that cause hazards. Molds produce poisonous mycotoxins, for example, aflatoxins (AFs), citrinin, ochratoxin, patulin, and trichothecenes. However, this mycotoxin production may vary according to the climatic conditions and temperature (Wu et al., 2017). The BIOHAZ are classified into two, macrobiological or microbiological hazards.

12.6.3.1 MACROBIOLOGICAL HAZARDS

Macrobiological hazards include the presence of insects or flies in foodstuffs. If the insects present in the food are poisonous, it may harbor pathogenic microorganisms and cause harmful effects. Usually, these microbiological issues are considered as external or physical substances and not as BIOHAZ (Raj et al., 2011).

12.6.3.2 MICROBIOLOGICAL HAZARDS

Microbes that cause diseases or infections will affect the consumers directly or indirectly:

- Microbes such as bacteria, viruses, and parasites/protozoa directly attack or infect body tissues
- The development of toxins (pre-formation of poisons in the body) by bacteria and mold (fungi) can affect the body indirectly (Singh et al., 2019)

12.7 FOOD ALLERGENS

Some of the proteins present in the food may be allergic to certain people up to an extent. The primary eight sources of the allergen protein are peanut, wheat, tree nuts, egg, soybean, fish, milk, and shellfish. These allergens may even have adverse reactions in terms of morbidity and fatality based upon the exposure and intensity of responses in the individual (Barlow et al., 2015).

12.8 HAZARD ANALYSIS

HACCP is a vital tool employed to achieve a superior level of consistency under the safety and quality of food and food commodities. It is also one of the most efficient methods with the lowest possible risk (Sikora and Strada, 2005). It is an organized and science-based method to analyze, assess, and regulate food hazards. Prerequisite programs (PRPs) like GMP, GAP (Good Agricultural Practices), and GHP (Good Hygienic Practices) are the primary support system for an efficient HACCP plan (Schmidt and Rodrick, 2003). Careful assessment and monitoring of each step involved in food manufacturing, preparation, and storage. Regulation of the critical levels is important for the safety of the food commodities through the implementation of the food safety programs. HACCP is a useful tool to produce safe food for consumers by emphasizing risk determination and control (Raj et al., 2011; Potter and Hotchkiss, 2012; Camino et al., 2017). HACCP systems usually involve 12 stages, of which 5 are necessary steps followed by 7 HACCP principles. Five preliminary steps of HACCP are as follows (Pierson, 2012):

- **Step 1**: Gather people and make a HACCP group
- **Step 2**: Describe the product and its properties
- **Step 3**: Describe the proposed use of the product to consumers
- **Step 4**: Create a flow chart to depict the process
- **Step 5**: Check the flow chart on site

Seven principles of HACCP are (Mortimore and Wallace, 2013):

- Carry out hazard analysis
- Ascertain the CCPs
- Lay down the critical limits
- Constitute the monitoring procedure
- Set up the corrective actions
- Frame verification procedures
- Follow record-keeping and documentation procedures (NACMCF, 1992)

12.9 RISK ANALYSIS

Analysis of risk is a method for evaluating the possibility of existence and severity of known or potential harm to humans that occurs due to the hazards caused by food. Implement risk analyzes during every stage of operation or phase in the system. Do the risk analysis to satisfy the requirements of regulatory standards, and a risk analysis can be either quantitative or qualitative (Lammerding, 1997). Execute risk analysis to:

- Provide a clear picture of risk
- Analyze various opportunities and results
- Determine causes and various elements corresponding to risk
- Expose the consequences of different measures of risks

There is fair discrimination between risk analysis, risk estimation, and risk interpretation: Risk assessment is a sum of risk evaluation and risk analysis. The basic framework of risk analysis process involves the following three main elements:

- Outlining the nature of risk
- Evaluation of risk, and
- Treatment of risk (Aven, 2015)

$$\text{Risk} = \text{Hazard} \times \text{Exposure} \times \text{Vulnerability (Pagliacci, 2019)}$$

12.9.1 RISK ASSESSMENT

Risk assessment is the estimation of the possibility of the uncertainty of the existence of detrimental events. It implements the groundwork for making decisions and evaluates the probability and impact of harmful health effects (Schmidt and Rodrick, 2003). A necessary organized procedure for assessing risk usually includes interpretation, forecasting, and unpredictability. The basic framework of risk assessment involves four elements: identification of hazard, assessment of exposure, the response of dose, and risk identification (Schmidt and Rodrick, 2003). Identification of hazard defines the nature of the issue that includes the determination of biological agents usually found in food responsible for causing harmful effects or illness. Exposure assessment involves the evaluation of approximate measurement of expected input of any biological agent inside the foodstuff. Dose-response assessment usually refers to the relation between the number of microbes consumed and the likelihood of prevalence of detrimental effects due to consumption. Lastly, the characterization of risk summarizes all the reports giving an approximation of the possibility of events and severity of harmful effects in a community (Lammerding, 1997).

12.9.2 RISK MANAGEMENT

Risk management can relate to activities and conditions carried out to explore opportunities as well as minimize losses, accidents, and hazards. To manage risks, the risk manager examines all the results from risk assessment and other causes that include economic, commercial, and industrial conditions to handle risks. Risk managers usually do decision making based on the safety of consumers and preferences of the community. Conduction of neither risk management nor risk assessment occurs in a vacuum. Establishment of regulatory standards to reflect the degree of protection of a prescribed hazard is mandatory (Aven, 2015). Policy strategies that are required to control one foodborne hazard might generate other new hazards (Aven, 2015). The management process usually includes bilateral communication between customers and the parties involved in each aspect of the process. Thus, the risk management is classified into three elements such as strategic risk, financial risk, and operational risk (Aven, 2015).

12.9.3 RISK COMMUNICATION

Risk communication refers to operations involving the interaction of stakeholders (all the interested parties, consumers, producers, scientists, industrialists, and government professionals) about analyzing risk and its management (Schmidt and Rodrick, 2003). Risk assessors explain the nontechnical terms in the data, models, and results of the risk analysis. The risk managers give numerous risk management strategies based on risk analysis (risk assessment) (Schmidt and Rodrick, 2003).

12.10 MICROBIAL FOODBORNE DISEASES (FBDs)

Foodborne diseases continue to be a global challenge to public health because of various reasons such as comprehensive handling or transportation of foods globally, leading to the transmission of pathogenic microbes into new terrestrial regions (Cliver and Riemann, 2002). More than 200 identified diseases have been transmitted by chemical, physical, and biological agents to the human body through food (Gupta, 2017). Biological agents are known to be the most infectious and essential in causing a higher number of foodborne diseases. Foodborne diseases include a range of analytical and etiologic surroundings and refer to a subset of enteric disorders (Cliver and Riemann, 2002). Food is a means of transmitting microbes, growing actively, may produce toxins and acting as a vehicle in the communication of diseases. The major clinical manifestations that usually occur due to foodborne conditions are as follows (Gupta, 2017):

- Vomiting and nausea
- Non-inflammatory diarrhea
- Inflammatory diarrhea
- Neurological signs

The incubation period between the time of ingestion and the outbreak of symptoms differs for different organism. The incubation time is usually one or two days when the infecting agent multiplies in the gut by ingestion. But in contrast, the incubation period is often only a few hours when toxins are already present in the food (Cliver and Riemann, 2002; Gupta, 2017).

12.10.1 FOOD INFECTION

The microbes usually ingested with food cause foodborne infectious diseases. The dosage required to produce infection varies with the nature and category of microorganisms. These kinds of diseases caused by microbes present in food are called food infection (Mohammad et al., 2018).

12.10.1.1 PATHOGENS

The dominant microbes that usually produce infection include bacteria, viruses, protozoa, and parasites. *Salmonella* sp., *Campylobacter* sp., and *Clostridium perfringens* are the most common bacteria involved. Viruses infrequently engage as causative agents such as Norwalk gastroenteritis virus and hepatitis A virus (HAV) through foods (Banwart, 2012).

12.10.1.2 PATHOGENESIS

1. **Site of Reaction**: Ingestion of microbes while eating the food may pass through the Gastrointestinal tract (GI) and cause associated syndromes before they reach other tissues (Banwart, 2012). Therefore, depending on various factors such as dosage, type of organism, and other related causes, the microbe will attack any organ or tissue in the body.
2. **Symptoms**: The organisms usually reproduce in the GI tract or infected tissues. Thus, there is a delay in the outbreak of manifestations by eight hours to several days that depends on the initial dosage and nature of pathogenic microorganisms. These attributes help to characterize food infections from intoxication (Cliver et al., 2011).
3. **Characteristics of Pathogenic Microbes (Cliver et al., 2011)**: (i) They should be viable *in vivo*; (ii) they should permeate through the mucosal barrier and reach the site of action; and (iii) should be able to multiply rapidly.
4. **Non-Invasive Infections**: Non-invasive bacteria cannot pass through the cells or tissues and are impotent to multiply, but they can cause illness. These infectious bacteria should be capable of sustaining and reproducing in the intestinal lumen. The most common non-invasive bacteria include *Vibrio cholerae* and enterotoxigenic *Escherichia coli* that divide and produce enterotoxins by adhering to the internal surface (Cliver et al., 2011).

12.10.2 FOOD INTOXICATION

Intoxication is a type of foodborne disease induced by an unsafe dosage of a toxicant present in food which is usually produced by bacterial pathogens. The agents causing food intoxications may be viable or non-viable, and their toxicity or detrimental reactions depend upon the dosage and vulnerability of toxins (Cliver et al., 2011).

12.10.2.1 TYPES OF TOXICANTS

The toxicants can be basically of two types, natural and synthetic. Natural toxicants are the necessary components of diverse animals, plants, and microbial species. Natural toxicants that commonly occurring in food are emerge from molds or bacteria which contaminate food (Cliver et al., 2011). Synthetic toxicants are human-made chemicals that intentionally or unintentionally enter foodstuffs. The most common examples are food additives and agricultural pesticides. Synthetic toxicants produce adverse effects at a higher dosage, which arises because of irregular management (Cliver et al., 2011). Foodborne toxicants can be further differentiated based on their mechanism. The most common examples are neurotoxin and aflatoxin.

12.10.2.2 PATHOGENESIS

There are various sites of action of toxicants, where some affect many tissues or organs, while few are tissue-specific. Various symptoms are produced because of food intoxications depends upon the dosage of toxicant, mechanism of action, organs or tissues targeted, age, and sex (Cliver et al., 2011). Exposing toxins to humans cause illness that can arise within a few minutes of ingestion (acute), or a few hours after ingestion (chronic) that requires a long time to develop. The significant distinction between foodborne infection and intoxication is that symptoms of acute poisoning usually occur within a few minutes to hours after the ingestion of contaminated food. In contrast, food infections take a long time of eight hours to several days (Cliver et al., 2011). Characteristics of food toxins are:

- Should retain their toxicity over all the stages of food processing until they reach the site of action
- Should be heat stable

- Should withstand digestive processes
- It should permeate through the intestinal barrier to reach the target tissue or organ

Toxins are entrapped in fibers and eliminate through faces. Another means of reducing noxious substance is by vomiting. Harmful effects are prone to occur when the dosage of chemicals surpasses the body's potential to eliminate toxins (Cliver et al., 2011).

12.11 TECHNOLOGIES IN FOOD SAFETY AND QUALITY (FSQ) MANAGEMENT

12.11.1 PROTEOMICS

Proteomics is the analysis of protein's structure, function, and the modification that occurs when subjected to any changes. Proteomics is a tool majorly used to assure the quality and safety of protein rich food sources such as dairy products, marine/aquaculture products, protein rich, meat, and poultry (Piras et al., 2016). Proteomics is also used in nutritional studies. The amount of essential amino acid present in the food and the biological value related to the protein in the food is studied using proteomics tools. For example, meat, egg, fish, milk, and other dairy products contain protein of higher biological value (Ortea et al., 2016). Bioactive peptides are peptides with some biological function and peptidomics is the study of endogenous peptides. Milk has bioactive compounds as well as their precursors. Proteomics has the upper hand as it can provide vital information that a genome or transcriptome cannot offer and helps to understand metabolome. Changes in protein conformation while processing may affect the bioavailability and enzyme activity. Identification of cellular biomarkers leads to the early prediction of the pathogens in the food (Giacometti and Josić, 2013). Biomarkers specific to foodborne pathogens (FBP) create a relationship between the proteome and the pathogenicity, and detection of proteome helps to prevent food spoilage.

12.11.1.1 PROTEOMICS TOOLS IN FOOD ANALYSIS

The authenticity, traceability, processing methods, quality, and safety control of the foodstuffs are analyzed using proteomics technology. The study of glycoprotein allergens present in the food is called glycomics (Andjelković and Josić, 2018). The importance of phosphorylation protein translational modifications (PTM) leads to another discipline called phosphoproteomics (Andjelković and Josić, 2018). They help in understanding the cellular response to various stimuli. Detection and assessment of PTMs became more important as the study of enzymatic and non-enzymatic PTM after processing and storage revealed that they could affect the food quality. Mass spectroscopy plays a vital role in proteomics study due to its high specificity, which is an essential feature required as the protein structure is highly complex, and the amount present in the food may be less to detect (Andjelković and Josić, 2018). Information obtained from MS may complement the arrangement derived from nuclear magnetic resonance (NMR), X-ray crystallography, electron microscopy, and other spectroscopic methods (Andjelković and Josić, 2018). The quantification of toxic substances may not necessarily determine the safety of the protein, as it can be active or inactive. Hence active based protein profiling (ABPP) can be done by MS, which provides information on the active component. Sample preparation, sampling, sample handling also contributes to the accuracy of the result. Data-dependent, targeted, and data-independent acquisitions are the three means of acquiring spectra from MS (Andjelković and Josić, 2018). Quantitative proteomics related to MS uses label-free strategy, label-based strategy, and isotope labeling. Proteomics is not only used for biological hazard detection but also for the exposure of xenobiotic contaminants. Moreover, proteomics is used to study food quality traits such as taste, flavor, consistency, and Nutritional properties. It is also used as a tool in food safety-related studies such as pathogens detection, allergens detection, and product adulteration (Piras et al., 2016).

12.11.2 BIOINFORMATICS

Bioinformatics is an interdisciplinary field consisting of biology, maths, and computation. It aids in the development of omics technology. It serves as the screening technique as it is a low cost and high-throughput technique (Wang et al., 2019). NCBI, UniProt, and BLAST are some of the database providers for the amino acid sequence (Tu et al., 2018). The software used for proteome analysis can be of two types: open access and the commercial. Open source helps in understanding the software with limited support, whereas the commercial ones will have technical support. Oxford Nanopore, Illumina, Thermo Fisher Scientific, PacBio are some of the sources used for analyzing the proteome (Mataragas et al., 2018).

12.14.1 EFFECT OF ENVIRONMENTAL POLLUTION ON FOOD SAFETY

Globally, the two significant issues are food safety and environmental pollution. Food crops can be polluted due to the accumulation of various heavy metals such as Cadmium, Mercury, Arsenic, Lead, and so on in the soil (soil pollution) that can become an environmental concern globally. These problems arise due to unrestrained industrialization, civilization, and exhaustive farming systems, and can have adverse effects on quality of soil, safety of food, and public well-being (Khan et al., 2017).

Water pollution poses another challenging situation on the dangers it causes to the safety of food and human health. Shortage of water (droughts), overdosage of pesticides, and chemical contaminants are the driving forces that affect food safety and cause health effects. Thereby residues of pesticides and heavy metals accumulation cause extensive threats to people. Thus, the main impacts of water and soil pollution on the safety aspects of food are the development of foodborne and carcinogenic diseases in humans (Lu et al., 2015).

12.15 CONCLUSION

This chapter outlined the attention given to the quality and safety of food worldwide owing to the increasing occurrence of foodborne diseases and numerous outbreaks and recalls at unacceptable levels. Consumers question the quality and safety of food products as they are concerned about the hazards (physical, chemical, and biological) and the risk related with the production of food that requires rigorous management at every phase of the food chain. The utilization of risk analysis, HACCP, and other food safety operations such as GMP and GHP from plant to the plate is an efficient means to limit and lower the adulteration of food. In addition to this, rapidly developing technologies such as WGS, NGS, Nanopore sequencing, and DNA metabarcoding applied in quality and safety systems of food are discussed. Over the last few decades, private standards have become an essential factor for the safety of food throughout the worldwide food supply chains. Adulteration in food products is taking place either for profit or to fake the quality of the product. Hence this chapter takes an opportunity to spread knowledge about food quality, food safety, analysis methods, and other topics related to food safety.

12.11.3 GENOMICS

Majorly used and rapidly revolutionizing tools used for the management of quality and safety of food are genomics tools and precisely whole-genome sequencing (WGS) that detect outbreaks worldwide. Over the past few years, omic approaches have also become essential tools in food safety and management apart from genomics and WGS due to the improved analytical methods and find its application in various fields (Stasiewicz et al., 2015; Prakash et al., 2020).

12.11.3.1 NEXT-GENERATION SEQUENCING (NGS)

Next-generation sequencing (NGS) is used to determine both metagenomics as well as whole-genome sequencing of a unique culture (Haynes et al., 2019). Metagenomics is the genome sequencing of multiple organisms present in the same sample (Jagadeesan et al., 2019). The study of microorganisms within the sample is less biased. Metagenomics, along with metaproteomics, has an enormous potential to assess FSQ. The depiction of microbial standards uses genome sequences of various food chains collected by CSFSC. A comparison of the results obtained from metagenomic sequencing with the criteria indicates the origin of the contamination. The incomplete microbial standard database and its inaccessible state act as a hindrance to this technique. Ion PGM, MiniSeq, MiSeq, Ion Proton, Ion torrent, NextSeq 500, HiSeq 2500, HiSeq 3000, HiSeq 4000, PaciBio RSII, PaciBio Sequel, MinION, GridION X5, PromethION are some of the sequencers used for metagenomic sequencing (Liu et al., 2012). GenBank, INDSC, DDBJ, EMBL-EBI, SRA are some of the database providers. They keep upgrading their database system as it is not complete yet. NGS can sequence complex genome efficiently than WGS. It can also detect non-culturable pathogens (Haynes et al., 2019).

12.11.3.2 WHOLE GENOME SEQUENCING (WGS)

Molecular typing methods have broad usage in recent times for various food safety applications (EFSA, 2015). WGS is one such technology that plays a vital part in the fields of the safety of food and security by providing genomic information to the food industry. WGS can rapidly detect a broad range of FBP, supervise the supply of raw materials, risk assessment, outbreak investigation, and sanitization regulation (King et al., 2017). This tool helps to ascertain the complete DNA sequence of an organism's genome in a distinct test. It is a significant test that is rapid, accurate, reduction, and prohibiting food safety-related problems. To assure a great degree of security for consumer health, continuous advancements in the understanding of molecular characteristics and genetics have led to the use of WGS methods. The potential benefits of WGS methods are that it provides more specific or accurate information on pathogens than other methods (King et al., 2017). WGS analyzes are less costly and provide results within a few days with higher sensitivity for agricultural sectors. This method does not benefit metagenomics but used for assessing the source ascription of foodborne illness (FAO, 2016).

12.11.3.3 NANOPORE SEQUENCING

Food duplicity is one of the major problems that has been increasing nowadays as it has a significant impact and effect on consumer's confidence. Food and food products have been contaminated purposely for economic profits, causing enormous consequences for the food industry. Developing rapidly transforming technologies such as MinION DNA sequencer from Oxford nanopore technologies (ONT) reduce the hazards related to food safety and detect the contamination in food. This technology can produce input within minutes in real-time (Voorhuijzen-Harink et al., 2019). In nanopore sequencing technology, the target DNA is placed on a thin membrane of the nanopore, and current is passed across the nanopore. The analysis of food samples using the metagenomics technology helps in determining and isolating bacterial pathogens that cause foodborne disease in a single system (Urban et al., 2020).

12.11.3.4 DNA METABARCODING

Sequencing of short section of DNA and comparing with the standardized references to find out the adulterations in food called DNA metabarcoding. In the case of complex samples, NGS of DNA barcodes can be done to detect the multiple species simultaneously. After NGS, they are subjected to filtering and clustering analysis to improve the throughput. Metabarcoding can also be applied to the degraded DNA containing samples. US Food and Drug Administration (FDA) has replaced the protein isoelectric focusing technique with this molecular technique for fish and fish products due to their efficiency in time (Staats et al., 2016).

12.12 FOOD SAFETY OPERATIONS

12.12.1 GMP

Good manufacturing practice (GMP) and standard operating procedures (SOP) constitute the foundation for HACCP implementation. GMP is not only applicable to the food industry but also drugs and other medical equipment industries. It concerns with the elision and riddance of unwanted foreign matter and inhibition and destruction of undesirable microorganisms (De Oliveira et al., 2016). GMP plays a vital role in food safety and aims to reduce foodborne disease, increasing the consumer acceptance level of the product, and meeting the quality standards. It covers most aspects of the industry such as:

- Premises (personnel and material flow, walls, and ceiling, ventilation, utilities, water system)
- Equipment (maintenance, equipment ID, logbooks, calibration, and cleaning)
- Personnel (qualification, job responsibilities, training, health, and hygiene)
- Quality assurance
- Sanitation programs
- Operations
- Transportation
- Storage
- Distribution
- Specifications
- Stabilities
- Samples
- Records
- Recall (Blanchfield, 2005)

PDCA (Plan, Do, Check, Act) cycle, which is used for continuous improvement, deals with diagnosing and correcting the issues and evaluating the corrective action. This PDCA cycle should be a continuous process for improving product safety and quality (De Oliveira et al., 2016).

12.12.2 GHP

Good Hygiene Practice (GHP) is an essential requirement of a food industry that enhances the safety and quality of the products (Ababio et al., 2016).

GHP is to prevent continual personnel unhygienic conditions and unhygienic environment conditions. Hence, it protects from the transmission of other foodborne diseases. Food handlers consider the carriers of pathogens which can survive on their mouth, skin, hands, bruises, hair, and clothes, and other surfaces (Margas and Holah, 2014). Hence the following necessary precautions should be taken:

- Washing hands (before and after eating, after using washroom)
- Sanitizing hands (before entering the food process area)
- Covering mouth with a mask
- Covering hair with a hairnet
- Wearing aprons inside the premises
- Wearing gloves
- Prohibition of smoking and eating inside the processing area
- Prohibition of workers with a disease such as flu inside the processing area
- Bruises should be adequately covered
- Medical examination of the workers
- Cleaning and sanitization of the equipment and floors regularly
- Training on GHP

Carrying out the above-mentioned processes is easy in large scale industries, but in the case of small and medium businesses, they have practical problems such as lack of qualification, time, training (Wambui et al., 2017). Hence these industries should be trained on GHP and its impact on food safety.

12.12.3 TRAINING

Training of workers and other employees in the industry or other food business is necessary for the maintenance of the safety and quality of the product (Seaman, 2010). The ethnicities, different cultures, and educational levels pose a challenge to the training. The training program should be chosen based on training needs (Reynolds and Dolasinski, 2019). For training to be practical, it should address the language barrier (Seaman, 2010). This increases the skills and the cultural behavior of the food handlers, thereby leading to a positive outcome. A review of the training should be conducted as there may be a slight change in it from time to time. The analysis should focus on the following essential factors. When food safety is concerned, the description of the mode of training conducted and the report on the methods of training evaluation is very important (Reynolds and Dolasinski, 2019).

The food safety topics to be discussed depend upon the guideline standards and regulations. The training can be in the form of lecture-style, videos, posters, photographs, brochures, and info sheets. The assessment can be done based upon pre/post questionnaire or post-intervention practical testing such as microbial/biochemical analysis (Gruenfeldova et al., 2019). During training pro gramme the actual skill and the workers' reaction to the training should be recorded (Seaman, 2010). Most people prefer onsite face-to-face training when compared to online training (Soon, 2020). The training should also be able to give in the required language as it is the most common factor that prevents small and medium businesses from being trained. All the food handlers should provide the basic training program.

12.12.4 INSPECTION

The demand for inspection has increased due to the considerations of adulteration, food safety, and food quality. Inspection at every part of the food industry, from raw materials to finished goods, storage conditions, and transporting vehicles is essential (Ali et al., 2020). The raw material inspection will be done based upon the standards prescribed by the regulation for the ingredients. Proximate analysis such as analysis of color, taste, smell, flavor, and nutritional value determines the quality of the product. Methods such as the electronic nose, have been developed for testing the same. An internal inspection is conducted regularly to check the equipment, walls, floors, lighting, and GMPs. Maintenance of records of the inspected documents for further reference is also mandatory (Ali et al., 2020).

12.13 FOOD SAFETY REGULATIONS AND FOOD QUALITY STANDARDS

Globally, food systems have undergone a rapid revolution with private standards, modern retailers, and advancements in integrated supply chains due to changes in consumer preferences. They are brought by growing living standards and concern about FSQ assurance (Swinnen, 2007). Standards for food quality and safety and the capability for tracing food within the supply chain has become more critical for both the producers and consumers. Guarantee of quality is a mandatory requisite at each step of food production and processing chain, to assure the safety of food in compliance with administrative and consumer requirements (Swinnen, 2007).

12.13.1 PUBLIC AND PRIVATE STANDARDS

Expectations of consumers regarding food risk and food quality have been enormously increasing. So, both the public and private sectors have acknowledged that it has resulted in plenty of food standards, certifications, and regulations. This has led the firms and other businesses to expand their market by gaining consumer confidence and trust (Chiputwa et al., 2015). Public and private standards can be distinguished based on their response to new challenges and demands. But a significant problem both the sectors face is harmonization (Henson and Humphrey, 2009).

12.13.1.1 GROWTH IN PUBLIC FOOD SECTOR

Recently, the execution of government regulations on food has started as consumers become concerned about FSQ, resource sustainability, and other growing environmental issues. Consumers have become aware and are keen on knowing about all the fundamental data or information regarding the origin, manufacture, and production of the food and food products they eat. Thereby, the government has taken measures by setting up stringent regulations and providing detailed information by creating accountability for the food industry, preventing immortal practices (Swinnen, 2007). Furthermore, public standards can expand the market and make it efficient by correcting market failures. Customers also gain confidence in quality and benefit due to standardization. Thus, governments have dominance in setting up public standards in comparison to private standards, and regulators can establish a single standard inevitably and punish the violators under criminal law. For example, establishment of a science-based approach recognized by the Codex Alimentarius Commission (CAC), which is known as the HACCP. It is a food safety program for the prevention of hazards at each stage of food production (Trienekens and Zuurbier, 2008). Internationally, there is rising propagation of quality meta-systems such as the ISO 9000 series of quality management standards. Other important meta-methods include GMP and GHP. A considerable drawback of public standards is that they are not flexible and do not adjust to changes as per consumer perceptions or technology (Swinnen, 2007).

12.13.1.2 GROWTH IN THE PRIVATE FOOD SECTOR

Private standards established by non-governmental entities such as manufacturers, retailers, or producers in the food industry contradict public standards set by regulatory authorities. Self-regulation is one form of standard-setting for private sectors to uphold customer confidence (Wouters and Geraets, 2012). Risk management is the main objective of specific standards related to the safety of food. The various aim, purpose, and constituents differ from sector to sector and according to standards types, but their legal status is a matter of debate. The primary objectives of private standards are:

- To enhance the standards and consistency of the supply chain to eliminate product failure
- To exclude numerous audits of food manufacturers and suppliers
- To back up retailer's and customer's objectives
- To provide data regarding production processes when food accidents occur (Trienekens and Zuurbier, 2008)

Over the past few years, these guidelines are becoming much more common components of the administration of global agri-food supply chains. They have been continuously monitored and imposed by third-party certification. Besides reducing risks, private standards also differentiate products by their extended attributes that indicate how they are formed (Henson and Humphrey, 2009). The privatization of standards is significant for both consumers and retailers as they are crucial to cost control and reputation, which is concerned with overall competition (Reardon and Farina, 2001). Ensuring customer satisfaction by setting up private standards as consumer preferences change with time. Various forces work in the food chain to ensure customer satisfaction. These significant forces include (1) An increase in consumers between grocery store chains, (2) Internal growth into retail grocery stores and retail discount stores, and (3) Technology and innovation. The primary reason for the success of this food distribution system is its flexibility and adaptability (Swinnen, 2007).

There is an increase in the need for private standards due to the following reasons:

- Due to development in recognized risks related to food, they have established standards in response to it
- Due to the increasing interests of customers in the processing of food and change in the perceptions of safety and quality of foodstuffs
- Due to the transfer in the responsibility of assuring food safety from the public to the private sector
- Due to the globalization of integrated supply chains (Henson and Humphrey, 2009)

But there are few concerns that the private standards bring to small farmers as the demanding nature of these standards are too complicated and expensive for them to meet along the supply chain. Also, they challenge the authority of public conditions of governance in the areas that have been historically following federal regulations (Henson and Reardon, 2005). The terms 'voluntary standards' and 'private standards' can be used conversely. Private rules usually set high standards for distinct food product characteristics, and it is more definite on how to accomplish specific goals and operationalize process guidelines (Swinnen, 2007; Henson and Humphrey, 2009).

Thereby, to respond to the increasingly globalized and complex food chain, many standards have been established. The ultimate role of standards, be it private or public, is basically to ensure and coordinate the agri-food value chains between the producers and provide information about the conditions of production, processing, and transportation of products (Henson and Humphrey, 2009).

12.13.2 GLOBAL AND INTERNATIONAL REGULATIONS ON FOOD SAFETY AND QUALITY (FSQ)

Globally, food safety issues are dealt with by the Food and Agricultural Organization (FAO), the World Health Organization (WHO), both UN organizations, and the World Trade Organization (WTO) (Trienekens and Zuurbier, 2008). The CAC (World's exceptional international food-standard-setting body) was established by FAO and WHO in 1963 to organize decisions regarding policymaking about food worldwide (FAO and WTO, 2017). The primary objective of the Codex is to ensure fair trade practices and assure the health of consumers. The international standards are established and regulated due to this objective. The Codex Alimentarius issues food standards ranging from characterization of raw materials and processing to hygienic practices, residual contamination, labeling, analysis, and inspection methodologies. Thus, Codex has immense importance in international food trade (Henson and Humphrey, 2009; FAO and WTO, 2017). WTO plays a significant role in solving trade disputes as they deal exclusively with international trade rules. Thereby, it helps in maintaining smooth, predictable, and free trade flow. A set of rules was established by the Codex Alimentarius and WTO together to provide safe foodstuffs, anticipated quality of food, and fair trade (Trienekens and Zuurbier, 2008). The SPS (sanitary and phytosanitary) and TBT (technical barriers to trade) affiliation strike a balance between rights to govern legal objectives such as food safety and public protection. At the same time, they assure regulations cannot become biased obstacles to trade. The SPS lists out rules that are required for the safety of the food and welfare of humans, animals, and plants. While TBT agreements cover a wide range of standards and regulations to protect and safeguard consumer health by ensuring the quality of the product (FAO and WTO, 2017). The benefits of having consistent food standards for the safety of customers are distinct due to rising demand in the market worldwide. National food standards and requirements are rated or estimated inside the legal framework constituting Codex standards as benchmark standards. Various countries and regions have their primary legislative assessment regarding FSQ (Wouters and Geraets, 2012).

12.13.3 GENERIC FOOD QUALITY AND SAFETY STANDARDS

ISO 9000 serves as a universal food quality management standard, ISO 22000 serves as a generic food safety management standard, and ISO 14000 serves as generic environment standards. These ISO standards are the requisites for any organization in the food cycle (ISO 22000 Standards, 2018). The generic food safety management standards elaborate on the scope of the standard, definitions, context of the organization, leadership, planning of operations, resources needed competence, support by awareness, communication and documented information, operation traceability, PRPs, HACCP, emergency preparedness, verification, control measures, corrective actions, performance evaluation, audit, management, and improvement. Any organization involved in the food chain should follow all these standards (Chen et al., 2020).

12.14 EFFECT OF CLIMATE CHANGE ON FOOD SAFETY

Food and climate have a complicated relationship between them, and thus the impact of climate variations is ambiguous (Lake and Barker, 2018). Change in climate conditions is one of the significant factors that directly or indirectly affect food and food products. Climatic changes stimulate changes in the nature that leads to the development of hazards in food, which can take place at any phase of the food cycle, starting from manufacturing to utilization or consumption. The various routes through which associated climate factors affecting food safety include temperature variations and precipitation changes, raising the magnitude of severe weather occurrences (floods and droughts), global warming, and alteration in the transportation routes of pollutants (Tirado et al., 2010). The food organizations trigger as well as worldwide climate change will have rapid effects in the future (Fanzo et al., 2018). There are adverse effects in food manufacturing and production in tropical regions as they are the primary cause of increasing temperature fluctuations and severe weather occurrences. It has a greater impact on the farmers who are profoundly affected due to decreasing yields (Niles et al., 2017). Nowadays, there is growing concern and attention about the effect of climate change on the security and superiority of food due to its degree of impacts on the production of food and in the food chain (Milicevic et al., 2019). The primary driving force for the contamination of crops through food security hazards are these extreme weather conditions (toxins and fungi), which leads to harmful effects on human and animal consumption (Milicevic et al., 2019). Another major increasing concern is the release of greenhouse gases that leads to alterations in the climate and are likely to have a more substantial influence on foodborne diseases (Lake and Barker, 2018). Thus, climate change has adverse effects on the protection of food starting from farm to fork, and food security has become a complex challenge that is gaining the attention to human health globally (Cole et al., 2018).

KEYWORDS

- **barcoding**
- **climate change**
- **foodborne disease**
- **food regulations**
- **food safety**
- **genomics**
- **hazards**
- **molecular technologies**
- **pathogens**
- **proteomics**
- **risk assessment**

REFERENCES

Ababio, P. F., Taylor, K. D. A., Swainson, M., & Daramola, B. A., (2016). Effect of good hygiene practices intervention on food safety in senior secondary schools in Ghana. *Food Control*, *60*, 18–24.

Ali, M. M., Hashim, N., Abd, A. S., & Lasekan, O., (2020). Principles and recent advances in electronic nose for quality inspection of agricultural and food products. *Trends in Food Science and Technology*, *99*, 1–10.

Andjelković, U., & Josić, D., (2018). Mass spectrometry-based proteomics as food omics tool in research and assurance of food quality and safety. *Trends in Food Science and Technology*, *77*, 100–119.

Attrey, D. P., (2017). Detection of food adulterants/contaminants. In: *Food Safety in the 21st Century*, (pp. 129–143). India: Academic Press.

Aven, T., (2015). *Risk Analysis* (p. 216). Norway: John Wiley and Sons.

Banerjee, D., Chowdhary, S., Chakraborty, S., & Bhattacharyya, R., (2017). Recent advances in detection of food adulteration. In: *Food Safety in the 21st Century* (pp. 145–160). India: Academic Press.

Banwart, G., (2012). *Basic Food Microbiology* (p. 774). New York: Springer Science and Business Media.

Barlow, S. M., Boobis, A. R., Bridges, J., et al., (2015). The role of hazard-and risk-based approaches in ensuring food safety. *Trends in Food Science and Technology*, *46*(2), 176–188.

Blanchfield, J. R., (2005). Good manufacturing practice (GMP) in the food industry. In: *Handbook of Hygiene Control in the Food Industry* (pp. 324–347). United Kingdom: Woodhead Publishing.

Bouzembrak, Y., & Marvin, H. J., (2019). Impact of drivers of change, including climatic factors, on the occurrence of chemical food safety hazards in fruits and vegetables: A Bayesian network approach. *Food Control*, *97*, 67–76.

Camino, F. M. M., Arisseto-Bragotto, A. P., & Block, J. M., (2017). Food quality, foodborne diseases, and food safety in the Brazilian food industry. *Food Quality and Safety*, *1*(1), 13–27.

Chen, H., Liu, S., Chen, Y., Chen, C., Yang, H., & Chen, Y., (2020). Food safety management systems based on ISO 22000:2018 methodology of hazard analysis compared to ISO 22000:2005. *Accreditation and Quality Assurance*, *25*(1), 23–37.

Chiputwa, B., Spielman, D. J., & Qaim, M., (2015). Food standards, certification, and poverty among coffee farmers in Uganda. *World Development*, *66*, 400–412.

Cliver, D. O., & Riemann, H. P., (2002). *Foodborne Diseases* (p. 411). San Diego, CA: Academic Press.

Cliver, D. O., Potter, M., & Riemann, H. P., (2011). *Foodborne Infections and Intoxications*. California, USA: Elsevier.

Cole, M. B., Augustin, M. A., Robertson, M. J., & Manners, J. M., (2018). The science of food security. *NPJ Science of Food*, *2*(1), 1–8.

De Oliveira, C. A. F., Da Cruz, A. G., Tavolaro, P., & Corassin, C. H., (2016). Food safety: Good manufacturing practices (GMP), sanitation standard operating procedures (SSOP), hazard analysis and critical control point (HACCP). In: *Antimicrobial Food Packaging* (pp. 129–139). Cambridge Academic Press.

European Food Safety Authority (EFSA). (2015). EFSA's 20th Scientific Colloquium on Whole Genome Sequencing of foodborne pathogens for public health protection. *EFSA Supporting Publications*, *12*(2), 743E.

Fanzo, J., Davis, C., McLaren, R., & Choufani, J., (2018). The effect of climate change across food systems: Implications for nutrition outcomes. *Global Food Security*, *18*, 12–19.

FAO and WTO. (2017). *Trade and Food Standards* (p. 71). Rome: Food and Agriculture Organization of the United Nations and the World Trade Organization.

FAO/WHO (Food and Agriculture Organization and World Health Organization). (2003). *Assuring Food Safety and Quality: Guidelines for Strengthening National Food Control Systems*. http://www.fao.org/3/y8705e/y8705e00.htm (accessed on 11 February 2022).

Food and Agricultural Organization (FAO). (2016). *Applications of Whole Genome Sequencing in Food Safety Management*. http://www.fao.org/fileadmin/templates/agns/WGS/FAO_Tech_Paper_WGS_Food_Safety_2016.pdf (accessed on 11 February 2022).

Giacometti, J., & Josić, D., (2013). Microbial proteomics for food safety. In: Toldra, F., & Nollet, L., (eds.), *Proteomics in Foods* (pp. 515–545). Boston, MA: Springer.

Girardon, P., Gabard, F., & Peyer, H., (2019). Food Safety Management System-HACCP-Risk Assessment. In: *Gases in Agro-Food Processes* (pp. 105–107). Cambridge Academic Press.

Gruenfeldova, J., Domijan, K., & Walsh, C., (2019). A study of food safety knowledge, practice and training among food handlers in Ireland. *Food Control*, *105*, 131–140.

Grunert, K. G., (2005). Food quality and safety: Consumer perception and demand. *European Review of Agricultural Economics*, *32*(3), 369–391.

Gupta, R. K., (2017). Foodborne infectious diseases. In: *Food Safety in the 21st Century* (pp. 13–28). Cambridge Academic Press.

Haynes, E., Jimenez, E., Pardo, M. A., & Helyar, S. J., (2019). The future of NGS (Next Generation Sequencing) analysis in testing food authenticity. *Food Control*, *101*, 134–143.

Henson, S., & Caswell, J., (1999). Food safety regulation: An overview of contemporary issues. *Food Policy, 24*(6), 589–603.

Henson, S., & Humphrey, J., (2009). *The Impacts of Private Food Safety Standards on the Food Chain and on Public Standard-Setting Processes*. http://www.fao.org/3/i1132e/i1132e.pdf (accessed on 11 February 2022).

Henson, S., & Reardon, T., (2005). Private agri-food standards: Implications for food policy and the agri-food system. *Food Policy, 30*(3), 241–253.

Hoffmann, V., Moser, C., & Saak, A., (2019). Food safety in low and middle-income countries: The evidence through an economic lens. *World Development, 123*, 104–611.

ISO 22000 Standards. (2018). *Food Safety Management Systems – Requirements for Any Organization in the Food Chain*. https://www.iso.org/standard/65464.html (accessed on 11 February 2022).

Jagadeesan, B., Gerner-Smidt, P., Allard, M. W., et al., (2019). The use of next-generation sequencing for improving food safety: Translation into practice. *Food Microbiology, 79*, 96–115.

Jia, C., & Jukes, D., (2013). The national food safety control system of China – a systematic review. *Food Control, 32*(1), 236–245.

Karippacheril, T. G., Rios, L. D., & Srivastava, L., (2017). *Global Markets, Global Challenges: Improving Food Safety and Traceability while Empowering Smallholders through ICT*. https://www.semanticscholar.org/paper/Global-Markets%2C-Global-Challenges%3A-Improving-Food-Karippacheril-R%C3%ADos/d92dec72b4253fd6a6c91d2f5ce362d-f8ae36a87 (accessed on 11 February 2022).

Khan, M. A., Khan, S., Khan, A., & Alam, M., (2017). Soil contamination with cadmium, consequences and remediation using organic amendments. *Science of the Total Environment, 601*, 1591–1605.

King, T., Cole, M., Farber, J. M., et al., (2017). Food safety for food security: Relationship between global megatrends and developments in food safety. *Trends in Food Science and Technology, 68*, 160–175.

Lake, I. R., & Barker, G. C., (2018). Climate change, foodborne pathogens and illness in higher-income countries. *Current Environmental Health Reports, 5*(1), 187–196.

Lammerding, A. M., (1997). An overview of microbial food safety risk assessment. *Journal of Food Protection, 60*(11), 1420–1425.

Liu, L., Li, Y., Li, S., et al., (2012). Comparison of next-generation sequencing systems. *BioMed Research International, 2012*. doi: https://doi.org/10.1155/2012/251364.

Lu, Y., Song, S., Wang, R., et al., (2015). Impacts of soil and water pollution on food safety and health risks in China. *Environment International, 77*, 5–15.

Margas, E., & Holah, J. T., (2014). Personal hygiene in the food industry. In: *Hygiene in Food Processing* (pp. 408–440). UK: Woodhead Publishing.

Mataragas, M., Alessandria, V., Ferrocino, I., Rantsiou, K., & Cocolin, L., (2018). A bioinformatics pipeline integrating predictive metagenomics profiling for the analysis of 16S rDNA/rRNA sequencing data originated from foods. *Food Microbiology, 76*, 279–286.

Milicevic, D., Lakicevic, B., Petronijevic, R., et al., (2019). Climate change: Impact on mycotoxins incidence and food safety. *Theory and Practice of Meat Processing, 4*(1), 9–16.

Mohammad, A. M., Chowdhury, T., Biswas, B., & Absar, N., (2018). Food poisoning and intoxication: A global leading concern for human health. In: *Food Safety and Preservation* (pp. 307–352). Cambridge, MA, USA: Academic Press.

Mortimore, S., & Wallace, C., (2013). *HACCP: A Practical Approach* (p. 475). New York: Springer Science and Business Media.

Moskowitz, H. R., (1995). Food quality: Conceptual and sensory aspects. *Food Quality and Preference*, *6*(3), 157–162.

Niles, M. T., Ahuja, R., Esquivel, J. M., et al., (2017). *Climate Change and Food Systems: Assessing Impacts and Opportunities* (p. 57). Washington, DC: Meridian Institute.

Ortea, I., O'Connor, G., & Maquet, A., (2016). Review on proteomics for food authentication. *Journal of Proteomics*, *147*, 212–225.

Pagliacci, F., (2019). Agri-food activities in jeopardy – A territorial analysis of risks from natural hazards. *Land Use Policy*, *87*, 104–141.

Pierson, M. D., (2012). *HACCP: Principles and Applications* (p. 223). New York: Springer Science and Business Media.

Piras, C., Roncada, P., Rodrigues, P. M., Bonizzi, L., & Soggiu, A., (2016). Proteomics in food: Quality, safety, microbes, and allergens. *Proteomics*, *16*(5), 799–815.

Potter, N. N., & Hotchkiss, J. H., (2012). *Food Science* (p. 608). New York: Springer Science and Business Media.

Prakash, B., Singh, P. P., Kumar, A., & Gupta, V., (2020). Prospects of omics technologies and bioinformatics approaches in food science. In: *Functional and Preservative Properties of Phytochemicals* (pp. 317–340). Uttar Pradesh, India: Academic Press.

Raj, D., Sharma, R., & Joshi, V. K., (2011). *Quality Control for Value Addition in Food Processing* (p. 348). India: New India Publishing Agency.

Ramaswamy, R., Ahn, J., Balasubramaniam, V. M., Saona, L. R., & Yousef, A. E., (2019). Food safety engineering. In: *Handbook of Farm, Dairy and Food Machinery Engineering* (pp. 91–113). United States: Academic Press.

Reardon, T., & Farina, E., (2001). The rise of private food quality and safety standards: Illustrations from Brazil. *The International Food and Agribusiness Management Review*, *4*(4), 413–421.

Reynolds, J., & Dolasinski, M. J., (2019). Systematic review of industry food safety training topics and modalities. *Food Control*, *105*, 1–7.

Scalco, A. R., Ganga, G. M. D., De Oliveira, S. C., & Baker, G., (2020). Development and validation of a scale for identification of quality attributes of agri-food products in short chains. *Geoforum*, *111*, 165–175.

Schmidt, R. H., & Rodrick, G. E., (2003). *Food Safety Handbook* (p. 805). Chichester: John Wiley and Sons.

Seaman, P., (2010). Food hygiene training: Introducing the food hygiene training model. *Food Control*, *21*(4), 381–387.

Sikora, T., & Strada, A., (2005). Safety and quality assurance and management systems in food industry: An overview. In: *The Food Industry in Europe* (pp. 85–95). Athens: University of Athens.

Singh, P. K., Singh, R. P., Singh, P., & Singh, R. L., (2019). Food hazards: Physical, chemical, and biological. In: *Food Safety and Human Health* (pp. 15–65). India: Academic Press.

Soon, J. M., (2020). 'Food allergy? Ask before you eat': Current food allergy training and future training needs in food services. *Food Control*, *112*, 107–129.

Staats, M., Arulandhu, A. J., Gravendeel, B., et al., (2016). Advances in DNA metabarcoding for food and wildlife forensic species identification. *Analytical and Bioanalytical Chemistry*, *408*(17), 4615–4630.

Stasiewicz, M. J., Den, B. H. C., & Wiedmann, M., (2015). Genomics tools in microbial food safety. *Current Opinion in Food Science*, *4*, 105–110.

Swinnen, J. F., (2007). *Global Supply Chains, Standards and the Poor: How the Globalization of Food Systems and Standards Affects Rural Development and Poverty* (p. 340). UK: CABI.

Taulo, S., Wetlesen, A., Abrahamsen, R., Kululanga, G., Mkakosya, R., & Grimson, A., (2008). Microbiological hazard identification and exposure assessment of food prepared and served in rural households of Lungwena, Malawi. *International Journal of Food Microbiology*, *125*(2), 111–116.

The National Advisory Committee on Microbiological Criteria for Food (NACMCF), (1992). Hazard analysis and critical control point system. *International Journal of Food Microbiology*, *16*(1), 1–23.

Tirado, M. C., Clarke, R., Jaykus, L. A., McQuatters-Gollop, A., & Frank, J. M., (2010). Climate change and food safety: A review. *Food Research International*, *43*(7), 1745–1765.

Trienekens, J., & Zuurbier, P., (2008). Quality and safety standards in the food industry, developments, and challenges. *International Journal of Production Economics*, *113*(1), 107–122.

Tu, M., Cheng, S., Lu, W., & Du, M., (2018). Advancement and prospects of bioinformatics analysis for studying bioactive peptides from food-derived protein: Sequence, structure, and functions. *Trends in Analytical Chemistry*, *105*, 7–17.

Urban, L., Holzer, A., Baronas, J. J., et al., (2020). Freshwater monitoring by nanopore sequencing. *Elife*, *10*, e61504. doi: 10.7554/eLife.61504.

Van, R. W., & Frewer, L. J., (2008). Consumer perceptions of food quality and safety and their relation to traceability. *British Food Journal*, *110*(10), 1034–046.

Voorhuijzen-Harink, M. M., Hagelaar, R., Van, D. J. P., Prins, T. W., Kok, E. J., & Staats, M., (2019). Toward on-site food authentication using nanopore sequencing. *Food Chemistry*, *10*, 2.

Wambui, J., Karuri, E., Lamuka, P., & Matofari, J., (2017). Good hygiene practices among meat handlers in small and medium enterprise slaughterhouses in Kenya. *Food Control*, *81*, 34–39.

Wang, C., Wang, Y., Liu, G., & Fu, L., (2019). Food allergomics based on high-throughput and bioinformatics technologies. *Food Research International*, *130*, 108–942.

Weng, X., & Neethirajan, S., (2017). Ensuring food safety: Quality monitoring using microfluidics. *Trends in Food Science and Technology*, *65*, 10–22.

Wester, P. A., (2018). Hazards, the hazard analysis, and the food safety plan. In: *Hazard Analysis and Risk Based Preventative Controls* (pp. 43–65). United States: Academic Press.

Wouters, J., & Geraets, D., (2012). Private food standards and the World Trade Organization: Some legal considerations. *World Trade Review*, *11*(3), 479–489.

Wu, S., Ricke, S. C., Schneider, K. R., & Ahn, S., (2017). Food safety hazards associated with ready-to-bake cookie dough and its ingredients. *Food Control*, *73*, 986–993.

CHAPTER 13

Sequencing, Bioinformatics Tools, and Databases in Food Microbiology

SUSHMA PRADEEP, ANISHA S. JAIN, CHANDAN DHARMASHEKARA, ASHWINI PRASAD, KOLLUR SHIVA PRASAD, CHANDRASHEKAR SRINIVASA, and CHANDAN SHIVAMALLU

ABSTRACT

Food is an essential component of our everyday lives. Before reaching the market, many food products are processed in some way, ranging from fermentation to packaging. Microorganisms play an important role in all of these processes, whether they are converting the food or spoiling or contaminating it. Bioinformatics is becoming more important in predicting and evaluating the desirable and unwanted effects of microorganisms on food. It combined with laboratory confirmation of selected findings is especially effective. This chapter discusses about how bioinformatics plays an important role in predicting and assessing the desired and undesired effects of microorganisms on food, genomics, and proteomics study to meet the requirements of food production, food processing, improving the quality and nutritive value of food sources and many others.

13.1 INTRODUCTION

Food consists of ample microbial populations composed of fungi, bacteria, and viruses. It might even contain detrimental microorganisms, which can reduce the quality of foods (spoilage microorganisms) or adversely affect the food safety (Mayo et al., 2019). Thus, food microbiology involves the study of microbes that are present or involved in the food, which may have

Food Microbial and Molecular Biology: From Fundamentals to Applications. Saher Islam, Devarajan Thangadurai, Jeyabalan Sangeetha, Zaira Zaman Chowdhury (Eds.)

both desirable and undesirable effects on the safety and quality of the food production (Mari et al., 2006).

Culturing method is the most common or the outmoded style of determining the composition of food-related microbiota, which relay on the cultivation and isolation of the microorganisms before typing and sequencing, but 98% of the environmental bacteria cannot be cultured under the routine laboratory techniques. Nevertheless, culturing technique is not always used for the overall microbial characterization of ecosystems, including foods (Liu et al., 2008). The isolation and cultivation of particular microorganisms require unknown growth factor and condition.

Foods may be stored or kept under harsh conditions, or it may have a low pH level, etc., due to which the microbes leave in a physiologically viable state than in cultivable state (Abee et al., 2004). Over the decades to overcome these problems, many culture-independent, molecular approaches have been identified; most of which are extensively used in food systems. These techniques help us in identifying, quantifying the food linked microbes, and offer sensitive and quick methods to determine the diversity and composition of microbial populations (Lemay et al., 2007). Culture independent methods are usually based on the microbial genetic material sequence (DNA and/or RNA) analysis (Figure 13.1).

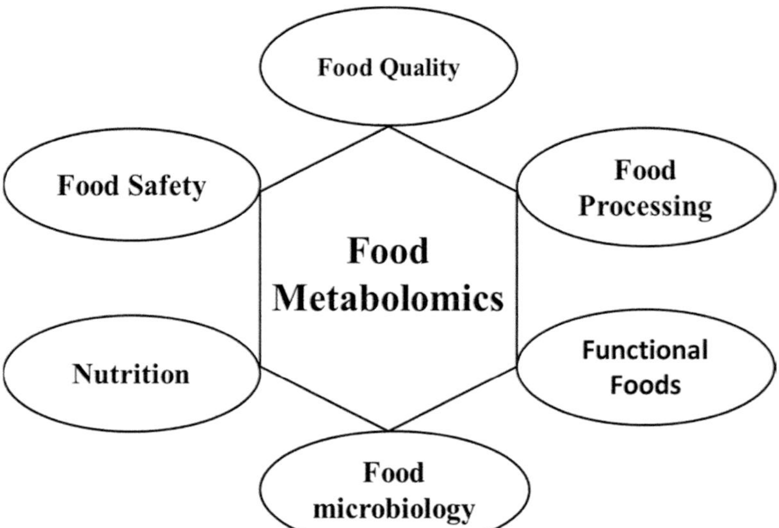

FIGURE 13.1 The sequential aspects involved in the food microbiology.

13.2 PCR GENE AMPLIFICATION TECHNIQUE

Most of which rely on the amplification of the DNA or RNA using the polymerase chain reaction (PCR) technique (Liu et al., 2008). These nucleic acids amplifying techniques majorly target the rRNA or rDNA genes. By matching or comparing the sequences obtained to that held in databases, establishes the phylogenetic relationships between the microbes (Mari et al., 2006) (Figure 13.2).

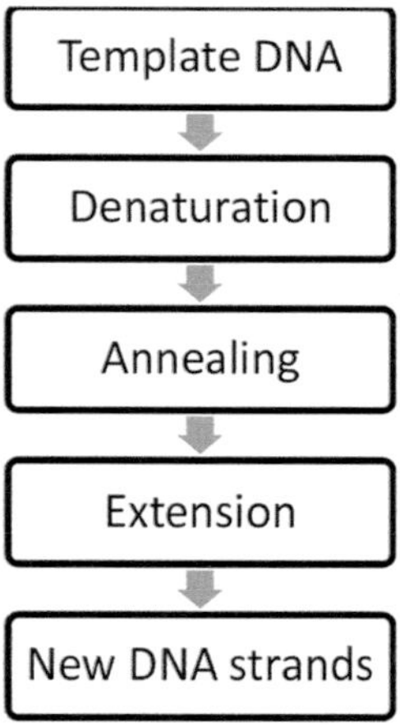

FIGURE 13.2 Flow chart representing one complete cycle of PCR.

The PCR-based techniques such as DGGE (denaturing gradient gel electrophoresis), TTGE (temporal temperature gradient gel electrophoresis), qPCR (real-time quantitative PCR), TRFLP (terminal restriction fragment length polymorphism), as well as single-stranded conformation polymorphism (SSCP) are available to date for analyzing the quality of food (Kumar et al., 2017). Out of which, qPCR (quantitative analysis), DGGE (qualitative/ semi-quantitative analysis) and have been extensively used in microbial characterize food backgrounds and in the field of food microbiology.

13.3 NEXT GENERATION SEQUENCING (NGS) TECHNIQUE

The characterization, identification, and typing of the microbial species or mixed microbial communities can be performed using whole-genome sequencing (WGS) with the high level of precision (Gomes et al., 2015). The relevant knowledge about the entire microbial genome can be extracted using microbial WGS followed by determining the single nucleotide polymorphisms (SNPs) provided with the highest resolution. NGS is the other technique that comes under the culture-independent method, the amount of nucleic acid sequence analysis using NGS method is extremely higher, which allows to have a deep description about the microbial constituents in the ecosystems. Alongside, NGS has aided the development of new genome-aided methods for linking genotype and phenotype (Kumar et al., 2017). NGS involves many high throughput "omics" techniques such as glycomics, proteomics, metatranscriptomics, genomics, metagenomics, and transcriptomics, etc., that are the new and high-throughput techniques used for sequencing. This technique has two different ways of analyzing the nucleic acid sequences, shotgun sequencing technique which considers the total nucleic acids of the microbe and targeted sequencing technique that does gene-specific sequencing. All of these techniques involved in NGS generate a large, genome-scale datasets that needs to be further analyzed (Abee et al., 2004) (Figure 13.3).

In a very short time frame, the NGS technique is proficient in processing millions of sequences without lowering its accuracy level; instead, this technique relies on automation and high throughput technologies. To do this, it might require a very few instruments runs to undergo a sequencing experiment (Desiere et al., 2001).

13.4 NGS INVOLVED IN FOOD MICROBIOLOGY

NGS in food microbiology is not only used to predict the prevalence of microbes in food production but has been widely used even to interpret the molecular level organization of these microbes that respond to different food-related conditions, which further gives the knowledge as to control or prevent the undesirable microorganisms (Gomes et al., 2015). Testing this technique in various platforms of food microbiology has been devoted a lot of work, usually used for the descriptive studies.

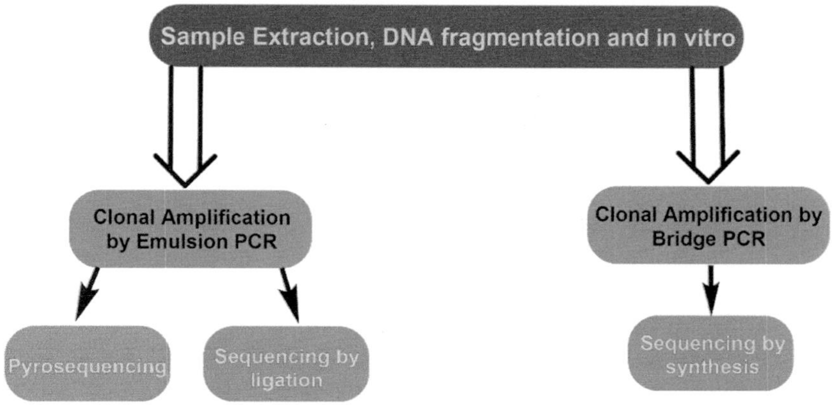

FIGURE 13.3 Next-generation sequencing workflow.

The data's obtained from the omic techniques would be useful to correlate between microbial diversity and food evolution that leads to the contribution of quality of the food products. New microbial genes and their species have been linked with some food processes. Based on this knowledge, during food production, the identification of the critical microbial variables would be easier, and hence control food safety and quality (FSQ). In this regard, NGS technique might help in gathering the knowledge about the relationships among the microbial safety and diversity properties (Abee et al., 2004). The safety and quality issues of the food can be directly related to the presence of unexpected microbe which is referred as contaminants (Figure 13.4).

Scientific hypothetical testing using NGS have been initiated, such as exploring the challenges, study of the deep interactions occurring between the microbial species and their strains, essential in the production of aroma and taste components, developing the metabolic models. When we consider foods, it is important to differentiate between inactive, active, and viable cells. NGS techniques focus on the active microbial groups by producing the cDNA obtained through the reverse transcription from RNA (Brul et al., 2006).

13.5 FUTURE PROSPECTS OF NGS IN FOOD MICROBIOLOGY

Next-generation sequencing (NGS) techniques will eventually reform food microbiology. Detecting the pathogen, profiling the microbial populations, data mining, phenotype, and genotype linking, determining the pathogens over food production and shelf life of the food products are the realm of this NGS techniques (high-throughput techniques) (Carrasco et al., 2012). Even the NGS technique must overcome some drawbacks and limitations that need to be addressed to improve the present understanding of data in food physiology and microbial diversity in the future.

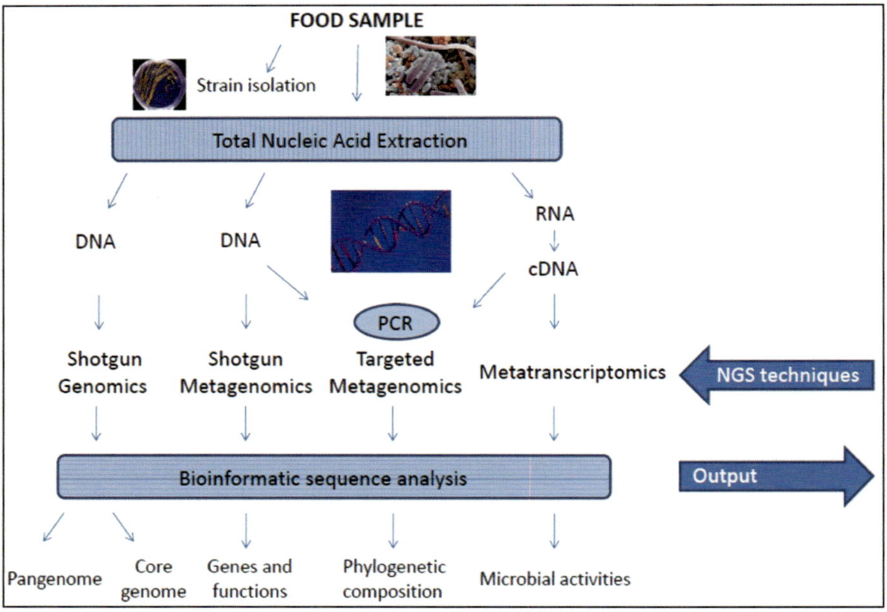

FIGURE 13.4 Applications of next-generation sequencing (NGS) in food microbiology.

13.6 WHAT IS BIOINFORMATICS?

Bioinformatics is a branch of science with the integration of both information technology and computer science into the field of Biology (Desiere et al., 2001). This field utilizes the resources of computer science to mainly deal with the biological data related to the nucleic acid sequences (DNA, RNA, and protein). This field of science actually develops software, tools, and algorithms that can record and analyze the data related to biology (Mayo et al., 2019). The biological data will always be in the raw format, and thus this sort of raw data requires a certain type of storage house in which the data can be organized, stored, and manipulated. These databases and software provide the users an opportunity to easily extract the data from the databases and can be further used for any experiment (Figure 13.5).

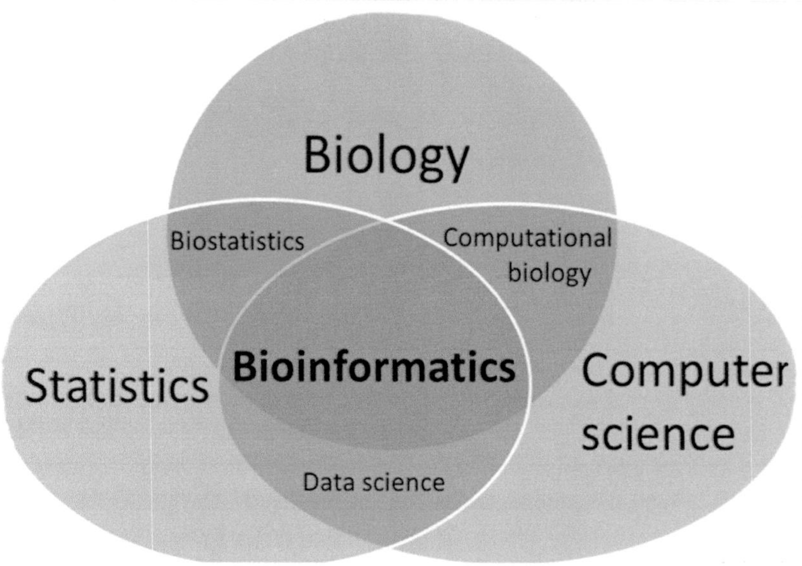

FIGURE 13.5 The involvement of various fields in bioinformatics.

13.6.1 APPLICATION OF BIOINFORMATICS

To solve the complex biological problems, the bioinformatics integrates several other fields into it, such as statistics, information science and technology, mathematics, and computer science (Bukin et al., 2019). Bioinformatics tools and software can be applied to various research areas like health and drug discovery, prediction of protein structure, comparative genomics, sequence analysis, genome annotation, and much more (Figure 13.6).

13.6.2 BIOINFORMATICS INVOLVED IN FOOD SCIENCE

Food sciences is described to be the combination of biological, physical, and engineering studies to know the nature of the foods involved during the food processing and for food quality improvement (Carrasco et al., 2012). Comparative genomic sequencing techniques of bioinformatics have been used in food-related pathogens and their production (Alkema et al., 2016). Using the computational algorithms and bioinformatics software the molecular modeling and dynamic simulation can be possibly performed to the taste receptors and also such simulations will possibly help to develop more efficient tasting components as additives for the food.

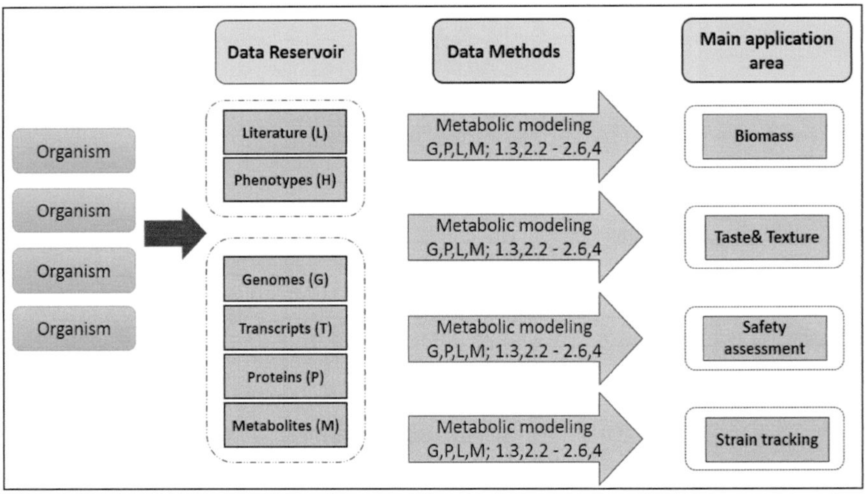

FIGURE 13.6 Applied bioinformatics in food microbiology.

13.6.3 MICROBIAL INFORMATICS

Bioinformatics plays a major role in determining desirable and undesirable effects of microbes on food. Food nutrition has growing importance in the growth of various chronic diseases; a lot of efforts have been taken to increase the nutritional composition and food quality (Alkema et al., 2015). The omic techniques have been enumerated in the area of food sciences. The advancement of bioinformatics in food microbiology is providing several ways for the researchers to accept new area or discipline. Microbial bioinformatics is used to optimize the biomass yield to improve the food safety, quality, and production (Lemay et al., 2007). To do this, sequencing has been carried at the genome level of the microorganism and metabolic models are designed for many microbes including food-associated microorganisms (Bukin et al., 2019).

The bioinformatics tools, comparative metabolomics and software can be applied to understand the genetic constitution or the organization of the pathogenic microbes. This genomic information can be further sequenced to know the protein function, gene elucidation and it also enriches the knowledge about the microbe and its disease mechanism. Microbial genomic information thus obtained forms the basis to study the microbe related protein and genes and synthesized toxins. Thus, bioinformatics approaches quicken and facilitate the analysis of cellular processes in order to control or treat the microbes (Brul et al., 2006).

13.6.4 BIOINFORMATICS INVOLVED IN SECURING THE SAFETY OF FOOD

Bioinformatics databases, tools, and software are playing a major role in securing the quality of food, taste, flavor and also maintaining food safety. The molecular evolution of the food taste receptors can also be tested using the bioinformatics tools (Chibuike et al., 2014). Many other studies have been carried out to focus primarily on the food taste receptors that links between the receptors that taste bitter and glucose regulation. Recently the chemical properties of many compounds associated with the food flavor and taste were included in the electronic database (Lemay et al., 2007). Because of the importance of bioinformatics in the field of food microbiology, various projects related to the genomic sequencing are concentrating on the foodborne pathogens, due to which now bioinformatics has started an innovative way to determine the cause of the foodborne illness (Bukin et al., 2019).

13.7 FOOD SCIENCE DATABASES

13.7.1 FOODOMICS DATABASE

It is the database that involves molecular profiles of the food. Twenty-eight standard references of USDA National Nutrient databases were released that contain the data for about 150 food-related molecules. The concurrent nutrient knowledge of less than 15 nutrients is provided by the nutrition facts labels (NFLs) (Mayo et al., 2019). The Foodomic profiles help us in implementing the precise medicine for every person; this can be done only when there is integration between NFLs and SR28. This is the precise way to give food molecules in the right amount at the right time to the right person (Gomes et al., 2015).

13.7.2 FOODWIKI DATABASE

It is the source of nutritive value and food information (Kumar et al., 2017). The immense amount of information or the data stored in this database possibly utilizes the bioinformatics strategies and protocols. Such source of information about the food and nutrition helps in enhancing the nutritive value, safety, and quality of the food (Mayo et al., 2019). This database plays a crucial role in the field of food sciences, which in turn leads to the progression of bioinformatics in the area of food microbiology.

13.7.3 EUROFIR-BASIS

This database mainly contains the information about the plant-based food compounds and uniquely pools biological activity and composition of the food. It comprises of 330 main edible parts of the food plants and various compound classes (Mochida et al., 2010). This database plays a very major role in providing a valuable source of information in the field of development of products in the food industry and diet and health-related research activities.

13.7.4 FooDB

This database is the major resource that provides the knowledge about the biology, chemistry related to the food constituents (Chibuike et al., 2014). It contains the information about macro and micronutrients of the compounds involved in food taste, color, aroma, texture, and flavor (Desiere et al., 2001). Every chemical entry of this database consists of more than 100 different data sets associated with precise physiological, compositional, and biochemical information (Mochida et al., 2010).

13.8 FOOD BIOINFORMATICS CONCEPTS AND TECHNIQUES

13.8.1 GENOME SEQUENCING

When considered a microbe of interest its complete genome sequence can be determined (Solieri et al., 2013). The high-quality and high-throughput sequencing data can be obtained by using NGS technique (Mochida et al., 2010).

13.8.2 COMPARATIVE GENOMICS

As the name indicates, the genome content or the sequences of various organisms are compared (Solieri et al., 2013). This genomic content must include nucleic acid sequences (DNA or RNA), gene order, genes, and regulatory sequences.

13.8.3 METABOLIC MODELING

By using the genome sequencing method, the metabolic pathways requirement as well as the growth prediction of the microbes can be carried out (Mochida et al., 2010). The construction of metabolic models in Genome-scale is done by using comparative or automated genomics (Solieri et al., 2013). With the help of these constructed metabolic models the growth simulation can be carried out. The bioinformatics tools used for the modeling are Vanted (https://omicstools.com/vanted-tool), PySces (https://omictools.com/pysces-tool) and Seed (https://pubseed.theseed.org/).

13.8.4 MICROBIOME ANALYSIS

The overall microbe content in a specific niche are said to be microbiome. This microbiome content is usually analyzed using sequence-based techniques like next-generation sequencing (Solieri et al., 2013).

13.8.5 16S rRNA SEQUENCING

Sequence generation from the 16S gene is termed as 16S rRNA sequencing (Ronholm et al., 2018). This technique is well established and cheap when compared to other sequencing techniques and plays a very major role in the characterization of compound cultures which has a limited prior data is available.

13.8.6 FUNCTIONAL PREDICTION

The result of the data generated from the 16S rRNA sequencing can be further used to determine the gene functions of a particular taxa using bioinformatics tool like PICRUSt (https://omictools.com/picrust-tool). This tool utilizes the already known data of the taxa where the genomes are already sequenced and generates the gene functions (Ronholm et al., 2018).

13.8.7 SHOTGUN METAGENOMICS AND META-TRANSCRIPTOMICS

These techniques are very powerful as the circumventing of culturing microbes is allowed while still determining their gene expression and gene content (Yi et al., 2015). Due to which it provides the knowledge about the functions of the encoded DNA at the molecular level and also for the expressed mRNAs.

13.8.8 ASSEMBLY

The sequence derived from the DNA/RNA can be arranged into large contigs, for these obtained large fragments the functional annotations can be effectively carried out.

13.8.9 ANNOTATION

The metagenome sequences can be taxonomically and functionally annotated by predicted or compared gene sequences using the known functions (Yi et al., 2015). Bioinformatics tools such as PhymmBL, MG-RAST, and MetaPHlAn.

13.8.10 PREDICTING PHENOTYPES

Statistical techniques, machine learning, and gene-matching methods are used to determine the bacterial strain phenotype based on the pathways, gene particulars or depending upon the classification (Wingender et al., 1996).

13.8.11 METABOLOMICS

Several metabolites in the biological samples are simultaneously measured. The reaction products of various microorganisms in the food samples and in the culture media can be determined using metabolomics technique (Yi et al., 2015).

13.9 CONCLUSION

NGS techniques will eventually reform food microbiology. Detecting the pathogen, profiling the microbial populations, data mining, phenotype, and genotype linking, determining the pathogens over food production and shelf life of the food products are the realm of this NGS techniques. There is an increasing success rate of bioinformatics approaches related to the food sciences, not only food sciences but it is also expanding the research by becoming a very essential tool in the field of life sciences. In the area of food sciences, it helps in protein sequence analysis of the food products to determine the function and allergenicity of it. Comparative genomic sequencing techniques of bioinformatics have been used in food related pathogens and their production. In concern to secure the FSQ, using bioinformatics, a lot of advancement has been made which allows automatic food property detection.

KEYWORDS

- **bioinformatics**
- **fermentation**
- **genomics**
- **microorganisms**
- **next-generation sequencing**
- **proteomics**

REFERENCES

Abee, T., Van, S. W., & Siezen, R. J., (2004). Impact of genomics on microbial food safety. *Trends in Biotechnology*, *22*, 653–660.

Alkema, W., Boekhorst, J., Wels, M., & van Hijum, S. A. F. T., (2016). *Microbial Bioinformatics for Food Safety and Production*, *17*, 283–292.

Alkema, W., Boekhorst, J., Wels, M., & Van, H., (2015). Microbial bioinformatics for food safety and production. *Briefings in Bioinformatics*.

Brul, S., Schuren, F., Montijn, R., Keijser, B. J. F., Van, D. S. H., & Oomes, S. J. C. M., (2006). The impact of functional genomics on microbiological food quality and safety. *International Journal of Food Microbiology*, *112*, 195–199.

Bukin, Y. S., Galachyants, Y. P., Morozov, I. V., Bukin, S. V., & Zakharenko, A. S., (2019). *Data Descriptor: The Effect of 16S rRNA Region Choice on Bacterial Community Metabarcoding Results*, 1–14.

Carrasco-Castilla, J., Hernandez-Alvarez, A. J., Jimenez-Martınez, C., Gutierrez-Lopez, G. F., & Davila-Ortiz, G., (2012). Use of proteomics and peptidomics methods in food bioactive peptide science and engineering. *Food Engineering Reviews*, *4*, 224–243.

Chibuike, C. U., (2014). Bioinformatics approaches prospects and challenges of food bioactive peptide research. *Trends in Food Science and Technology*, *36*(2), 137–143.

Desiere, F., German, B., Watzke, H., Pfeifer, A., & Saguy, S., (2001). Bioinformatics and data knowledge: The new frontiers for nutrition and foods. *Trends in Food Science and Technology*, *12*(7), 215–229.

Gomes, L.. H. F., Alves-Ferreira, M., & Carels, N., (2015). Functional genomics. In: *Plant Biology and Biotechnology, Vol. II: Plant Genomics and Biotechnology*, Bahadur, B., Rajam, M. V., Sahijram, L., & Krishnamurthy, K.V., (eds), Springer, New Delhi, pp. 223–246.

Kumar, A., & Chordia, N., (2017). Bioinformatics approaches in food sciences. *Journal of Food: Microbiology, Safety and Hygiene*, *2*, 2–5.

Lemay, D., Zivkovic, A. M., & German, J. B., (2007). Building the bridges to bioinformatics in nutrition research. *The American Journal of Clinical Nutrition*, *86*, 1261–1269.

Liu, M., Nauta, A., Francke, C., & Siezen, R. J., (2008). Comparative genomics of enzymes in flavor-forming pathways from amino acids in lactic acid bacteria. *Applied and Environmental Microbiology*, *74*, 4590–4600.

Mari, A., Scala, E., Palazzo, P., Ridolfi, S., Zennaro, D., & Carabella, G., (2006). Bioinformatics applied to allergy: Allergen databases from collecting sequence information to data integration. The allergome platform as a model. *Cellular Immunology*, *244*, 97–100.

Mayo, B., et al., (2019). *Impact of Next Generation Sequencing Techniques in Food Microbiology*, 293–309.

Mochida, K., & Shinozaki, K., (2010). Genomics and bioinformatics resources for crop improvement. *Plant and Cell Physiology*, *51*, 497–523.

Ronholm, J., (2018). *Editorial: Game Changer – Next Generation Sequencing and Its Impact on Food Microbiology, Front. Microbiol.*, *9*, 1–3. https://doi.org/10.3389/fmicb.2018.00363.

Solieri, L., Dakal, T. C., & Giudici, P., (2013). Next-generation sequencing and its potential impact on food microbial genomics. *Ann Microbiol. 63*, 21–37. https://doi.org/10.1007/s13213-012-0478-8.

Wingender, E., Dietze, P., Karas, H., & Knuppel, R., (1996). TRANSFAC: A database on transcription factors and their DNA binding sites. *Nucleic Acid Res*., *24*, 238–241.

Yi, L. S., Chin, T. L., Mohamad, M. S., Deris, S., Subair, S., & Ibrahim, Z., (2015) A review on metabolic pathway analysis in biological production. *Mini-Reviews in Organic Chemistry 12*(6), 506–523. doi: 10.2174/1570193X13666151218191358.

Index

B

C

G

H

Q

R